## UAV Networks and Communications

The first book to focus on the communications and networking aspects of UAVs, this unique resource provides the fundamental knowledge needed to pursue research in the field.

The team of authors covers the foundational concepts of the topic, as well as offering a detailed insight into the state-of-the-art in UAVs and UAV networks, discussing the regulations, policies, and procedures for deployment (including analysis of risks and rewards) along with demonstrations, test-beds, and practical real-world applications in areas such as wildlife detection and emergency communications.

This is essential reading for graduate students, researchers, and professionals in communications and networking.

**Kamesh Namuduri** is a Professor in the Electrical Engineering Department at the University of North Texas.

**Serge Chaumette** is a Professor of Computer Science at the University of Bordeaux, France, and leads UAVs research and activities at Bordeaux Computer Science Research Laboratory (LaBRI).

**Jae H. Kim** is an Executive and Senior Technical Fellow of Boeing Research and Technology, and an Affiliate Professor at the University of Washington, Seattle.

**James P.G. Sterbenz** is a Professor of Electrical Engineering and Computer Science and Director of the Networking Systems Laboratory in the Information and Telecommunication Technology Center at The University of Kansas.

# UAV Networks and Communications

Edited by

KAMESH NAMUDURI
University of North Texas

SERGE CHAUMETTE
University of Bordeaux

JAE H. KIM
Boeing Research and Technology

JAMES P.G. STERBENZ
University of Kansas

CAMBRIDGE
UNIVERSITY PRESS

**CAMBRIDGE**
**UNIVERSITY PRESS**

University Printing House, Cambridge CB2 8BS, United Kingdom

One Liberty Plaza, 20th Floor, New York, NY 10006, USA

477 Williamstown Road, Port Melbourne, VIC 3207, Australia

314-321, 3rd Floor, Plot 3, Splendor Forum, Jasola District Centre, New Delhi - 110025, India

79 Anson Road, #06-04/06, Singapore 079906

Cambridge University Press is part of the University of Cambridge.

It furthers the University's mission by disseminating knowledge in the pursuit of education, learning and research at the highest international levels of excellence.

www.cambridge.org
Information on this title: www.cambridge.org/9781107115309
DOI: 10.1017/9781316335765



First published 2018

*A catalogue record for this publication is available from the British Library*

*Library of Congress Cataloging in Publication data*
Names: Namuduri, Kamesh, editor. | Chaumette, Serge, editor. | Kim, Jae H. (Jae Hoon), editor. | Sterbenz, James P. G., editor.
Title: UAV networks and communications / edited by Kamesh Namuduri, University of North Texas, Serge Chaumette, University of Bordeaux, Jae H. Kim, Boeing Research and Technology, James P.G. Sterbenz, University of Kansas.
Description: Cambridge, United Kingdom ; New York, NY, USA : Cambridge University Press, 2018. | Includes bibliographical references and index.
Identifiers: LCCN 2017044885 | ISBN 9781107115309 (hardback : alk. paper)
Subjects: LCSH: Drone aircraft–Control systems. | Aeronautics–Communication systems. | Wireless communication systems.
Classification: LCC TL589.4 U27 2017 | DDC 629.135–dc23 LC record available at https://lccn.loc.gov/2017044885

ISBN 978-1-107-11530-9 Hardback

Additional resources for this title are available at www.cambridge.org/namuduri

Dedicated to
Our Teachers

# Contents

# Preface

Aviation authorities around the world have been making progress towards integrating UAVs (unmanned aerial vehicles) into their national airspaces. In parallel, private industry has been developing innovative UAV-based applications, such as drone-based package delivery, medicine delivery, pipeline monitoring systems, and disaster-area aerial surveys. However, before UAVs can become integrated into the civilian airspace and such real-world applications become reality, there are several technical, societal, and regulatory challenges that need to be addressed by the research community. The most important among them is the need for enhanced situational awareness of UAVs in the airspace.

Three different, yet complementary, paradigms emerged to address enhanced situational awareness of UAVs: satellite communications, cellular-communications, and aerial communications and networks. This book focuses on the third strategy, i.e., enhanced situational awareness through self-organized aerial networking of UAVs. It provides the necessary knowledge for students, researchers, and professionals to gain an understanding of the research challenges in UAV networks and communications. Collaborating with several eminent research scholars and subject matter experts, the editors developed nine chapters that take the reader from the foundations to active research topics in this exciting domain.

The first chapter, "Introduction to UAV Systems," introduces the reader to UAV types and missions. It provides the background and context for UAVs and UAV networks with a focus on their civilian applications. It also discusses the state-of-the-art in engineering and technology aspects of UAV networks and the benefits of deploying such networks.

The second chapter, "Air-to-Ground and Air-to-Air Data Link Communication," provides the background on wireless communication used in manned aviation. It discusses the technologies proposed for L-band Digital Aeronautical Communication. It provides the fundamental insights relevant for aerial communication on unmanned and small UAVs, learned from experience with the advanced terrestrial mobile broadband communication extrapolated to the aerial case.

The third chapter, "Aerial Wi-Fi Networks," provides the characteristics of aerial links in three-dimensional space (3D). Aerial networks differ from other wireless networks, such as mobile ad hoc networks or vehicular ad hoc networks. It discusses the communication requirements for aerial network applications in terms of throughput, delay, data exchange frequency, etc. It defines different levels of autonomy in aerial networks from the perspective of communication needs.

The fourth chapter, "Disruption-Tolerant Airborne Networks and Protocols," presents an architecture and protocol suite suitable for the aeronautical environment: highly dynamic, high-velocity multi-hop networks that require the greatest change to past networking architectures, with comparisons to traditional end-to-end transport and routing protocols.

The fifth chapter, "UAV Systems and Networks: Emulation and Field Demonstration," discusses the design, implementation, and deployment of a UAV network for the purpose of transmitting video surveillance data amongst nodes. It presents a heterogeneous network consisting of multiple stationary and mobile ground-based nodes, as well as multiple autonomous aerial vehicles. It presents a design process to enable emergent system safety through appropriate integration of critical subsystems and collaboration across multiple UAVs.

The sixth chapter, "Integrating UAVs with NAS – Regulatory, Technical, and Research Challenges," provides the background and context for integrating UAVs within a civilian airspace system. This chapter will cover regulatory concerns, social issues, and technical challenges with respect to integration efforts for UAVs.

The seventh chapter, "Safety, Security, and Privacy Aspects in UAV Networks," discusses the challenges in terms of safety, security, and privacy, which are the three dimensions for integrating UAVs in the civilian airspace and for designing real-world applications.

The eighth chapter, "Collaboration Between Autonomous Drones and Swarming," addresses some of the major issues that should be solved if swarms are to be used in the field. It explains why swarming can significantly increase the possibilities of a mission. It then outlines and dives into a number of challenges and research directions that need to be explored further.

The ninth chapter, "Real-World Applications," reviews two of the many applications of UAVs and UAV networks that researchers are currently pursuing: (1) wildlife detection and (2) emergency communications. These examples showcase the unique value and innovation that UAVs can bring to the real-world applications.

The editors sincerely believe that these nine chapters will guide aspiring students, researchers, and professionals to gain a broad understanding of this emerging topic of UAV networks and communications.

Kamesh Namuduri
Serge Chaumette
Jae H. Kim
James P. G. Sterbenz

# Contributors

**Mohammed J.F. Alenazi**
University of Kansas

**Egemen K. Çetinkaya**
University of Kansas

**Serge Chaumette**
University of Bordeaux

**Claudiu Danilov**
Boeing Research and Technology

**Leanne Hanson**
U.S. Geological Survey

**Samira Hayat**
Alpen-Adria–Universität at Klagenfurt

**Jae H. Kim**
Boeing Research and Technology

**Jean-Marc Moschetta**
Institut Supérieur de l'Aéronautique et de l'Espace

**Kamesh Namuduri**
University of North Texas

**Hemanth Narra**
University of Kansas

**Natasha Neogi**
NASA

**Truc Anh N. Nguyen**
University of Kansas

**Andres Ortiz**
AeroVironment

**Kamakshi S. Pathapati**
University of Kansas

**Kevin Peters**
University of Kansas

**Sofie Pollin**
KU Leuven

**Justin P. Rohrer**
Naval Postgraduate School

**Damien Sauveron**
University of North Texas

**Arunabha Sen**
Arizona State University

**James P. G. Sterbenz**
University of Kansas

**Betrold Van den Bergh**
KU Leuven

**Evşen Yanmaz**
Alpen-Adria-Universität Klagenfurt

# 1 Introduction to UAV Systems

Jean-Marc Moschetta and Kamesh Namuduri

This chapter provides the background and context for unmanned aerial vehicles (UAVs) and UAV networks with a focus on their civilian applications. It discusses, for example, the types of UAVs, fuel, payload capacity, speed, and endurance. It will also discuss the state-of-the-art in engineering and technology aspects of UAVs and UAV networks and the advantages of UAV networks, including enhanced situational awareness and reduced latency in communications among the UAVs. It presents the applications of UAV networks, research opportunities, and challenges involved in designing, developing, and deploying UAV networks, and the roadmap for research in UAV networks.

Over recent decades, many different terms have been used to refer to UAVs, the most recent of which being remotely piloted aerial system (RPAS), which insists that the system is somehow always operated by somebody on the ground who is responsible for it. The term is very much like the old name for UAVs of the 1980s, that is remotely piloted vehicle (RPV). The RPAS puts emphasis on the fact that the aerial system includes not only the flying vehicle but also, for example, a ground control station, data link, and antenna. It also provides room for the case where several aircraft belonging to the same system may be remotely operated as a whole by a single human operator. In that case, it is not possible for the operator to actually control each flying vehicle as if he or she was an RC pilot.

Yet, in aeronautics, piloting an aircraft basically means flying an aircraft. It has a very precise meaning which is related to the capability to control the attitude of the vehicle with respect to its center of gravity. While most UAVs are remotely operated, they almost all have an on-board autopilot in charge of flying the aircraft. Therefore, it is not a remotely piloted vehicle but only a remotely *operated* vehicle where navigation commands are sent to the aircraft. Furthermore, navigation orders such as waypoints, routes, and decision algorithms may even be included in the on-board computer in order to complete the mission without human action along the way. In this way, human judgment is devoted to actions at higher levels, such as decision making or strategy definition. The term "remotely operated aircraft system" (ROAS) would therefore make more sense to the current scientific community.

Nevertheless, in the present book, the classical terms UAV or UAS have been chosen to refer either to the aerial vehicle itself (UAV), or to the whole system (UAS), which classically includes a set of UAVs (or possibly one), a control station, data links, a support equipment, and human operators.

## 1.1 Introduction to UAV Types and Missions

Many authors have already proposed various classifications for the different kinds of UAS. One may classify UAS by vehicle types, sizes, mass, mission range, altitude, endurance, etc. Each kind of classification is a way to point out a particular feature, but is doomed to hide another important aspect of UAS. Most lectures given on UAS start with a classification of UAVs based on some sort of conventional typology including: high altitude long endurance (HALE), medium altitude long endurance (MALE), tactical UAVs, vertical take-off and landing (VTOL) UAVs, and mini- and micro-UAVs. The main drawback of such descriptions is that they are basically based on existing systems, mixing mission capabilities (VTOL, long endurance), size (mini or micro), and other features such as altitude (high or medium altitude). Such a classification does not provide a comprehensive outlook of the various choices as applied to missions and vehicle configurations. Furthermore, it makes it very difficult to anticipate future UAS since it is rooted on the existing UAS market segmentation.

A more appropriate way to classify the different kinds of possible UAS would be a double-entry matrix to combine typical mission profiles and the major vehicle configurations.

Mission profiles may include:

1. Recognition missions (outdoor/indoor) requiring VTOL capabilities,
2. Surveillance missions (close range/long range) requiring long endurance capabilities,
3. Other specific missions such as delivering goods, monitoring special facilities ranging from wind turbines to nuclear plants, some tactical missions in the military domain requiring covertness (low acoustic and radar signature), and robust transmission.

In terms of mission profiles, it should be pointed out that most end-users have difficulty in actually defining their mission requirements without resorting to the prior definition of a configuration at the same time. Yet, it is very important in the UAS design process to properly distinguish between mission requirements and the payload/vehicle definition. For instance, in order to survey a remote area in the ocean, one may specify the size of the area, the distance between the launch zone and the area of interest, the maximum time allowed to get the required piece of information, additional practical constraints related to logistics, regulations, operating costs, etc. If the remote area is far from the launch zone, one has to select a long-range vehicle. If the remote area is not that far but permanent surveillance is required, the system may consist of either a single long-endurance vehicle or a fleet of smaller vehicles, each vehicle having a limited endurance but providing almost unlimited surveillance capability by taking turns between vehicles. The latter option may represent a much better trade-off between cost and mission performance than the former option. Indeed, a small vehicle, which is easier to deploy than a larger one, may also be equipped with a cheaper payload since it is devoted to a much smaller surveillance area.

Vehicle configurations are typically split into three main categories: fixed-wing, flapping-wing, and rotary-wing configurations. One should add a fourth category which combines any of the first three categories. The fourth category would mainly include convertible vehicles, either tilt-rotor, tilt-wing, or tilt-body platforms. It would also include most of the existing ornithopters, which usually combine flapping wings and a fixed-wing control surface, which plays the role of a tail or elevator. Other vehicle configurations, such as airships and paragliders, may be considered as a separate category, although they represent a smaller portion of current and future UAS.

### 1.1.1 Fixed-wing UAVs

Fixed-wing UAVs may typically range from micro-sized UAVs, also called micro air vehicles (MAVs), up to UAVs almost larger than any existing conventional aircraft. An example of a small fixed-wing MAV is given by the Wasp from AeroVironment, a 41-cm span electrically powered flying wing of 275 grams. Even smaller fixed-wing MAVs may be designed, such as the 10-cm span flexible-wing MAV developed by Professor Peter Ifju from the University of Florida in 2005 (see Figure 1.1) [26].

As opposed to extremely small-scale fixed-wing UAVs, the Boeing "SolarEagle" (Figure 1.2) is supposed to be a "satellite-drone" which can fly virtually 24/7 thanks to its solar cells covering the upper part of its wings and the very stringent constraints on the airframe fabrication to make it as light as possible. The 130m span fixed-wing

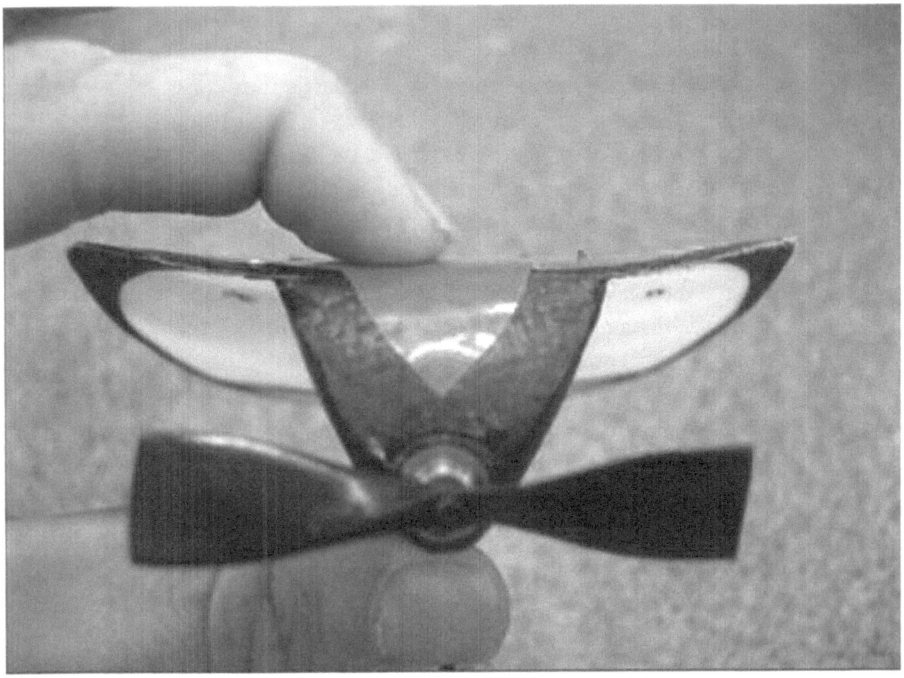

**Figure 1.1** A 10cm-span fixed-wing MAV (Courtesy of Michall Sytsma)

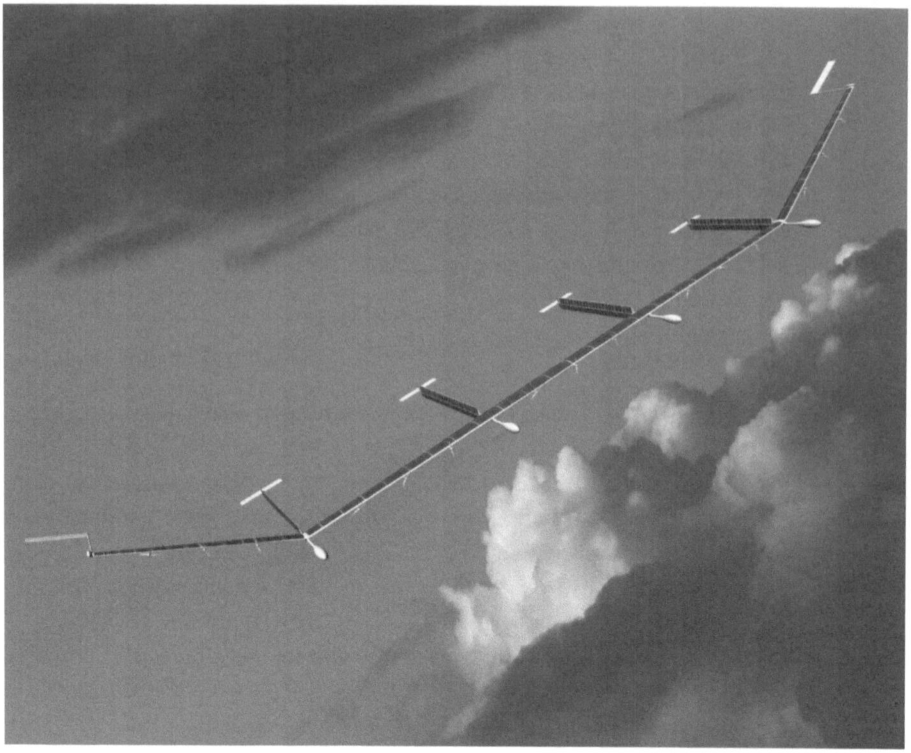

**Figure 1.2** Boeing SolarEagle concept in flight (photo credit: Boeing)

solar-powered UAV has to struggle against the famous square-cube law, which states that mass increases quicker than wing surface. As a consequence, solar-cell UAVs may be more appropriate at smaller sizes since a greater portion of the power needed to supply the motor may be obtained from the sun as compared to larger aircraft.

As an example, a 50cm-span fixed-wing covered with thin flexible solar cells, called Solar-Storm, has been designed and fabricated in order to extend the endurance of an existing version entirely powered with standard batteries. On sunny days, the Solar-Storm (see Figure 1.3) [7] was able to extract up to 45% of the total power needed to fly. From a practical point of view, it should be noted that such small solar-powered vehicles do not require a battery charger which needs to be plugged into some electrical source. While one mini-UAV is airborne, an identical model may recharge itself on the ground.

Although fixed-wing UAVs intrinsically suffer from difficulty to hover, they remain very good candidates for long-range or long-endurance surveillance missions as compared to rotary-wing UAVs. Even hand-launched medium-sized fixed-wing UAVs (less than 10kg) may stay airborne for up to 8 hours a day, which is usually more than enough for a typical surveillance mission. Although airplane design has become a well-known engineering technique for conventional airplanes, it is still poorly documented for mini- or micro-UAVs because of the low Reynolds effects degrading the aerodynamic and propulsive performance. It should be pointed out that careful design and fabrication techniques should be specifically applied and adapted to the field of mini-UAVs in

**Figure 1.3** A 50cm-span fixed-wing solar-powered UAV (Solar-Storm from Murat Bronz)

order to achieve good performances. Furthermore, long-endurance requirements rely on high values of the ratio $C_L^{3/2}/C_D$, where $C_L$ denotes the lift coefficient and $C_D$ denotes the drag coefficient. As a consequence, long-endurance fixed-wing UAVs correspond to fairly high values of $C_L$ and may lead to cruise conditions close to wing stall. Designing a long-endurance fixed-wing UAV should therefore include the requirement of minimum load factor, take-off, and landing performances. Specific wind tunnel tests and optimization process should then be conducted as illustrated in Figure 1.4, which shows the fixed-wing mini-UAV DT18 developed by Delair-Tech in the ISAE-SUPAERO low-speed wind tunnel [133].

Beyond some recent progress in the miniaturization of fuel cells, one interesting way to dramatically enhance mini-UAVs endurance is to extract energy from the atmosphere. Energy harvesting may be realized using thermals, such as in the case of gliders, or wind gradients. The best example of such a mechanism in nature is given by the albatross flight, which benefits from wind gradients created by the atmospheric boundary layer above the sea surface. That phenomenon, also known as dynamic soaring, is now better understood and can be mathematically simulated. Some authors have suggested that the principles of dynamic soaring could be exploited to create an unmanned aerial vehicle that could be used for surveillance, monitoring, and search and rescue missions over the ocean (see Figure 1.5) [1].

### 1.1.2 Flapping-wing UAVs

From the very beginning of aviation, some authors have argued that engineers should get inspiration from existing flying animals, either birds or insects. The idea underlying such a view being that animals have been gradually optimized over the centuries.

**Figure 1.4** A 1.8m-span fixed-wing long-endurance UAV (DT18 from Delair-Tech)

**Figure 1.5** Long-endurance mini-UAV concept inspired from the albatross flight (Courtesy Philip Richardson, 2012)

**Figure 1.6** AeroVironment with left-hand image courtesy of Lavvy Keller, Litiz Ba/Getty Images. Right-hand image courtesy of Coral von Zumwalt

Fascinating examples of small and large flying animals include various species ranging from the fairyfly, the smallest known flying insect at only 0.15mm long (0.0059in.), up to the famous pteranodon, a flying dinosaur with up to 7m wing span although its actual weight is still a matter of debate [450]. A special mention should be made about the hummingbird, which represents a source of inspiration for a nano air vehicle recently developed by AeroVironment (Figure 1.6) [254].

Understanding the aerodynamics of flapping wings is still, to a large extent, an open question due to the intrinsic flow field complexity and the unsteadiness involved. Over the past 40 years, it has been the focus of many research groups, involving various experimental and numerical techniques [382].

It has not yet been clearly established whether flapping flight is actually more efficient than rotary-wing systems, although it has been shown that existing birds and insects do not display a very efficient way to hover [294] as compared to conventional rotors, even at very low Reynolds numbers. Furthermore, recent studies have revealed that flapping flight might be much less efficient for some insects than previously thought [308]. The reason for such poor aerodynamic performance could be related to the fact that the begin and end positions of the flapping motion have very limited aerodynamic efficiency because the relative air speed becomes very low at those points. In contrast, a rotary wing can provide almost constant lift along its revolution.

Another limitation of flapping wings is their intrinsic technological complexity. In flight, flapping wings have to simultaneously provide lift and thrust, and also contribute to the control in pitch, roll, and yaw, which makes an autopilot extremely difficult to design. Finally, the fact that rotary wings have not emerged from the biological evolution of natural systems should not prevent engineers from considering rotary-wing UAVs as valuable candidates for VTOL missions. Indeed, neither wheels nor propellers or rotors, although highly efficient, have been produced by the natural process of evolution. Some authors point out that there are a few exceptions to this lack of imagination from nature, such as maple seeds or the bacteria flagellum. However, the maple seed is only a passive rotary-wing glider, which benefits from its increased lift-to-drag ratio to reach remote places when dropped by the parent tree. Yet, the SAMARAI monowing nano air vehicle [463] is inspired by the maple seed flight and powered by a micro jet located at the wing tip with a total mass of only 10 grams.

In the long run, flapping-wing UAVs might become very useful in specific recognition missions requiring covertness because of their ability to mimic birds or insects and to easily disappear from the human sight. Flapping-wing UAVs may also benefit from new materials such as electroactive polymers associated with different kinds of MEMS [176]. Furthermore, the development of recent microfabrication technologies has enabled complex articulated mechanisms at small scales that open the way towards insect-like resonant thorax [452].

### 1.1.3 Rotary-wing UAVs

Beside the limitation of fixed-wing UAVs and the complexity of flapping-wing UAVs, rotary-wing UAVs have attracted a good deal of attention from the scientific community. According to recent figures, among the 3000 to 4000 UAVs flying in France and currently registered by the French authorities, about 80% are rotorcraft, that is multi-rotors. A first reason for this attention is related to the fact that rotary-wing configurations provide the capability of hovering, which is essential to guarantee clear identification. Hovering is also a way to easily take off and land without a complex procedure, such as a prepared airfield or a specific landing device. Furthermore, multi-rotors are easy to fabricate and fairly straightforward to fly indoors. As quad-rotors were almost the only multi-rotors available 10 years ago, more recent multi-rotor aircraft now include hexa-rotors, octo-rotors, and various combinations of coaxial multi-rotors. Increasing the number of rotors is generally considered to be a good way to enhance security since if a motor fails, the other motors can immediately compensate. Usually, the different rotors are equally distributed in the azimuthal direction. Yet, some designers have chosen to adopt different configurations in order to allow for a better field of view ahead of the vehicle. Such an example is given by the ASTEC Falcon 8, which has been very popular over the past two years (Figure 1.7).

While helicopters consist of combining a main rotor and an anti-torque rotor, they also rely on a cyclic-pitch swash-plate to allow for flight control. Therefore, designing a helicopter requires a lot more experience and expertise than designing multi-rotors. When reducing the rotor diameter, Reynolds effects start degrading the propulsion efficiency. For a given overall maximum dimension, it is more efficient to use a single rotor rather than many rotors of a smaller diameter, which would cover the same disk area. However, in order to cancel the resulting torque, one can either resort to an anti-torque rotor as in conventional helicopters or add a counter-rotating rotor underneath. Such a coaxial rotor allows for altitude hold and control around the vertical axis. A recent example of a portable coaxial UAV has been given by the Sprite, a 1.2kg coaxial drone equipped with a two-axis gimballed camera (Figure 1.8). The rotorcraft can fly up to 10–12 minutes and can easily be backpacked after folding the blades.

**Figure 1.7** An eight-rotor mini-UAV developed by Ascending Technologies (photo credit: Lakeside Labs GmbH)

It should be mentioned that coaxial rotors suffer from a loss in propulsion efficiency due to the fact that the lower rotor is blown by the propeller slipstream produced by the upper rotor instead of being blown by a uniform freestream flow. As a consequence, the overall efficiency loss is generally considered to be around 30% with respect to a pair of isolated counter-rotating rotors. Nevertheless, the interaction penalty is compensated by the benefit of using a larger disk area.

Because of apparent rotating parts, rotorcraft may have difficulty coping with obstacles. Consequently, rotorcraft UAVs are often equipped with a crashproof outer structure, which protects the rotors. Such protections involve a significant weight penalty and may not perform very efficiently if they are not capable of absorbing energy during crashes. EPP foam associated with carbon rods or rubber bands may be used to offer various forms of bumpers or "mechanical fuses." As an example of such a "mechanical fuse," propellers may be mounted on the motor shaft using a simple rubber O-ring, which will help avoid the propeller and the shaft being damaged in case of a collision between the rotor blades and an obstacle. In terms of general UAV design, it is advisable to think in terms of lightness and flexibility rather than in terms of stiffness and weight. A soft and light aircraft will recover from a crash much better than a stiff and heavy vehicle.

One good design option that improves the robustness of rotorcraft consists of adding a duct around the rotor. Ducted rotors are more efficient than unducted rotors because they almost completely cancel out the blade tip losses. As a consequence, propulsion efficiency at a given disk area is increased. Furthermore, long ducts may contribute

**Figure 1.8** A 1.2kg coaxial rotor mini-UAV developed by Ascent AeroSystems (courtesy of Ascent AeroSystems)

an extra lift, mainly due to the design of a diverging nozzle. By combining a proper inlet and nozzle design with optimized rotor blades with almost no blade tip losses, one can obtain a shroud with additional lift and propulsion efficiency, which completely compensates for the weight penalty. The Br2C is an example of a vehicle that takes advantage of a protecting outer structure with full weight compensation due to the extra lift and propulsion efficiency provided by the shroud effect (Figure 1.9). As opposed to the Sprite coaxial UAV, the Br2C is controlled by a pair of flaps located within the rotor slipstream. A disadvantage of long-ducted rotorcraft is the difficulty to withstand strong cross winds due to the bluff body effect.

### 1.1.4 Convertible UAVs

The success of multi-rotors is somewhat plagued by their difficulty to perform adequately in windy outdoor conditions. High-speed forward flight is limited by various aerodynamic side effects, such as a poor rotor efficiency when the incoming freestream is dramatically tilted with respect to the rotation axis. While fixed-wing UAVs fail to properly achieve hover flight, rotorcraft are limited to low-speed forward flights and are usually much less efficient in fast-speed flight phases. Therefore, some UAV designs aim at combining the advantages of fixed-wing and those of rotary-wing configurations.

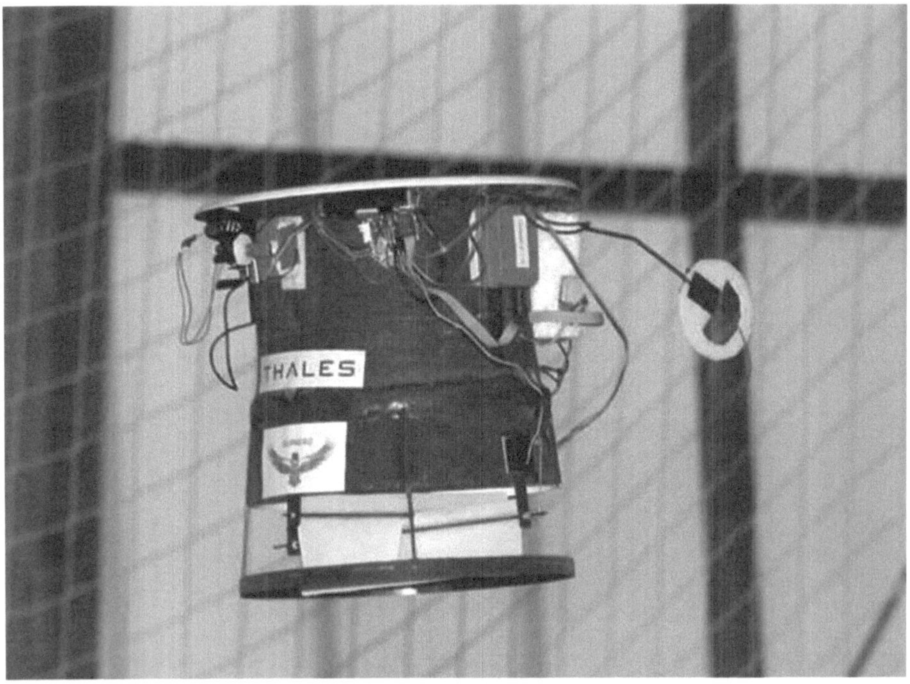


**Figure 1.9** A 500-gram ducted coaxial rotor micro-UAV developed by ISAE-SUPAERO (copyright Aéroland. Reproduced with permission from Sylviane & Christian Veyssiere)

These design combinations are called convertible UAVs. Combining the advantages of fixed-wing and rotary-wing configurations may be done following two different design strategies. One is to start from an airplane configuration and to modify it so as to achieve vertical flight. The other strategy consists of starting from a rotorcraft configuration and modifying it in order to achieve horizontal flight. As an example of the first strategy, one can mention the 20kg *Flexrotor* UAV developed by the Aerovel Corporation, USA. It basically consists of a regular 3-meter span airplane with an oversized propeller and two small anti-torque rotors located at either wing tip (Figure 1.10).

The *Flexrotor* belongs to the family of tilt-body UAVs or tail sitter UAVs, which means that they may take off and land vertically and perform cruise flight horizontally. When flying in airplane mode, the folding blades in the wing tip rotors allow limitation of the drag penalty. The aircraft is powered by a large propeller, which also plays the role of a main rotor when flying in helicopter mode. Therefore, the pitch may be varied so as to adjust itself with the flight phase: low pitch in hover and high pitch in cruise flight.

An example of the second strategy is given by the convertible biplane concept, which consists of combining a standard multi-rotor configuration with a set of lifting surfaces added underneath [215]. Again, the key point is that when flying horizontally, the whole body is tilted at an angle of 90 degrees. In airplane mode, the vehicle behaves as a biplane flying wing with a good aerodynamic efficiency. While the flying wing may not

**Figure 1.10** A 20-kg convertible UAV developed by Aerovel Corporation (copyright Aerovel. Reproduced with permission)

be statically stable because of the absence of a horizontal tail, the motors placed along the wings may be used to maintain control in pitch. Recently, a commercial version of such a biplane tilt-body UAV concept has been proposed by the French UAV company Parrot with the *Swing*, which makes use of an X-wing rather than a regular biplane wing (Figure 1.11).

Other examples of convertible configurations include tilt-rotor and tilt-wing UAVs. Tilt-rotor UAVs consist of mounting the rotors on a rotating axis, which allows the main body to remain horizontal while transitioning from cruise to hover. On tilt-wing UAVs, a portion of the wing located in the propeller slipstream is physically connected to the rotor so that both the rotor and that part of the wing rotate during transition flight. In both cases, such convertible aircraft require an additional tilting mechanism, which means additional weight and complexity. Also, with movable parts, including motors, the location of the overall center of gravity will vary during transition, which adds some complexity when developing the autopilot. An example of a tilt-wing UAV is given by the *AVIGLE* developed by RWTH Aachen University [328]. The *AVIGLE* UAV looks like a regular airplane except that its wing can be tilted vertically while the fuselage stays horizontal. It should be noticed that an additional vertical rotor has to be added near the tail in order to maintain control in pitch during transition (Figure 1.12).

A good example of an efficient tilt-rotor UAV is given by the *Skate* developed by Aurora Flight Sciences. The *Skate* is a rectangular flying wing powered by a pair of electric motors mounted on independent tilting mechanisms which allow for control in roll and pitch. Yaw control is supplied by differential throttle. No other moving parts,

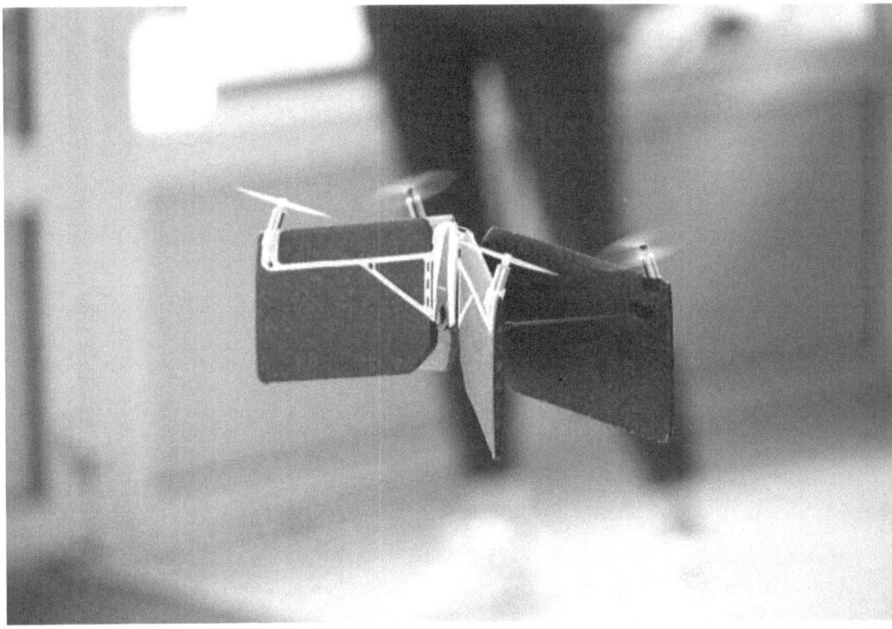

**Figure 1.11** A tilt-body 4-rotor mini-UAV developed by Parrot (photo credit: ISAE-SUPAERO)

**Figure 1.12** Left: a tilt-wing UAV developed by RWTH Aachen University (reproduced with permission from the Institute of Flight System Dynamics); right: a tilt-rotor UAV developed by Aurora Flight Sciences (reproduced with permission from UAVGlobal.com)

such as flaps or elevators, are therefore needed to control the vehicle. In cruise flight as well as in hover, the rotors are almost aligned with the wing chord since the vehicle is tilted vertically in hover. Only the transition flight requires the rotor axis to tilt with respect to the wing. Some tilt-rotor configurations, however, require the fuselage to remain horizontal as in the case of the Osprey V-22. The main advantage of such a tilt-rotor configuration is that the arrangement of the embedded system, antennas, and payload does not need to be modified to take into account a change in attitude when hovering. Yet, tilting the rotors usually implies a download force due to the propeller slipstream impinging on a portion of the wing.

**Figure 1.13** Left: a paraglider UAV developed by Flying Robots, Switzerland (photo credit: Flying Robots, Switzerland); right: a lighter-than-air UAV developed by Ride Engineering (photo credit: Ride Engineering, Russia)

As a final note, one should mention two additional configurations which play a limited role in the field of UAV designs. The first configuration is the paraglider. A paraglider consists of a fuselage usually equipped with a propeller in pusher position and a parachute which plays the role of a flying wing. An example of such a UAV is given by the *Swan* developed by Flying Robots, Switzerland (Figure 1.13). The main advantage of a paraglider UAV is its capability to fly very slowly and to be packed in a very compact way. Because they can be deployed and dropped from an airplane, paragliders are good candidates for search and rescue missions that cover large areas. The second configuration is the airship. An example of such a solution is given by the *Sphere-P2* project developed by Ride Engineering. While lighter-than-air UAVs are attractive because of their capability to stay airborne for a very long time, they suffer from two major drawbacks: (1) their sensitivity to winds, (2) the limited payload which can be lifted for a given airship volume. Some airships are tethered so as to be kept within a certain range for permanent surveillance of an area. In the *Sphere-P2* project, a coaxial rotor has been designed to provide altitude hold, while the horizontal control is obtained from a movable center of gravity.

### 1.1.5 Hybrid UAVs

Special attention should be paid nowadays to a new category of UAVs, which has recently emerged for very practical reasons. When flying in the vicinity of the ground, either in a forest or in an urban environment, a UAV which has to carry out a recognition mission cannot avoid hitting unpredictable obstacles of any kind: trees, electric wires, antennas, chimneys, roofs, etc. Also, some recognition missions may include building intrusion and the vehicle might be required to enter very narrow corridors or tunnels. In such mission profiles, the obstacles cannot be avoided. Using a conventional ground vehicle may still be very limited because jumping over obstacles is always difficult and risky. Also, in many cases, landing the UAV in the middle of the mission may be desirable. For instance, a police operation may suddenly require a totally silent UAV, which implies switching off the motors. Then, the UAV has to land or to cling to a surface but still be able to take off and continue the flight without human intervention. Hybrid UAVs are vehicles which aim at combining the capabilities of aerial and ground vehicles.

The main idea of hybrid UAVs is that obstacles are no longer considered as problems but as *opportunities* to add some new features. In terms of design, adding an outer crash-proof structure, such as a set of carbon rods, represents a weight penalty but may also bring a new capability on-board, such as rolling on the ground or hanging from the ceiling. A first example of such a hybrid vehicle is given by the *MAVion* "Roll & Fly," which is a rectangular flying wing powered by two counter-rotating propellers in tractor position equipped with a pair of free wheels located on either side of the wing (Figure 1.14). Far from obstacles, the vehicle can fly vertically, thanks to its two elevators located along the wing trailing edge, that is in the propeller slipstream. The MAVion can also fly horizontally as a conventional bimotor flying wing. In both situations, control in pitch and roll is provided by the elevators, which remain efficient over the whole flight domain. Control in yaw is provided by differential throttle. When hitting a flat surface, such as floor, ceiling, or walls, the wheels not only protect the propellers but also allow them to roll at a constant distance from the wall. Differential throttle can help the vehicle to "drive" when rolling on the ground.

Following the same idea of combining ground and aerial vehicles, two additional interesting concepts should be mentioned here. They are both based on the idea that the

**Figure 1.14** Left: the *MAVion* "Roll & Fly" rolling along a vertical wall (photo credit: ISAE-SUPAERO, France); right: the micro-UAV *Vision-Air* "Stick & Fly" clinging to a window, motors switched off (photo credit: ISAE-SUPAERO, France)

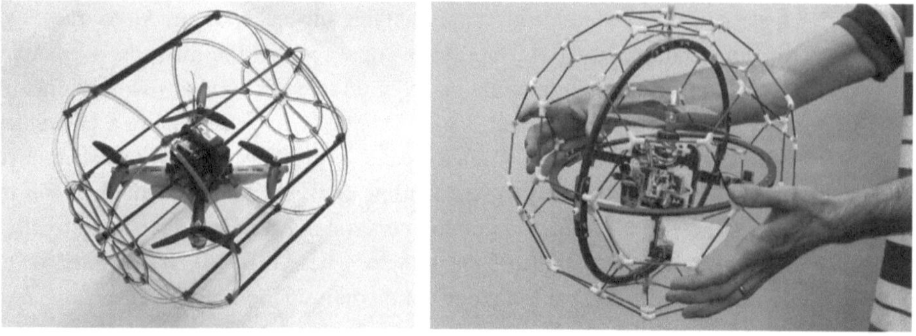

**Figure 1.15** Left: the *HyTAQ* quadrotor equipped with a rolling cage (reproduced with permission from Matthew Spenko, Illinois Institute of Technology); right: the mini-UAV *GimBall* with a double-axis rotating sphere (copyright Alain Herzog. Reproduced with permission)

outer crashproof structure may freely rotate around the flying vehicle. The first example is given by the HyTAQ project (Hybrid Terrestrial and Aerial Quadrotor) developed by the Illinois Institute of Technology (Figure 1.15, left). On the HyTAQ, a rolling cage has been added to an original quadrotor in order to make terrestrial locomotion possible. During terrestrial locomotion, the vehicle consumes much less energy compared to the aerial mode and can easily cope with an obstacle by simply flying over it. The second example is given by the GimBall developed by the Ecole Polytechnique Fédérale de Lausanne, Switzerland (Figure 1.15, right). In the GimBall, the aerial vehicle is fitted inside a sphere, which can freely rotate around a vertical axis and around a horizontal axis. As a consequence, the vehicle can cross very complex environments, such as a forest or a network of wires, without being stuck.

As a conclusion to the first section, it appears that cutting-edge technology has thoroughly reshaped the standard classification of UAVs so that they cannot be reduced to fixed-wing, rotary-wing, or flapping-wing UAVs. A general overview of UAV concepts requires inclusion of novel configurations, such as convertible and hybrid UAVs. The use of convertible and hybrid UAVs is believed to be of the utmost importance for the purpose of networking UAVs since they open the way to multi-tasking missions, which require cooperation and dynamic tasks allocation.

## 1.2 UAV Swarming and Miniaturization

There are many good practical reasons to develop unmanned aerial systems (UAS), one of which is purely economic. If one can achieve a given surveillance or recognition mission for less money, it will have a competitive advantage over conventional systems such as light aircraft. This also happens to be the case within UAS between larger UAVs and smaller ones in which the small size of each individual vehicle is compensated by a large number of such vehicles operating as a team.

Although networking UAVs can virtually be done using vehicles of any size, it only makes sense for mini or micro-sized UAVs. Indeed, only mini-UAVs can be launched in a short time since they require a very limited logistic footprint and few crew members. As a consequence, launching dozens of UAVs with the view of achieving a coordinated flight would simply not be possible if each UAV required more than a minute to be launched. Otherwise the first airborne UAV will have terminated its mission while the last one will not have taken off. Only small UAVs may be eligible for the networking of a large number of vehicles.

Operating a large UAV such as the *GlobalHawk* requires a large number of human operators, and conducting a multi-vehicle surveillance mission is only possible with mini- or micro-UAVs. Therefore, UAV swarming is basically a matter of the number of operators while increasing the number of vehicles. Instead of requiring many operators for controlling multiple UAVs, the idea of UAV networks would be to have a flock of vehicles controlled by a single operator. Having a fleet of UAVs controlled by a single operator not only requires a high level of autonomy for each flying vehicle, it also requires new control and navigation algorithms to efficiently drive the UAV network. These new algorithms will be further detailed in the following chapters. For the moment, it is important to look at the practical issue of launching dozens of flying vehicles in a row and manipulating a flock of UAVs heavily reliant on the capability to miniaturize each vehicle up to a point where crashing one vehicle does not represent a major technical or economical concern, and will still allow the mission to be fully completed. As a consequence, it is important to carefully examine to what extent UAVs may be miniaturized before going any further.

## 1.3 UAV Miniaturization: Challenges and Opportunities

If UAV networks consistently rely on the capability to miniaturize aerial vehicles, miniaturization itself implies several opportunities as well as new design challenges.

In terms of opportunities, miniaturizing UAVs offers a small visual and electromagnetic footprint. For some applications related to defense and security, smaller vehicles may therefore represent a great advantage in terms of covertness. Smaller vehicles also tend to produce less noise and to become barely noticeable if properly adapted to their environment by the well-known technique of camouflage. Another advantage of miniaturizing UAVs is that they can slot into highly confined environments, such as tunnels, collapsed buildings, ventilation pipes, pipelines, sewer pipes. In such tight spaces, ground vehicles are more likely to become stuck than flying vehicles. Finally, smaller vehicles usually means cheaper vehicles. Losing a 100-dollar flying robot, while hundreds are pursuing the recognition mission, is not a big issue, while losing a Predator-sized UAV is more likely to be of critical importance for the operator. Finally, combining a large number of vehicles flying in cooperation may, in some applications, represent a very efficient way to achieve complex and multi-tasking missions, while a single vehicle would require considerably more effort.

Although very desirable, miniaturizing UAVs faces major design challenges and technical bottlenecks, such as gust sensitivity, energy source, aerodynamic efficiency to name a few.

### 1.3.1 Gust Sensitivity

Designing mini-UAVs cannot be reduced merely to scaling down conventional aircraft configurations. There are several reasons for this. One is related to the gust sensitivity perceived by the vehicle. In order to illustrate that effect, let us consider a conventional fixed-wing aircraft. In level flight, the lift equation equates the vehicle weight and the lift force as

$$mg = \frac{1}{2}\rho S V^2 C_L \tag{1.1}$$

where the mass $m$ and the wing surface $S$ vary as $L^3$ and $L^2$ respectively according to the famous square-cube law, $L$ being the overall vehicle size. Provided that $C_L$ remains almost of the order of unity, the flight speed necessarily varies as

$$V \simeq \sqrt{L} \tag{1.2}$$

which shows that the flight speed needs to decrease when scaling down the vehicle. Now, if we consider the equation of motion along the pitching moment axis, we may write

$$J\ddot{\theta} = \frac{1}{2}\rho S L V^2 C_m \simeq L^4 \tag{1.3}$$

where $J$ represents the moment of inertia that is proportional to $L^5$. As a consequence, equation (1.3) reduces to

$$\ddot{\theta} \simeq L^{-1} \tag{1.4}$$

which simply means that the roll acceleration will tend to increase when down-sizing the vehicle. As a result, a smaller vehicle will be much more sensitive to wind gusts than a larger one. That effect comes in addition to the fact that by flying more slowly as a consequence of (1.2), the smaller vehicles will also encounter atmospheric perturbations where typical speeds become comparable to the vehicle speed. In other words, flying mini-UAVs goes back to flying a regular airplane through a storm.

### 1.3.2 Energy Density

Although the energy density of gasoline remains high compared to the best available batteries, thermal combustion engines fail to remain efficient when their size is drastically reduced. The reason for this phenomenon is that the heat produced in the combustion chamber is proportional to $L^3$ while the heat flux dissipated through its walls only reduces as $L^2$. Consequently, miniaturizing thermal engines will inevitably lead to poor thermodynamic efficiencies since most of the heat produced within the combustion chamber will rapidly evaporate through the walls. Increasing the rotation speed to compensate for heat losses will not bring a viable solution either because of the limitations on chamber residence time. Furthermore, poor pressure tightness and friction increase are additional problems, which also ruin the attractiveness of thermal combustion engines when reduced in size [381]. Mini-UAV designers are therefore left to the sole choice of electrically powered vehicles, which suffer from limited specific energy with a maximum value of about 200Wh/kg for a high quality lithium-polymer battery. In spite of rapid progress in the field of fuel cells and other new chemical alternatives to LiPo cells, energy density remains a strong limitation for further miniaturizing UAVs at the present stage.

### 1.3.3 Aerodynamic Efficiency

The Reynolds effect, which drives the importance of viscous effects in the flowfield surrounding a flying vehicle, varies as

$$Re = \frac{VL}{\nu} \simeq L^{3/2} \tag{1.5}$$

showing that the importance of viscosity dramatically increases when the vehicle size is reduced. When the Reynolds number is lower, laminar separation is likely to occur, which results in poor maximum lift capabilities and a high drag level even at low angles of attack. Since the aerodynamic performances of wing airfoils as well as the efficiency of the propeller blades drop miserably when the Reynolds number decreases, it follows that the aerodynamic efficiency of small aircraft represents a crucial design challenge which requires new aerodynamic ways of producing lift with limited drag.

### 1.3.4 Other Design Challenges

Miniaturizing UAVs is not only a difficulty for physical reasons related to aerodynamics, propulsion, and flight control, it also represents a technical challenge for other practical reasons – one of which is associated with electromagnetic interference. Indeed, when all electronic components are packed within a tight space, the electromagnetic field created by the motor tends to jam signals within the magnetometer or the GPS receiver. Also, the experience of miniaturizing UAVs has revealed that the weight of electric wires represents a significant portion of the overall mass for small UAVs. Integration is therefore needed in order to reduce the weight due to electrical connections between the various components.

## 1.4 UAV Networks and Their Advantages

A network of UAVs can be viewed as a flying wireless network in which each UAV serves as a node transmitting its own information to other nodes or receiving the information intended for it or relaying information meant for others in the network. The network could be ad hoc without any supporting infrastructure or it could be supported by ground-based and/or satellite-based communication infrastructures. The topology

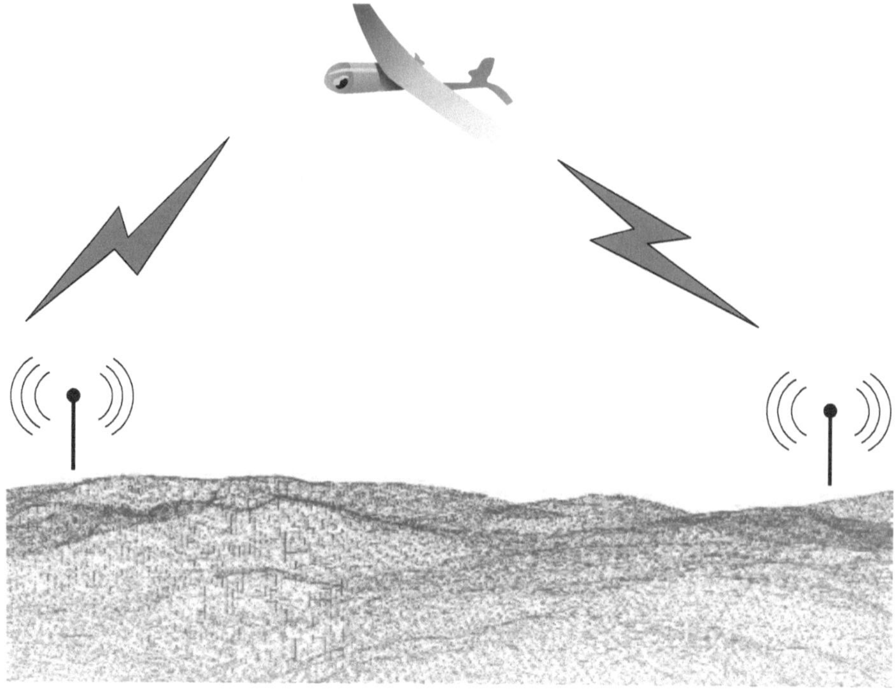

Figure 1.16 A UAV can serve as a relay node between a transmitter–receiver pair, extending the communication range between them

or configuration of the UAV network may take any form, including a mesh, star, or even a straight line, and it primarily depends on the application and use case scenario. First, let us understand why we need a UAV network. A single UAV just by being at a higher altitude offers several benefits, foremost among these being a clear line of sight between the transmitter on the ground (or in the air) and the receiver in the air (or on the ground). Indeed, this is the reason for placing antennas intended for cellular or broadcast communications on a tower, at a typical altitude of 50 to 200ft. A single UAV node could serve as a relay between a transmitter–receiver pair located on the ground extending the range of connectivity between them as shown in Figure 1.16. A UAV enabled communication infrastructure provides a better alternative to ground-based infrastructure, especially when a clear line of sight between a transmitter and receiver is not available due to uneven terrain or cluttered environment.

Figure 1.17 shows how two UAVs can work together relaying information from one radio to another on the ground. Multiple UAVs can serve as a chain of relay nodes, extending the range of communication. Figure 1.18 shows a group of UAVs forming an ad hoc network, as in a mobile ad hoc network or an aerial MANET. An aerial MANET is a multi-hop networking solution for delivering information over long distances. Each node in the aerial MANET acts as a terminal as well as a relay node or router carrying information within the network. In an ad hoc configuration, there is no need for any other infrastructure such as satellites or centralized servers to support the UAV swarm.

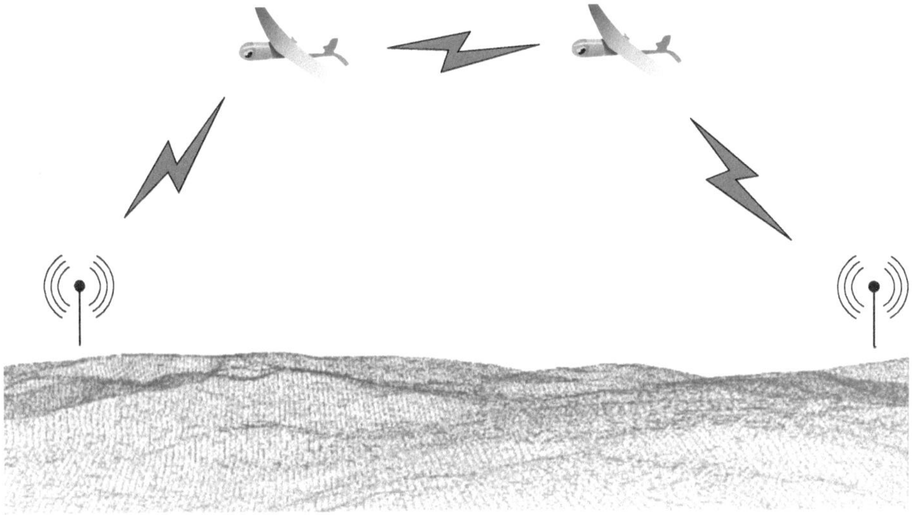

**Figure 1.17** Two UAVs working together as a simple relay network extending the range of coverage on the ground

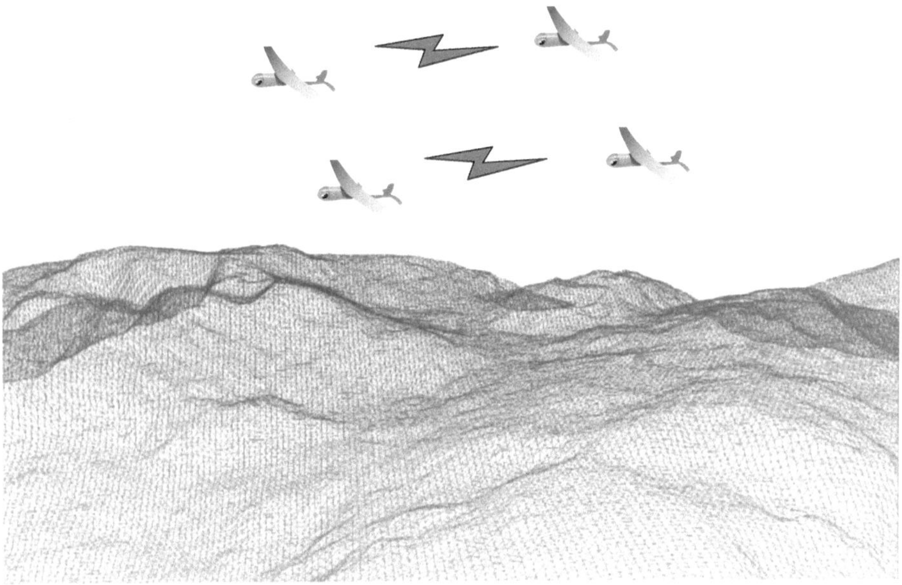

**Figure 1.18** Multiple UAVs forming an aerial mobile ad hoc network

However, in real-world applications, ground- and satellite-based services will improve the reliability and robustness of the UAV network. For example, a Global Positioning System (GPS) sensor helps to estimate and exchange geolocation information among the UAVs. A UAV network with ground- and satellite-based communication infrastructure is commonly known as an airborne network.

### 1.4.1 Unique Features of Airborne Networks

Since aerial nodes move much faster than nodes on the ground, the topology of an aerial network will be very dynamic. Figure 1.19 shows an example of an airborne network. The extremely changing dynamics require specific protocols for routing and secure information exchange. In addition, sense-and-avoid and situational awareness strategies are necessary to make sure that the nodes maintain a minimum safe distance during their flight.

Airborne networks are unique and significantly different from vehicular networks involving only ground vehicles in many perspectives. Classical mobility models and security strategies designed for MANETs and ground vehicular networks are not suitable for airborne networks. Mobility models that take into account the unique characteristics, such as smooth turns, and high-level information assurance, authentication, and integrity verification strategies that can meet the minimum latency requirements, are needed for airborne networks.

An airborne network is a cyber physical system (CPS) in which there is an intense interaction between its physical and cyber components. While computation, communication, and networking elements form the cyber components of the system, flight paths, maneuver geometries, and multi-mode resources, including ground-based nodes and control stations, form the physical components of the CPS. The fundamental challenge for airborne networks is to bring the synergistic interactivity between its cyber and physical components. This synergy, if successfully explored and exploited, will immensely benefit the next generation of air transportation systems; for example, predicting the trajectories of airborne vehicles within the neighborhood (say, 1000 square mile region), forming a trusted network with friendly nodes, reconfiguring the network as its topology changes, and sharing audio and video streaming data securely over the air among the pilots will significantly improve the situational awareness of an airborne vehicle and enhance the safety capabilities of the air transportation system. However, fundamental design principles, which are needed to explore this synergy between the cyber and physical dimensions, are yet to be developed. There is a great need for generating experimental datasets, which will lead to such design principles.

### 1.4.2 Mobility Models for UAV Networks

Mobility models provide a framework for connectivity studies, network performance evaluation, and eventually the design of reliable routing protocols. In particular, mobility models capture the random movement pattern of each network agent, based on which rich information related to the varying network structures can be estimated, such as node distribution and the statistics of link and path lifetime. In order to provide accurate predictions to facilitate airborne networking, it is crucial to develop realistic and tractable mobility models for Airborne Networks. Some mobility models have been studied extensively in the literature, such as random direction (RD), and random waypoint (RWP). The RWP model assumes that an agent chooses a random destination (waypoint) and traveling speed; upon arrival, it pauses before traveling to the next

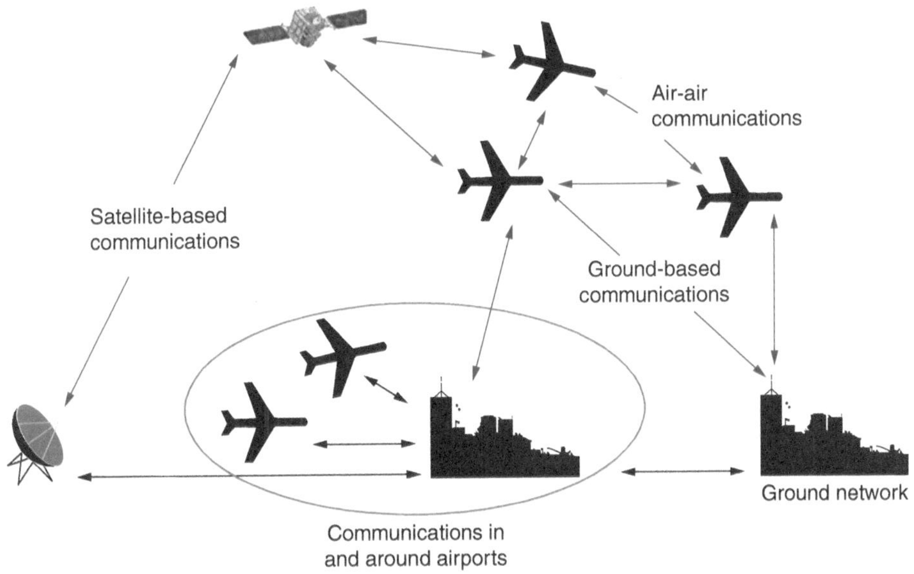


**Figure 1.19** A real-world airborne network consisting of unmanned aerial systems as well as the satellite- and ground-based communication infrastructure

destination. The extended version of the RD model assumes that an agent chooses a speed and direction randomly after a randomly selected traveling time. The stochastic properties of these common models, such as their spatial distributions, can be found in the literature. The widely used RWP and RD models are well suited to describe the random activity of mobile users in MANETs. However, they lack the capability to describe features that are unique to airborne vehicles. For example, it is easy for mobile users and vehicles on the ground to slow down, make sharp turns, and travel in the opposite direction (see an enhanced random mobility model that captures such movements). Aerial nodes are not capable of making such sharp turns or instantly reversing the direction of travel. Hence, there is a need to develop realistic models that capture features that are unique to airborne networks.

### 1.4.3 State of the art in UAV Networks

UAV networking and communications is an emerging field of research. Although there has been immense literature on the applications of small UAVs, the bulk of this research has been in theory and simulations only, with a limited number of real implementations in academic and research institutions. Below, we discuss some of the recent implementations of UAV networks and their outcomes.

#### AUGNet (University of Colorado, Denver, 2004)

AUGNet is an implementation of ad hoc UAV-ground networks consisting of ad hoc nodes on the ground and ad hoc nodes mounted on small UAVs [71]. This test-bed illustrates two use cases of AUGNet. The first use case is a relay scenario in which a UAV, with a better view of the ground nodes, enhances connectivity of an ad hoc network of ground nodes. In the second use case, ad hoc networking between UAVs increases the operational range and improves communications among the UAVs. Experimental results demonstrate that a UAV supported network generates shorter routes that have better throughput and improved connectivity over nodes at the edge of network coverage. More recent experiments on this test-bed have provided detailed data on network throughput, delay, range, and connectivity under different operating regimes. Such experiments are needed to understand the performance limits of UAV networks.

**UAV Networking with Commercial Off-the-shelf Components (Air Force Research Laboratory and Harvard University, 2006)**

Air Force Research Laboratory (AFRL) and Harvard University jointly pursued UAV networking with commercial off-the-shelf (COTS) communication equipment [198]. The availability of low-cost and yet highly capable COTS-based communications equipment and UAV platforms allowed the team to conduct two field experiments for UAV-based networks using communication equipment that supports 802.11 (at 2.4GHz and 5GHz) and 900MHz technology, respectively. The experiments were carried out to compare bandwidth and communication range and networking capabilities. The experimental data that were collected through these field experiments were more accurate and realistic than any simulated data available at that time.

**Robust Airborne Networking Extension (Boeing and the Naval Research Laboratory, 2009)**

The Robust Airborne Networking Extension (RANGE) research project was carried out by Boeing Research and Technology and the Naval Research Laboratory (NRL) with support from the Office of Naval Research (ONR). The team developed, tested, evaluated, and demonstrated protocols and techniques for resilient mobile inter-networking of UAVs and ground stations to extend surveillance range and battle-space connectivity [124]. The field test included an 802.11 ground–UAV network of 11 ground stations, a mobile vehicle, and two fixed-wing UAVs. The field tests demonstrated hybrid air/surface networking scenarios and mobile ad hoc networking (MANET) capabilities.

**UAVNET (University of Bern, 2012)**

UAVNET is a prototype implementation of a mobile wireless mesh network using UAVs [303]. Each UAV carries a lightweight wireless mesh node, which is directly connected to the flight electronics of the UAV using a serial interface. The flying wireless mesh nodes are interconnected, and communicate with each other over the IEEE 802.11s protocol. Every wireless mesh node works as an Access Point (AP), providing access for regular IEEE 802.11g wireless devices, such as notebooks, smartphones, and tablets. The prototype implementation is capable of autonomously interconnecting two communication peers by setting up an airborne relay, consisting of one or several flying wireless mesh nodes. Experimental results demonstrate that a multi-hop UAV relay network achieves significantly higher throughput compared to terrestrial-based relay network.

### Mobility Model for UAV Networks (2014)

Mobility models abstract the movement patterns of mobile nodes in MANETs. They are typically used for estimating the performance of network protocols in different application scenarios. Realistic mobility models are necessary to create a realistic simulation environment. A paparazzi mobility model for UAVs has been suggested in [65]. They developed the paparazzi mobility model, which is a stochastic model that imitates paparazzi UAV behavior based on the state machine in which the five states represent the five possible UAV movements: stay-at, waypoint, eight, scan, and oval. The mobility model is compared with the well-known random waypoint mobility model. In a more recent study, a smooth-turn mobility model has been suggested [438]. This model captures the tendency of airborne vehicles to make straight trajectories and smooth turns with large radii.

### SkyScanner (2015)

SkyScanner is a research project targeting the deployment of a fleet of fixed-wing mini-drones for studying the atmosphere [6]. This is a collaborative project with five partners, including the Laboratory for Analysis and Architecture of Systems (LAAS) at the Centre National de la Recherche Scientifique (CNRS), the Groupe d'étude de l'Atmosphère Météorologique (GAME) at the Centre National de Recherches Météorologiques (CNRM), the Department of Aerodynamics, Energetics and Propulsion (DAEP) at the Institut Supérieur de l'Aéronautique et de l'Espace (ISAE), Systems Control and Flight Dynamics at ONERA, and the UAV Laboratory at Ecole Nationale de l'Aviation Civile (ENAC). The scope of the SkyScanner project includes atmosphere sciences, aerodynamics of mini-drones, energy harvesting, and distributed fleet control. The project relies on strong cooperation between UAVs that collectively build a 3D map of atmospheric parameters and decide which areas to map further.

## 1.5 Summary

This chapter discussed UAVs in terms of their types and mission capabilities. It outlined UAV swarming, UAV miniaturization, and the opportunities and design challenges in UAV miniaturization. It outlined the advantages of UAV networks and presented several UAV networking projects that were demonstrated over the past few years. Mobility models for UAV networks were briefly discussed.

# 2 Air-to-Ground and Air-to-Air Data Link Communication

Bertold Van den Bergh and Sofie Pollin

Wireless communication has always been essential, even in manned aviation, to ensure ground control of airplanes and airspace. In addition to wireless communication, radars operating on various radio bands have always been an essential part of any flight control system as they can give accurate information about aircraft locations. Given the tremendous improvements that have been realized in wireless communication over recent decades, first from analog to digital systems and then the digital scaling enabled by Moore's law, we see today a huge range of different technologies in use and sharing the same aerial spectrum. In this chapter, we will start by giving the background on wireless communication used in manned aviation from the early days. This includes radar as well as the early digital communication. We will then discuss two novel technologies currently being proposed for L-band Digital Aeronautical Communication. Finally, we will conclude with some fundamental insights, learned from experience with advanced terrestrial mobile broadband communication extrapolated to the aerial case relevant for aerial communication on unmanned and small UAVs. As technology scales, more and more processing power is available on ever smaller chips, yet to ensure good link budget and long distance communication, we also have to rely on fundamental properties of the electromagnetic waves and the wireless channel. We analyze the use of multiple antenna techniques, the breakthrough technology for improving communication range and rate in all terrestrial mobile broadband technologies of the last decade, and comment on their usability for unmanned small aerial vehicles.

## 2.1 Air-to-Ground Communication for Manned Aviation

Wireless technologies are commonly used in manned aviation. This section gives an overview of both radar and communication solutions for manned aviation. Systems for unmanned aerial vehicles can be seen as a revolution of these systems, combining legacy manned communication technology as a basis, with disruptive innovations from the mobile broadband communication scene. Nevertheless, for this book, we believe a comprehensive overview is useful for the reader.

## 2.1.1 Radar for Ground-based Aircraft Identification

In controlled airspace it is very important to accurately know the location of all participating traffic. This is done by two types of radar, that is the primary and secondary surveillance radars. We will first introduce both radar systems, and will then zoom in on the antenna typically used for such systems.

### Primary Surveillance Radar

This system is a traditional radar where a large directional antenna is used to transmit a signal. The device listens for received echoes that are delayed as a function of the time of flight, or distance to the aircraft. This approach avoids the installation of equipment in the aircraft, which is a huge advantage when considering size and weight-constrained systems. However, the disadvantage is that any object that causes a strong enough reflection will be detected. This results in clutter caused by birds, clouds, and even landscape features. In addition to this, a main disadvantage of a radar is that it merely detects, and cannot identify, objects.

Another limitation is the fact that almost all Primary Surveillance Radar (PSR) systems are not able to determine the altitude of the aircraft. This is caused by the fact that the system can only pan the antenna in the horizontal plane. The vertical beam pattern is fixed by the physical shape of the antenna. Indeed, the horizontal beam is very narrow, resulting in a high bearing resolution, but the vertical pattern is broad, in order to illuminate and detect as many objects in the air as possible.

The most likely way to create a primary radar that can measure altitude would be using a phased-array with beam steering, which is more expensive than mechanical steering as one now needs an analog front-end for each antenna element in the phased-array, and front-end designs with good phase coherence. Alternatively, one could propose to mount the antenna on a pan-tilt positioning device and scan the entire 3D space this way. Unfortunately, the radar images are very time critical, requiring an update rate of 5–15s. The antenna has a very high mass so it cannot be moved physically at high velocity and acceleration. Some systems exist, mainly for defense purposes, employing multiple feeds or a single feed with a moving secondary reflector. Weather radars often use an antenna with two axis positioning, but they only have to deliver an image every few minutes. Figure 2.1 shows the radar station in Bertem, Belgium, operated by Belgocontrol. The large antenna is the Primary Surveillance Radar. The smaller antenna on top is used by the Secondary Surveillance Radar.

### Secondary Surveillance Radar

The main goals of the Secondary Surveillance Radar (SSR) are determining the altitude and identifying the aircraft. As it relies on a transponder that has to be installed in the aircraft, it cannot be used to detect and track non-cooperative targets.

The transponder receives interrogation requests sent by the ground station, which are answered with a reply depending on the transponder mode. The commonly used ones for civil aircraft are detailed in Table 2.1 below.

Note that modes A and C are almost always used together, so both the altitude and ID code are sent back. A problem with the basic modes A and C is that they lack

**Table 2.1** Secondary radar interrogation modes

| Mode | Description |
|---|---|
| A | Transmits a 4-digit aircraft ID code. This ID is not globally unique, but is instead assigned by the traffic controller. |
| C | Transmits the aircraft pressure altitude. |
| S | Transmits many types of information. Supports selective interrogation. Globally unique ID. |

**Figure 2.1** Primary and secondary radar in Bertem, Belgium

support for selective interrogation. If an area is covered by many radars, the transponders will transmit a significant amount of replies, leading to increased congestion and interference. Furthermore, if two aircraft are close together, their interrogation replies can collide, reducing radar system effectiveness. To resolve these issues, mode S was introduced. All transponders have a globally unique 24-bit ID code, which can be used for specific interrogation as well as data communication between the ground and air stations. Interrogations use 1030MHz, while replies are sent on 1090MHz.

### Cosecant Squared Antenna Pattern

Typical air surveillance radar antennas have a cosecant squared pattern. In this section, we will shortly explain what this specific pattern is, and why it is used. To do this we will first derive a formula calculating the power received by a radar.

The first step is determining the power density $S_{\mathrm{i}}$ produced by the radar transmitter at the target. Suppose our radar system has a transmitter power of $P_{\mathrm{tx}}$, which is spread over the surface of a sphere with radius $R$, then $S_i$ is given by

$$S_{\mathrm{i}} = \frac{P_{\mathrm{tx}}}{4 \cdot \pi R^2}.$$

Of course, any practical radar system will use an antenna with gain, which means that the power is predominantly sent in one direction. The parameter $G$ defines the height of the power density when compared to an isotropic radiator. The power density at the target ($S$), taking into account the directionality, is

$$S = G \cdot S_{\mathrm{i}}.$$

Not every object will reflect the radar signal similarly. A large passenger plane may be a very good reflector, while a wooden bi-plane might not be. As such, we define the radar cross section $\sigma$ as the area that intercepts the emitted radar radiation and isotropically scatters it back to the receiver. The power $P_{\mathrm{s}}$ scattered by the object is

$$P_{\mathrm{s}} = S \cdot \sigma,$$

which will again be exposed to losses when calculating the power density $S_{\mathrm{r}}$ at the radar receiver. Note that we assume a monostatic radar, which means that the distance from the transmitter to the target and the distance from the target to the receiver are equal. Since the radar cross section is defined for an isotropic radiator, the gain is equal to 1, which gives a received power density of

$$S_{\mathrm{r}} = \frac{P_{\mathrm{s}}}{4 \cdot \pi R^2}$$
$$S_{\mathrm{r}} = \frac{S \cdot \sigma}{4 \cdot \pi R^2}$$
$$S_{\mathrm{r}} = \frac{G \cdot P_{\mathrm{tx}} \cdot \sigma}{(4 \cdot \pi)^2 R^4}.$$

The actual power $P_{\mathrm{r}}$ at the antenna port depends on the antenna aperture $A$

$$P_{\mathrm{r}} = S_{\mathrm{r}} \cdot A. \tag{2.1}$$

The gain of an antenna is proportional to its aperture

$$G = \frac{4 \cdot \pi \cdot A}{\lambda^2}$$
$$A = \frac{G \cdot \lambda^2}{4 \cdot \pi}.$$

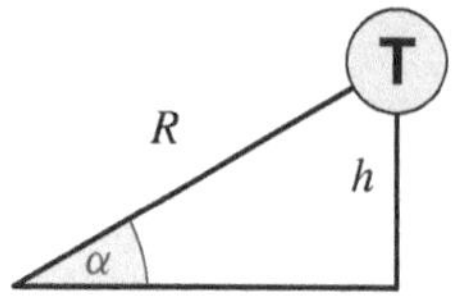


**Figure 2.2** The elevation angle $\alpha$ of a target at range R and altitude h

From Equation (2.1) and assuming the same antenna is used for transmit and receive gives

$$P_r = S_r \cdot \frac{G \cdot \lambda^2}{4 \cdot \pi}$$

$$P_r = \frac{G \cdot P_{tx} \cdot \sigma}{(4 \cdot \pi)^2 R^4} \cdot \frac{G \cdot \lambda^2}{4 \cdot \pi}.$$

The final received power can be computed as

$$P_r = \frac{G^2 \cdot P_{tx} \cdot \lambda^2 \cdot \sigma}{(4 \cdot \pi)^3 R^4}. \tag{2.2}$$

In order to provide uniform sector coverage to an air surveillance radar, we want an antenna that delivers a $P_r$ that is independent of the range for an airplane traveling at a certain altitude through the beam. As seen in (2.2), for a constant $P_r$

$$G^2 \sim R^4$$

$$G \sim R^2.$$

As seen in Figure 2.2, the range $R$ is dependent on the elevation angle $\alpha$ and the altitude $h$ of the plane

$$\sin \alpha = \frac{h}{R},$$

thus

$$R = \frac{h}{\sin \alpha} = h \cdot \csc \alpha.$$

Knowing that $G$ is proportional to $R^2$ and assuming constant altitude, we get

$$G \sim (h \cdot \csc \alpha)^2$$

$$G \sim (\csc \alpha)^2.$$

We hence determine that the ideal antenna has a gain that is proportional to the cosecant square of the elevation angle.

### 2.1.2 Distance and Direction Measurements Beyond Radar

Distance measuring equipment (DME) allows measurement of the slant range from the aircraft to a transponder station on the ground. While the purpose is similar to that of the radar, it is implemented by means of a two-way packet exchange. The interrogating aircraft will send a pulse coded message from the aircraft to the ground station, which is answered with a reply message by the ground station after a fixed delay. The measurement unit in the aircraft can then determine the range to the DME beacon by measuring the round-trip time. Note that the measured distance is slant range. Indeed, if one is 1km above the DME, the distance readout will indicate 1km.

The DME system uses 126 different channels with channel spacing of 1MHz. The downlink frequency range is 1025MHz to 1150MHz. The uplink band is 962MHz to 1213MHz. The typical peak transmitter output power is around 1kW, which is very high compared to mobile broadband systems such as Wi-Fi and will make coexistence challenging as will be discussed later.

DME stations are often co-located with VHF Omnidirectional Range (VOR) stations. The VOR is a fixed land-based transmitter. It transmits a signal that allows the aircraft to calculate the bearing to and from the site. This is often used together with DME to calculate the position of the aircraft. Alternatively, bearings to two VORs can be taken, intersecting these bearings will yield the position of the aircraft.

The VOR system works by transmitting a VHF signal in the 108MHz to 118MHz frequency range. This signal has three parts. First, there is an omnidirectional component. The second component is highly directional. It is transmitted by a phased array antenna rotation at 30Hz. By comparing the directional and omnidirectional component, the bearing can be calculated. The third signal is a Morse coded signal for station identification. Some VORs may also broadcast voice messages.

In addition to this, there are non-directional beacons that operate at very low frequencies, between 190kHz and 1750kHz. Their signal is a carrier modulated with a Morse code ID and voice information. They are received by a direction finding receiver in the aircraft.

### 2.1.3 Instrument Landing System for Precise Localization

The Instrument Landing System is a ground-based radio system. It helps airplanes perform a precise landing. The system has two parts, the localizer and the glide slope. The localizer is an antenna array installed beyond the end of the runway. By correctly delivering signals to the antennas, two signals are transmitted. One is modulated at 90Hz and directed right, the other at 150Hz and directed left. The receiver in the aircraft will measure the relative strength of the 150Hz and 90Hz components. For a correct approach on the center of the runway, the strengths should be equal.

The localizer system only provides horizontal guidance. To guide the aircraft vertically, a separate system, the glide slope, is used. It works on the same principle. A 90Hz modulated signal is radiated upwards, while a 150Hz signal is radiated downwards. The approach path is typically around $3^\circ$.

### 2.1.4 Voice Communication between Air and Ground

To allow aircraft to communicate using voice, the VHF airband is most commonly used. Its frequency range is between 118MHz and 137MHz. The actual band starts at

108MHz, but the first 10MHz are reserved for non-voice operation (VOR, ILS, ...). In many countries, the channel spacing is 25kHz, while Europe uses 8.33kHz channels. For long-range operation in remote areas, shortwave may be employed. Military aircraft have a UHF allocation between 225MHz and 400MHz.

The modulation employed on VHF is standard amplitude modulation. The big advantage of using AM is that when multiple signals are received at the same time, the user will hear a mix of both transmissions. This allows talking over a transmission that is already taking place. Frequency modulation has a very strong capture effect: if two signals are sent at the same time, only the strongest one is received.

## 2.2 Modernization of Aerial Communication for Future Growth

According to a report issued by Eurocontrol [154], it is predicted that Europe will have 14.4 million flights in 2035 under the most likely scenario "Regulated Growth." With this continuously growing demand for air travel, it is likely that the current air traffic management (ATM) systems in the United States and Europe will reach their capacity limits in the coming years. Two large-scale projects have been started to develop a next-generation air traffic management system that can cope with this demand:

- Single European Sky ATM Research (Europe);
- Next Generation National Airspace System (United States of America).

Both projects are harmonized under the framework of the ICAO. To enable these new air traffic management systems, it is necessary to develop improved communication, surveillance, and navigation techniques. We start with an overview of the surveillance and navigation techniques, and then conclude with the novel technologies for improved communication.

### 2.2.1 Modern Surveillance and Navigation

To enhance surveillance, it is likely that Automatic Dependent Surveillance–Broadcast (ADS-B) will be used. This is a cooperative surveillance technology where aircraft periodically broadcast their positions that are obtained via satellite navigation. The information is broadcast without an interrogator asking for it. Instead, many ground stations simply receive the periodic squitter, enabling the aircraft to be tracked. Additionally, the signal can be picked up by receivers on other aircraft to provide situational awareness. ADS-B is likely to replace radar as the typical method for civil airspace surveillance. Note that the system requires target cooperation, making it unsuitable for defense applications.

ADS-B consists of two services, ADS-B Out and ADS-B In. ADS-B Out is a transmitter that broadcasts the navigational information about the aircraft. ADS-B In is a receiver that can receive information broadcast by other aircraft or ground stations.

ADS-B provides many benefits:

- Situational awareness: Pilots of aircraft equipped with an ADS-B receiver (ADS-B In) can see the location and altitude of surrounding aircraft. Weather can also be shown on the cockpit screen. The view obtained in a full ADS-B In system is comparable to what is seen on the screen at air traffic control. This creates a form of distributed situational awareness, which greatly improves see-and-avoid.
- Improved accuracy of navigation: Since the position of the aircraft is known with a high accuracy, aircraft can be controlled with less separation, improving efficiency and reducing environmental impact. Additionally, more efficient diversions around weather and restricted airspace are possible.
- Identification: The data broadcast by the ADS-B transponder include a unique identification code. This allows each aircraft to be uniquely identified.
- Improved safety: As said above, ADS-B also allows transmitting data from the ground to the plane. Two services offered are Flight Information Service Broadcast (FIS-B) and Traffic Information Service Broadcast (TIS-B). FIS-B provides weather information and text-based advisories, such as NOTAMS. TIS-B delivers course information about aircraft flying in radar contact with the controllers.
- Search and rescue: Since the aircraft transmits a very accurate position approximately once a second, it is much easier to predict where an airplane may have crashed.
- Small footprint: An ADS-B ground radio is very small. Therefore, coverage gaps can be easily filled by adding another ground station, even in locations where it would be impossible or cost prohibitive to install a radar site.
- Cost: The cost of an ADS-B groundstation is significantly less than the cost of a traditional primary and secondary radar station.

There are two dominant air protocols for ADS-B traffic:

- 1090MHz Extended Squitter: This system transmits the data using a modified Mode S transponder on 1090MHz. The format for the extended squitter message is standardized by the ICAO [219]. This is the system used in Europe.
- Universal Access Transceiver: This 978MHz system is only used in the United States of America below 18000ft MSL. It is the most featured data link protocol for ADS-B.

### 2.2.2 Digital Communication for ATM

According to Schnell *et al.* [375], ATM modernization requires a paradigm shift away from voice toward digital data communications. Increased and more complex information exchange between controllers and pilots demands the use of modern communication technologies. As described in a previous section, almost all communication between air traffic control and the pilot is performed using a voice channel. Clearly, voice has a high overhead, making it unsuitable for efficiently transferring the information needed for future operational procedures. While voice works reliably, it also has an extremely low spectral efficiency, as the content of the message could typically be represented in just a few data bits. To resolve these problems, it is necessary to develop a digital data link. Ideally, it would work in the VHF spectrum band to obtain optimal coverage. In fact, ICAO has already standardized a VHF data link two decades ago: VDL. The link has a very low throughput and as such is not suited for future ATM scenarios. Therefore, it was decided to develop a new data link standard, operating in an L-band portion (960MHz to 1164MHz), which is already allocated on a secondary basis to aeronautical communications. Two systems are under development to cover these needs: LDACS1 and LDACS2. LDACS stands for L-band Digital Aeronautical Communications System. The communication range of both systems is designed to be 200 nautical miles (370km).

Note that a significant part of the proposed frequency range is shared with very high power pulsed DME (see Section 2.1.2) transmitters. Therefore, any data link that is deployed here must either operate in the small piece of spectrum that is not used by DME (non-inlay) or transmit only between DME signals (inlay). Fortunately, the DME signal has a spectrum as shown in Figure 2.3. As seen, around 500KHz of spectrum can be used between two DME transmitters in the inlay deployment scenario. Deployment between DME signals is of course preferred as this allows deployment of the new system without having to change the channels of any existing transmitters.

### LDACS1

LDACS type 1 [139] is the most promising candidate for the future air-to-ground communications standard LDACS. The LDACS1 mode aims to be deployed in the inlay scenario. Therefore, the total bandwidth is limited to around 500kHz. Of course, it is also possible to deploy the system in a non-inlay scenario. Figure 2.3 shows the LDACS1 signal in between two DME transponders. In some situations, this spectral shape alone is not enough. Therefore, the system also contains a blanking system for pulsed interference.

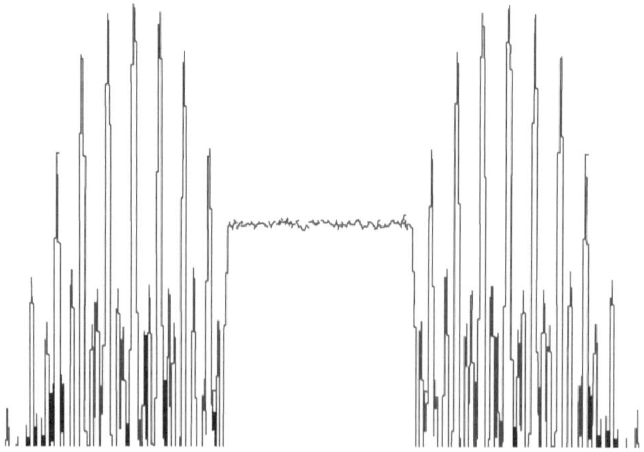

**Figure 2.3** LDACS1 deployed between two DME stations

**Table 2.2** LDACS1 system parameters

| Name | Value |
|---|---|
| Total bandwidth ($B$) | 625kHz |
| FFT size ($N_{\text{fft}}$) | 64 |
| Subcarrier spacing ($f_{\text{sc}} = B/N_{\text{fft}}$) | 9.77kHz |
| Active subcarriers ($N_{\text{a}}$) | 50 |
| Active bandwidth ($N_{\text{a}} \cdot (f_{\text{sc}} + 1)$) | 498.1kHz |
| OFDM symbol duration ($t_{\text{fft}} = 1/f_{\text{sc}}$) | 102.4$\mu$s |
| Cyclic prefix duration ($t_{\text{cp}}$) | 17.6$\mu$s |
| Total symbol duration ($t_{\text{s}} = t_{\text{fft}} + t_{\text{cp}}$) | 120$\mu$s |

**Table 2.3** Comparison of LDACS systems proposed

| Parameter | LDACS1 | LDACS2 |
|---|---|---|
| Duplex method | FDD | TDD |
| Modulation | OFDM | GMSK |
| Throughput | 561kbps–2.6mbps | 270.8kbps |
| Frequency range | 960–1164MHz | 960–975MHz |
| Coexistence | Spectral inlay and pulse blanking | None |

LDACS1 is an FDD mode based on OFDM modulation. While TDD approaches are easier to implement, they would incur large overheads due to the long ranges the system is designed to work with. Indeed, the guard interval would need to be very long. Another advantage is the fact that the forward (ground to air) and reverse (air to ground) links can be aligned with the DME up and downlink frequencies as described in Section 2.1.2. This will greatly reduce the coexistence constraints.

The LDACS1 OFDM parameters are given in Table 2.2. A complete overview of the frame structure is given in [386].

### LDACS2

LDACS2 [272] is the second standard for the described ground-to-air data link. Its physical layer is actually very similar to the second generation GSM system. Since it is not able to cope with DME interference, it can only be used in the 960MHz to 975MHz frequency range. Table 2.3 summarizes the most important parameters of LDACS1 and 2 [238].

## 2.3 Practical UAV and MUAV Data Links

In this section, a short overview of typical communication technologies used in commercial UAV systems, or micro UAV systems (MUAV), will be given. Typically the communication needs are twofold. First and foremost, one should be able to control and monitor the operating of the UAV. To be able to perform a useful function, the UAV should be outfitted with one or more payloads. These payloads often need to be able to deliver data to other UAVs or a ground control station. Below we discuss various existing technologies that could be used for this. In Section 2.4, we will then perform some further measurements and an analysis of these technologies

### 2.3.1 Control and Telemetry

The wireless link used for controlling and monitoring the UAV is typically designed to have a long range and high reliability. Bandwidth requirements are typically low. If the trajectory of the UAV is controlled by the pilot, the latency should also be as low as possible.

### 2.3.2 Payload or Application Data Communication

For high bandwidth UAV to UAV communications, needed for typical UAV applications that rely on videos, for, for example, surveillance, the following systems are often used:

**IEEE 802.11**

This standard is most commonly known for wireless networks in the home and enterprise. Popular versions of the standard such as IEEE 802.11a/n operate using an OFDM physical layer in the 2.4 and 5GHz range. The channel bandwidth is most commonly 20MHz or 40MHz. Two narrowband modes are defined, typically used in professional applications, offering a bandwidth of 5MHz and 10MHz. The latest version of the standard (IEEE 802.11ac) allows up to 160MHz channel width on the 5GHz band. A major benefit of this is that due to its wide application in consumer products, chips and wireless modems are available at very low cost. Another significant advantage from an integration point-of-view is that it emulates an ethernet cable. Indeed, standard networking applications can often be reused on the UAV without many changes.

**Cellular 3/4G**

The main advantage of using cellular systems for UAV payload communication is that the entire network side is already provided by the mobile operator. This has the advantage that the system is easy to use. Simply turn it on and an Internet connection is available, assuming that coverage is also sufficient in the air. This advantage is also a disadvantage – the user has no control over the network. There may not be coverage in the right areas,[1] or the quality of service may be too low. Beyond possible coverage issues, depending on the way the operator has configured the network, there can also be connectivity issues related to Network Address Translation and firewalls.

**Analog Systems**

This section is mostly applicable to the transmission of video signals. Many UAVs in the hobby and light commercial category actually transmit the camera images as a simple analog signal.

[1] As most base stations have antennas that tilt downward, to the ground, it is yet to be confirmed how good coverage is at high altitudes.

The systems are based on legacy PAL or NTSC video signals. This signal is then FM modulated on an RF carrier and transmitted. There is no real standard and the signal bandwidth and video equalization parameters are manufacturer dependent. A bandwidth of 20MHz is, however, typical.

Analog transmission has many disadvantages: poor image quality, low resolution, no possibility for secure encryption, susceptibility to noise, and very bad spectral efficiency.

Nevertheless, there are two clear advantages that are not easily achieved with modern digital communication systems.

The first advantage is that since the pixel stream from the image sensor more or less directly modulates the RF signal, the latency is extremely low (less than 15ms is achievable). This is often used in hobby systems to pilot the UAV using the video feed (FPV) while flying fully under manual control. Another advantage is the graceful degradation. When the distance increases more and more noise appears on the image, telling the user that the radio system is getting near its limits. This is in contrast to most digital systems that tend to work perfectly until a certain point, and then not at all.

Of course, as with DVB-based technologies, this system will be transmitting with a duty cycle of 100% prohibiting any coexistence with other communication technologies. In many countries, this will imply strong transmit power limitations or the need for licensed spectrum.

## 2.4 Analysis of Terrestrial Wireless Broadband Solutions for UAV Links

UAV aerial communication requirements will need an even stronger paradigm shift in aerial communication, and hence the earlier discussed LDACS1 and LDACS2 will most likely not be sufficient. First of all, UAV frames are typically a lot smaller, ranging from fixed wing medium- and small-sized airplanes, to even smaller octocopters and quadrocopters. The smallest quadrocopters that can be found today have a diameter of about 10cm. As a result, the weight of the communication module (including link layer analog and digital processing and the antenna) should be extremely small. Secondly, UAV applications are typically related to search and rescue or surveillance, for which video data are to be transmitted wirelessly from the UAV to the ground. As a result, data requirements are a lot higher than currently considered in LDACS1 or LDACS2. Thirdly, with the emergence of UAV systems for a broad range of military and civilian applications, we expect the density of UAV nodes in an aerial network to be large, that is a lot more simultaneous communication links will have to be supported than today.

To meet weight and data rate requirements, it is often believed that the technology widely available today for mobile terrestrial broadband communication would be suitable for UAV systems. Examples here are IEEE 802.11a, GPRS, or LTE cellular networks. For these nodes, chipsets are widely available, integrating sometimes both the analog and digital processing in a single chip. The resulting communication solution can hence be very light, assuming we can also make a good yet small antenna for the UAV frame. To illustrate the challenges here, we will measure the performance of some widely used antennas for IEEE 802.11 communication in the ISM band for some widely used small UAV frames in Section 2.4.1.

Data rate requirements in those wireless broadband communication technologies are typically met by increasing the number of spatial streams, which means that by relying on multiple antennas at both the transmitter and the receiver, it is possible to increase the link layer throughput between that transmitter and receiver. This is, however, only possible under specific conditions of the wireless propagation environment, and we will study in more detail if this is the case of an air-to-air link in Section 2.4.2.

For the air-to-ground link, we will consider two cases. First, we will study the air-to-ground link from an interference point of view. Each UAV in the air is interfered by multiple terrestrial stations that use the same spectrum. These interfering links reduce significantly the signal to noise and interference ratio seen by the UAV. By implementing beamforming, which is an alternative multiple antenna technique that does not rely on multi-path channel conditions, the impact of this terrestrial interference can then be significantly reduced. More details are given in Section 2.4.3.

Analysis of IEEE 802.11 beyond the link and physical layer fundamental properties is considered in more detail in the next chapter.

### 2.4.1 Single Antenna UAV System Analysis

While chipsets for analog and digital broadband communication are very light, the main bottleneck for these communication systems is still the antenna size and weight. At 2.4GHz, the ideal half-wavelength antenna has a length of about 6cm, which is relatively large for most small UAV frames.

In addition, it is likely that the UAV frame contains a significant amount of conductive materials. Since physical separation of the antenna and UAV is difficult to achieve, it is likely that antenna performance will be impacted. To confirm this, we have taken a measurement where a typical skew-panar-wheel omnidirectional 5GHz antenna is installed on a quadrotor with metal arms. It can be expected that these metal arms will work as a reflecting ground plane. As seen in Figure 2.4, the signal strength in the measurement direction is significantly reduced [41]. The same experiment has been repeated for the Swinglet, which is a plastic fixed-wing UAV. To verify that the reduction is indeed caused by the influence of the metal body, the UAV frame has been modified with plastic arms. Indeed, the signal level is now comparable to the free-space case.

Assuming the metal arms form a perfect groundplane, we can analyze the resulting radiation pattern analytically. Figure 2.5 shows the resulting radiation pattern for different antenna heights (top and bottom on one picture). As seen in Figure 2.4, the signal varies significantly with the angle of arrival. Figure 2.6 shows this more closely.

### 2.4.2 Multiple Antenna UAV Air-to-Air Link Analysis

All terrestrial mobile broadband communication technologies widely used today rely on the use of multiple antennas to boost link layer capacity. While IEEE 802.11n and LTE technology allow the use of up to four antennas at both transmitter and receiver,

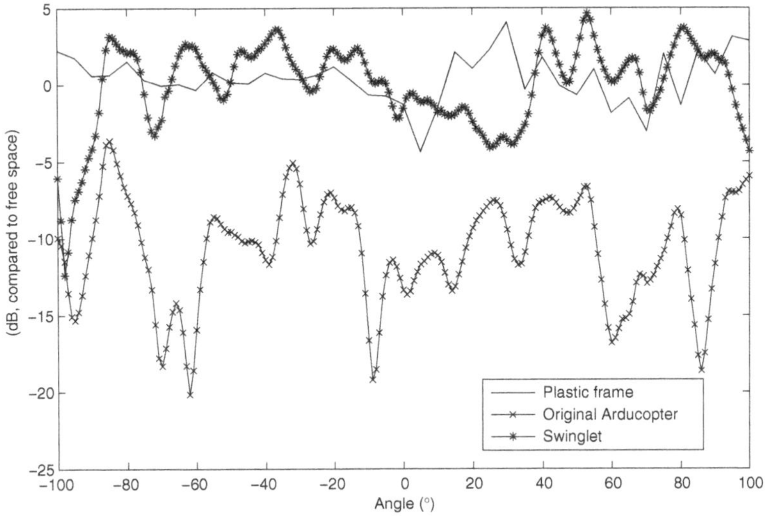


**Figure 2.4** Antenna test results

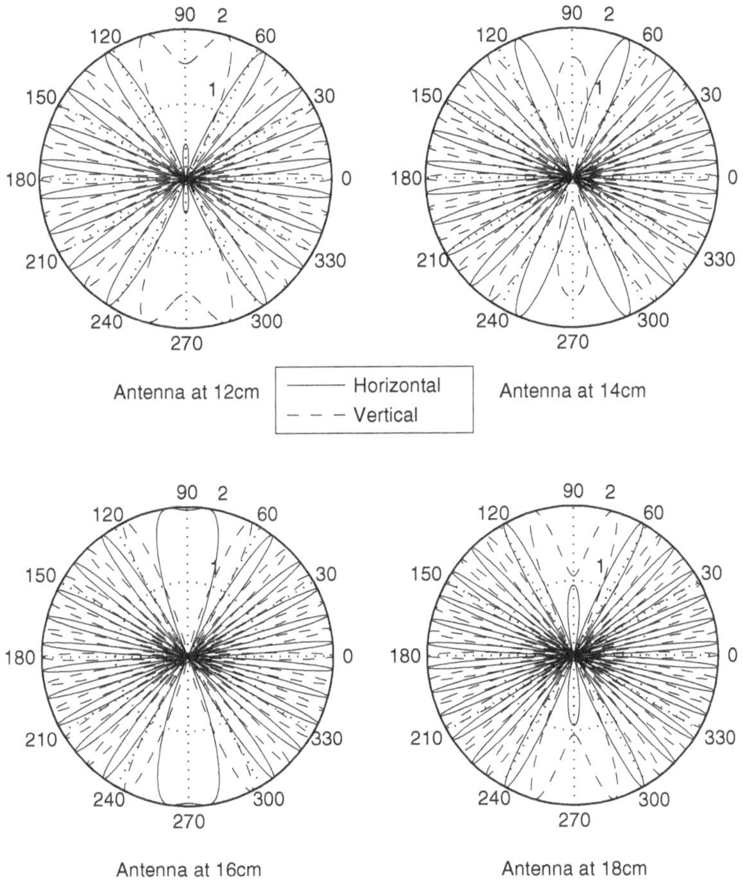


**Figure 2.5** Polar plot of total antenna gain

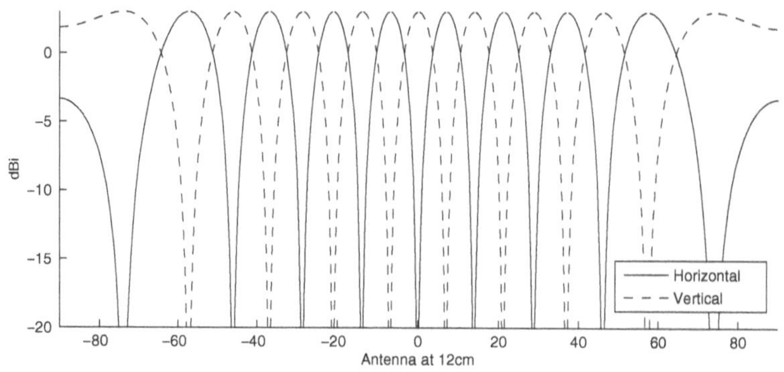


**Figure 2.6** Plot of a single antenna

the next generation technologies IEEE 802.11ac and LTE-Advanced are even considering eight antennas at the access point or base station. The motivation is this: when scaling up the number of antennas on both transmitter and receiver by a factor four, it is possible (in theory) to also scale up the link layer throughput with a factor four. This technology relies to a large extent on the spatial degrees of freedom present in typical terrestrial communication systems: every transmit and receive antenna pair sees a different channel, and if those channels are perfectly orthogonal, it is straightforward to send information on each channel. Yet, this orthogonality is only achieved when there are a lot of multi-path reflections in the environment, causing each location (or antenna) to see another bunch of frequencies, adding up either constructively or destructively.

The air-to-air communication link has, however, a very strong line-of-sight (LOS) component. This means that the direct ray between the transmitting UAV and the receiving UAV is very strong – orders-of-magnitude stronger than the reflections that are caused by the ground and other objects. While in cluttered environments there might indeed be multiple slightly different reflections (multi-path), these reflections will likely all have very low power at the receiving UAV compared to the LOS component. As a result, the LOS channel seen between the multiple transmit and multiple receive antennas is very correlated. Because of this correlation, it is very difficult to rely on multiple antenna techniques for spatial division multiplexing (SDM) on the air-to-air link.

To verify that MIMO is not effective in the UAV scenario, we measured the performance of an aerial IEEE 802.11n link. IEEE 802.11n enables the use of up to four antennas, but most widely available chipsets only support up to two. We mounted two antenna IEEE 802.11n cards on a pole with a height of 7m, to measure aerial link performance. By doing the measurement on a pole, we could eliminate other effects, as reported in Section 2.4.1. Both antennas were using the same polarization.

Despite the noisy measurement, Figure 2.7 shows that using MIMO does not give the expected factor two increase in throughput. Only at very short ranges does MIMO give some throughput improvement. Using antennas with different polarization could potentially increase the performance for two-stream MIMO since they will enable two uncorrelated channels.

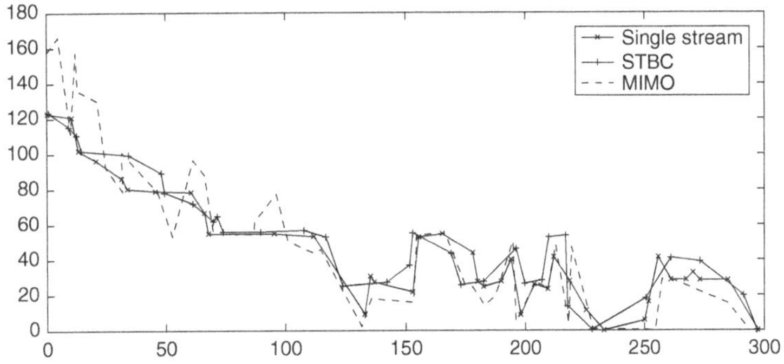


**Figure 2.7** Comparison of different 802.11 operating modes

## 2.4.3 Multiple Antenna UAV Air-to-Ground Link Analysis

From our above analysis, the use of multiple antenna techniques for small UAVs is challenging. First of all, the antenna weight and size often dominate the weight and size of the communication payload, and hence it is not trivial to put multiple antennas on a single (small) UAV frame as the impact would be considerable. Secondly, even if the antenna weight were not of concern, the LOS propagation environment inhibits the use of multiple antennas for spatial division multiplexing. Alternatively, the use of multiple antennas for beamforming can be considered. In this case, the antennas cooperate to strengthen the signal in the LOS direction, and ensure that the waves transmitted by both antennas add constructively at the receive antenna. Similarly, the multiple antennas at the receiving UAV can cooperate to receive signals coming from the LOS direction, neglecting as much as possible signals from other directions. In this section, we will first analyze the impact of harmful interference from the ground on the signal-to-noise and interference ratio at the receiving UAV. We will then show how beamforming can be used to improve the receiver's SINR significantly.

### Harmful Interference Measured

To verify our hypothesis, that interference levels go up with increasing altitude, we designed in [430] a very lightweight IEEE 802.11 packet sniffer that can be used on almost any UAV or even with helium balloons. With this sniffer, it is possible to (a) log RSSI power levels of an IEEE 802.11 transmitter to see how they vary with height and (b) track the number of IEEE 802.11 transmitters overheard. The measurement setup is detailed in Figure 2.8, where the sniffer is carried by a balloon so only the impact of ambient interference is measured.

Measurements carried out in an urban environment indeed point out that the number of detected networks versus altitude of the balloon increases significantly, as illustrated in Figure 2.9. The measurement was done on a large open field near a residential area in the city of Leuven (note that with a balloon, one can fly everywhere, there are no regulatory restrictions). The height of the balloon is approximated by the length of the rope. In case of wind, this is not accurate and often the balloon is blown away.

**Figure 2.8** The wireless scanner is carried by a helium balloon

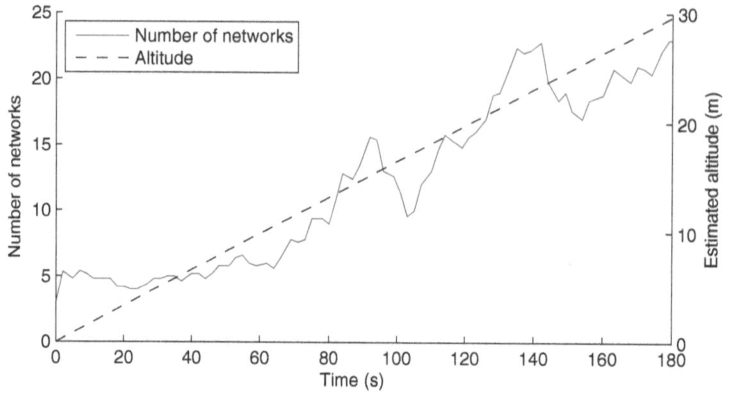


**Figure 2.9** The number of detected networks versus altitude

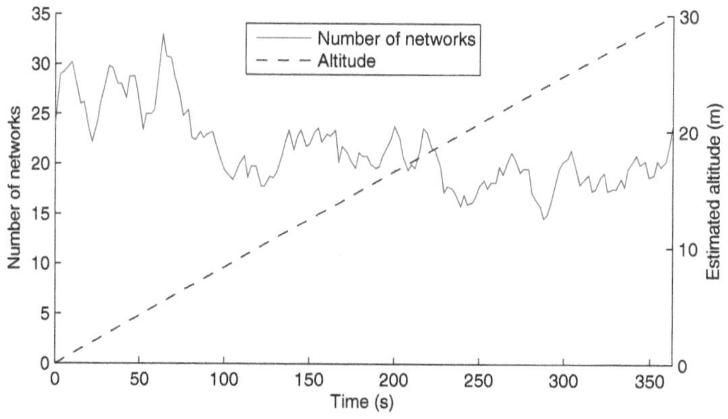


**Figure 2.10** Shadowing effect control case

This causes dips in the number of detected IEEE 802.11 networks, see Figure 2.9. To prove that this effect is because of limited shadowing of aerial networks compared to terrestrial networks, we did the same measurement in a highly shadowed environment. Results from Figure 2.10 indeed confirm that, in such scenarios, the number of detected networks does not increase with altitude.

### Harmful Interference Model

To model the effects measured above, we create a simple model. Since we are working in an air-to-ground scenario, it may be expected that the path will be dominated by the line-of-sight component. Therefore, the path loss is usefully estimated by the Friis equation

$$L_\mathrm{p} = G_\mathrm{t}G_\mathrm{r}\left(\frac{\lambda}{4\pi R}\right)^2$$

Yet, measurements confirmed very discrete shadowing effects should be considered. Shadowing may be expressed simply as an additional path loss per meter of penetration through the object

$$L_\mathrm{s} = G_\mathrm{s}^d$$

With this model, we can determine the signal-to-interference and noise ratio (SINR) of the receiver in the air, communicating with a ground base station and receiving interference from a large number of wireless access points randomly distributed on the ground. Assuming omnidirectional antennas and free-space propagation, a link spanning a distance of 300m and operating at 2.45GHz will have a path loss of 90dB. If we assume the airborne transmitter has an output power of 20dBm, this results in a received signal level of −70dBm. This level has more than enough margin to ensure an error-free

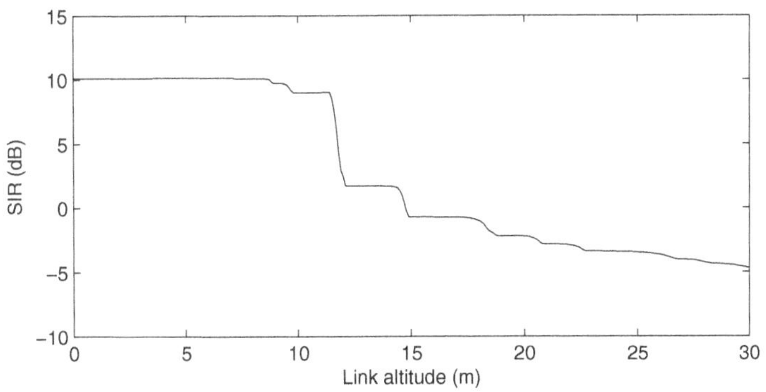


**Figure 2.11** Signal to interference ratio

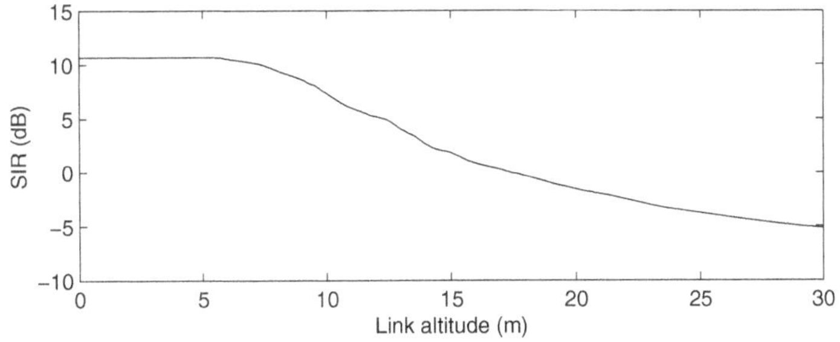


**Figure 2.12** Averaged signal to interference ratio

IEEE 802.11 link. One would assume this link to get better only when the UAV goes to a higher altitude, since the Fresnel zone will become entirely clear. However, taking into account the interference, we get a completely different picture. The high levels of background interference make the signal-to-interference ratio negative. The effect is demonstrated in Figure 2.12 for the average over 100 simulation runs (with different random locations of the buildings and terrestrial interferers).

## 2.5 Conclusions

In this chapter, we started with an overview of communication typically used between airplanes and ground stations. First-generation systems consisted of radar systems, or some simple communication technology suited mainly for voice communication. With the emergence of more and more data communication requirements, and the need for more throughput, modern digital communication technologies such as LDACS1 and LDACS2 were then introduced. With the emergence of small- and medium-sized unmanned aircraft, the aerial communication requirements will increase even further towards low-weight solutions, high capacity links for video transmission, and very dense networks. We have explored the feasibility of achieving low-weight aerial communication solutions, with high data rates and good coexistence with the ground communication systems. It was shown that low-weight communication is possible, but the main challenge is interference from the ground. High throughput and interference avoidance can be achieved in terrestrial communication by relying on multiple-antenna techniques, but this is challenging for aerial air-to-air links. As UAV communication is, however, very susceptible to interference from the ground, the use of multiple antennas for beamforming to minimize interference to and from the ground is a relevant technology. In the next chapter of this book, the use of IEEE 802.11 technology for UAV networking is further explored and discussed.

# 3 Aerial Wi-Fi Networks

Evşen Yanmaz and Samira Hayat

## 3.1 Introduction

Autonomous unmanned aerial vehicles (UAVs) are considered with increasing interest in a diverse set of civil and commercial applications, such as environmental and natural disaster monitoring, border surveillance, emergency assistance, search and rescue missions, delivery of goods, and construction [208, 132, 362, 265, 434, 110, 280, 163]. Use of single or multiple UAVs as communication relays or aerial base stations for network provisioning in emergency situations and for public safety communications has been of particular interest due to UAVs fast deployment and large coverage capabilities [331, 193, 206, 429]. The deployed aerial platform size and the number of necessary platforms are key factors in determining the on-board communication system in addition to the coverage and capacity requirements [73].

Small-scale multi-rotors are of particular interest in practice due to their ease of deployment, low acquisition and maintenance costs, high maneuverability, and ability to hover. Research and development of small-scale UAVs focused initially on control issues, such as flight stability, maneuverability, and robustness, followed by designing autonomous vehicles capable of waypoint flights with minimal user intervention. Considering the limited flight time and payload capacity of these vehicles and the area sizes the aforementioned applications might span, the interest is shifting more and more toward collaborative multiple UAV systems for efficient and successful missions. The number of UAVs might be from tens to hundreds traveling over tens of meters to kilometers for different applications (see Figure 3.1 for an illustration).

Wireless communication is an essential factor not only for network provisioning with UAVs but also for successful deployment of systems consisting of multiple small-scale UAVs. It is highly likely that high-performance links and connectivity in three-dimensional space will be required for several applications (e.g., monitoring certain areas) with data delivery meeting a certain quality of service demands [35]. The question arises as to which wireless technology should be employed for aerial networks that consist of air-to-ground and air-to-air links, and data needs to be delivered regardless of significant height and orientation differences. Taking into account the quality of service to be satisfied over diverse links and the high node mobility, it is not yet clear as to whether networking protocols developed for ground networks are readily deployable on UAV networks.

Several wireless technologies can be exploited for UAV networks such as IEEE 802.15.4, IEEE 802.11x, 3G/LTE, and infrared [70, 302, 460, 40, 31, 145, 207]. Though

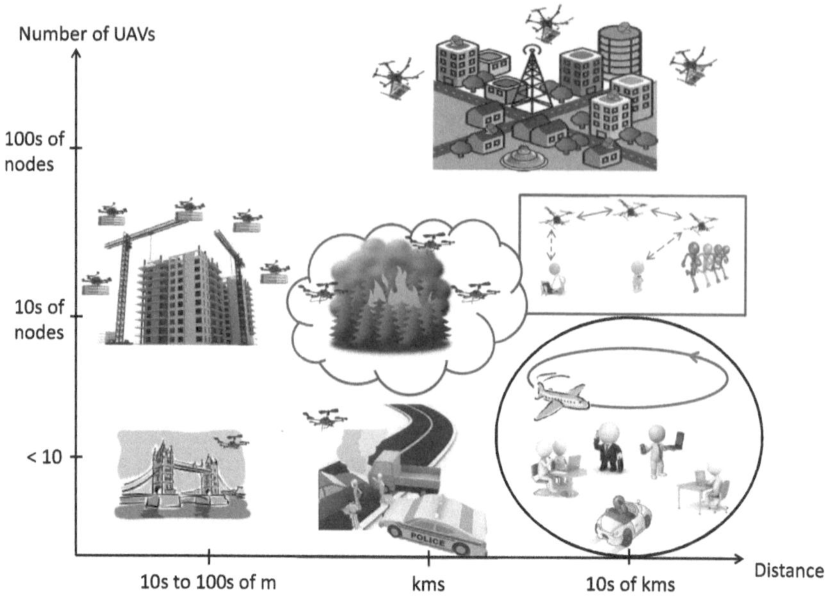


**Figure 3.1** Application areas over a range of distance vs. number of nodes

deployed generally assuming a two-dimensional communication plane, either for low mobility of people or high mobility of ground vehicles (i.e., IEEE 802.11p), a candidate that has been heavily investigated for aerial networks is IEEE 802.11, due to its broad availability and the support of both infrastructure and ad hoc modes for a wide range of services and data intensive applications.

In this chapter, we report on performance results for single and multi-hop aerial Wi-Fi networks. In particular, in Section 3.2, we first provide the characteristics of aerial links in three-dimensional space (3D) that differ from other wireless networks, such as mobile ad hoc networks or vehicular ad hoc networks. Then, in Section 3.3, we focus on demands from the communication viewpoint to enable autonomous aerial networks, and define different levels of autonomy in relation to communication needs. Section 3.4 provides communication requirements for potential aerial network applications in terms of throughput, delay, data exchange frequency, etc. Real-world measurements for multi-hop aerial Wi-Fi networks are reported in Section 3.5. These findings together with the quantitative demands provided in Section 3.4 can be used to determine the feasibility of Wi-Fi technology for the application of interest and to adapt the existing technologies to meet the necessary quality of service demanded. Finally, Section 3.6 summarizes our observations and provides a further outlook.

## 3.2 Aerial Network Characteristics

Due to the nature of the devices used, some characteristics specific to aerial networks arise that differ vastly from those of other networks such as mobile ad hoc networks (MANETs), vehicular ad hoc networks (VANETs), and wireless sensor networks

(WSNs) [49]. In the following, we analyze these differentiating characteristics, and the demands on the communication system imposed by these characteristics. Specifically, we focus on the *types of vehicles employed*, which might affect the range of communication; the *3D nature of aerial links*; the *high mobility* of aerial network devices, which might result in frequent topology changes; and the *payload and flight time constraints*, which have a direct impact on the deployed network size and network lifetime.

### 3.2.1 Vehicles

The vehicles used for aerial networking not only are different from the devices used in MANETs, VANETs, and traditional WSNs, but they come in various forms due to the requirements of the applications they are deployed for. A classification of aerial vehicles (i.e., balloons, airships, gliders, propeller and jet engines, helicopters, and multi-copters) based on their range, endurance, weather and wind dependency, maneuverability, and payload capacity is provided in [148]. The aerial network can encompass one or many of these vehicles, depending on the requirements of the network. In a scenario where there is a need to provide long-term connectivity to ground devices, such as network coverage provisioning via aerial base stations, balloons may be the devices of choice due to their high endurance [351, 429]. However, their altitude and range is limited due to the fact that they are tethered using ropes to an operator (person or ground vehicle). If large areas need to be covered, for example for monitoring and mapping applications, fixed-wing devices can be more favorable due to their longer flight times. However, in operations where the UAV is expected to hover close to objects with good command over flight maneuvers, for example for structural monitoring [152], rotary-wing systems may be the devices of choice [148].

The choice of the vehicles affects the range of operation and the number of required vehicles. Large unmanned devices with specialized radio transceivers usually offer a longer range of connectivity over a single link. If, on the other hand, cheaper, open source devices are used, for example, employing Wi-Fi compliant radios, the same range of connectivity can be expected using multiple devices with multi-hop ad hoc networking. The payload capabilities of small UAVs restrict heavy communication infrastructure to be installed on-board the UAVs. The antennas on-board should not only be light, but also be able to provide omnidirectional coverage. The design and performance of a lightweight antenna structure that enables omnidirectionality in 3D space is described in [460]. The choice of UAV thus affects the communication design of the aerial network, as the channel used varies depending on the device's transceiver characteristics.

### 3.2.2 3D Nature

Another important differentiating characteristic of aerial networks is its 3D nature. The capability of devices in an aerial network to move in a 3D space may render the network useful in situations where the devices have to move around obstacles, for example in urban environments, or disaster scenarios such as earthquakes. In other situations, such as monitoring and mapping of large areas from a high altitude, or mobility in a rural environment, the 3D nature of the network is still advantageous. For instance, in the case where multiple UAVs are employed to perform mission tasks, the spatial decorrelation offered by altitude differences can help avoid collisions amongst the devices [455].

The 3D nature of the network demands the support of various types of links. The links in an aerial network can be either air-to-air (A2A), air-to-ground (A2G), or ground-to-air (G2A). These links have been analyzed against each other as well as against ground-to-ground (G2G) links [31, 460, 302]. It has been stated that these links have to be modeled differently due to their distinct channel characteristics, which affects the supportable network-related quality-of-service (QoS), and hence the sustainable traffic on each type of link. The wireless channel is also affected by elements in the 3D space, which corresponds to the terrain over which the UAV is flying, along with the number of obstacles in the space.

### 3.2.3 Mobility

In many application scenarios, the aerial devices facilitate time efficiency due to their high mobility. For example, [184, 435] emphasize that search and rescue can be performed in a more timely fashion with the use of UAVs. High mobility also makes the aerial network different from other types of networks, such as MANETs and WSNs. Due to this high mobility, the terrain over which the UAVs are flying is expected to change very frequently, for instance, from woodlands to lakes to buildings during a single flight. Not only do terrain-induced blind spots affect the wireless channel, but they may also introduce frequent topology changes amongst multiple devices that require connectivity (UAVs, ground clients, and base stations). High mobility is also a characteristic of vehicles in VANETs. However, the VANET mobility model follows restricted routes in 2D, for example, highways and roads, whereas aerial devices are characterized by the demand for mobility in 3D space. Thus, not only may the terrain over which the UAVs are flying change frequently, but also the altitude of flight may have to be varied to avoid obstacles/collisions.

Considering these characteristics, the communication protocols established for an aerial network should not only allow robust networking of the highly mobile UAVs, but should also be more flexible than VANET protocols in terms of mobility modeling. As in other networks, mobility can be used as an advantage in aerial networks, where the network may not be fully connected all the time. In this case, the highly mobile devices can be positioned at optimized locations in a time efficient manner such that some network QoS can be supported [40]. Also, the controlled mobility in 3D space can be used to enhance range using directional antennas [32].

### 3.2.4 Payload and Flight Time Constraints

The demand for commercially available UAVs has increased due to their cost effectiveness and usability. However, commercially available vehicles that are finding utility in such a vast variety of application areas, though cheap, are constrained in their payload carrying and flight time capacities.

There is an inverse relation between the payload and flight time capacity of the UAVs [13]. This means that by reducing the payload of the UAV, the flight time can be improved. Using the current technology, these improvements in flight times are achieved by allowing multiple UAVs to share the payload [295]. In such a case, a tight synchronization between the closely located UAVs is a requirement for safety. Highly accurate imaging/position sensors along with real-time communication is a necessity for collision avoidance.

## 3.3 Communication Demands of Autonomous Aerial Networks

Now that we have specified the general characteristics of aerial networks, in this section we shift our focus to *autonomous* aerial networks. For several of the envisioned UAV applications, it is expected that the system of UAVs work autonomously toward a desired goal. The level of autonomy depends on the constraints of the system at hand and the requirements of the application. It also determines the communication needs of the aerial network. With more communication capabilities, e.g., over long distances or with high capacity links, more autonomy can be brought into a system of multiple UAVs. The following discussion illustrates this relationship between autonomy and communication. As a first step, it is important to classify the autonomy related to UAV systems. Thus, we define "Device autonomy" and "Mission autonomy" in the following subsections as two categories of autonomy in aerial networks.

### 3.3.1 Device Autonomy

Device autonomy relates to the control of the UAVs and can be used to specify whether a UAV can fly autonomously or needs remote controlled (RC) navigation by a human pilot. It is important to note that to ensure safety, UAVs are obligated by law to stay in RC range for human intervention in case of an emergency. UAVs can fly autonomously following pre-computed or adaptive waypoints. These waypoints can be decided by a central processing entity, like a base station, and then sent over a communication link to the UAV. The UAV can also decide its path on-the-fly by using the information collected from the environment (terrain, obstacles, as well as presence of other UAVs) via on-board sensors. Communication demands vary based on the methodologies employed, but always increase as the level of autonomy is enhanced [406].

Considering device autonomy, we will classify transmitted traffic into control traffic (RC data exchange), coordination traffic (waypoint or mission plan exchange), and sensed traffic. In the following, we describe the varying traffic exchange requirements considering different levels of device autonomy. For instance, when there is no device autonomy, and a human operator is responsible for the control and navigation of the UAV through an RC, control data need to be exchanged between the UAV and RC unit. If we go a step further in the level of device autonomy, where a central entity provides waypoints to the UAV to fly autonomously, the data exchange requirement changes and also includes support for coordination traffic. For fully autonomous devices, where the next waypoint to fly to is decided on-board the UAV itself, the UAV is required to be equipped with some sensors to locate itself related to obstacles and other UAVs in the vicinity. In this case, the RC traffic exchange is accompanied by the demand for sensed data exchange. These sensed data need to be provided to the on-board processing unit, and may also be provided to a central entity (for decision making by ground personnel in case of, for example, a disaster situation, or for providing a higher level of redundancy to ensure safety). Additionally, some coordination traffic exchange may be required amongst the decision-making UAVs; to acquire knowledge of the individual path plans of the neighboring UAVs.

For a failsafe communication system design, it is important to consider the basic control information exchange requirements that enable device autonomy. This can help in determining whether currently available technologies are capable of supporting such information exchange. Currently, the autopilot control for UAVs includes tasks such as pitch attitude hold, altitude hold, speed hold, automatic takeoff and landing, roll-angle hold, turn coordination, and heading hold [87]. This demands that system states be provided to the autopilot at a rate of 20Hz [31]. Current technology promises support for such rates [87]. As an example, with AR Drone, the control loop maintains a connection using the watchdog command every 30ms. The control commands are 20–60 bytes. The device performs emergency landing if no command is received in a duration of 2 seconds.[1] The requirements for coordination and sensed data traffic exchange are detailed in Section 3.4 for various applications.

### 3.3.2 Mission Autonomy

Mission autonomy relates to the coordination between the entities in the network, including UAVs, base stations, and other devices forming part of the network. Having a central decision-making entity (DME) offers a simpler solution than a distributed system of decision-making devices, in terms of the design and processing power required on-board each device. However, distributed decision making may offer superior solutions that avoid single points of failure. Also, as mentioned previously, in aerial networks, which suffer from payload and flight time constraints, parallel processing on-board multiple devices may be a desirable attribute to increase time efficiency.

For an aerial network, we define mission autonomy and corresponding traffic requirement depending on the decision-making entities and the decision-making process, as shown in Table 3.1 using a two-dimensional decision matrix. We define decision-making entities as either *centralized* or *distributed*, represented by the rows of the matrix. The columns stand for the decision-making process, which, according to our definition, can be either *offline* or *online*. The elements of the matrix describe the methods that can be adapted for the mission completion. The level of autonomy in the network depends on a combination of entities and processes. For instance, an offline, centralized decision may provide the least amount of mission autonomy, while an online decision, made in a distributed manner, ensures a higher level of mission autonomy.

[1] AR Drone SDK 1.6 Developer Guide. Parrot USA, 2011

**Table 3.1** Decision matrix for level of mission autonomy

| | | Online | |
|---|---|---|---|
| | Offline | *Min Info* | *Max Info* |
| Distributed individual | Telemetry | Telemetry | Telemetry sensed |
| Distributed consensus Centralized | Telemetry | Telemetry coordination | Telemetry coordination sensed |

Observe that the communication demands do not depend entirely on the mission autonomy. These demands increase with the amount of information exchanged during online processing. They also depend on whether the distributed decision making is done on an individual basis or through consensus between the network entities (e.g., UAVs). Consensus-based decision making [51] is expected to pose higher demands on the design of the communication component than individual based [458]. This is because distributed individual decision making does not require coordinated information exchange between entities in the network. The data that do need to be supported by the communication module considering such classification are shown in Table 3.1. We divide the exchanged mission data into telemetry, coordination, and sensed data. The classification is based on the functionality each type of data provides, as their names indicate. To be more precise, telemetry data include the IMU (inertial measurement unit) and GPS (global positioning system) information. Coordination data are any data that need to be exchanged for coordinating the entities in the network. This may include synchronization information, flight path decisions, routing information, etc. Lastly, sensed data encompass anything that is used to measure the physical environment. The information exchange before the mission starts (offline decision dissemination) and RC data exchange are not considered here.

## 3.4 Quantitative Communication Requirements

In this section, we elaborate on communication-related quantitative requirements derived from aerial applications found in the literature. Table 3.2 summarizes the required data exchange frequency, throughput, delay, and traffic type for expected applications of UAV networks. We classify the types of traffic as real-time, periodic, and delay tolerant (DT). We present the values for the data types *coordination* and *sensed* as defined in Section 3.3. Coordination traffic may include telemetry, some sensed traffic, and decision making and task-assignment commands. Sensed data include any traffic generated by the sensors on-board the UAVs [35]. The control data that consist of telemetry data on the downlink and control commands on the uplink are common to all applications. The standard frequency of telemetry data exchange is 4–5Hz or less [214]. RC traffic also needs to be supported due to safety requirements, forcing the UAV to be in the RC range at all times. Hence, control data are expected to be more frequent (20–50Hz) to enable real-time system response to RC commands [87]. Regarding throughput, telemetry expects approx. 24kbps. For RC data, as the packet size is very small (8 channel RC packet = 11 bytes), the throughput expected is very low ($\sim$ 5kbps).

The applications provided in Table 3.2 are very diverse in terms of the number of nodes, the area sizes the UAVs need to span, the terrain the aerial network is deployed for, etc. Therefore, the quality of service requirements vary significantly, which imposes constraints on the communication technology that can be used. We envision that the provided values can help in designing a robust network for the application scenario at hand. When the size of the area is known along with the throughput requirements for reliable data transfers, it is also easier to estimate whether using multiple hops to achieve full connectivity over the area satisfies the throughput requirements, or if it is more desirable to use DTN [40]. Another important issue to mention is that for the same network, if we employ multiple hops instead of a single-hop, the links that are to be supported in the network include not only A2G and G2A, but also A2A. Moving from single-hop to multi-hop, the throughput supported in the network will also reduce [459]. That is why all the requirement values provided in Table 3.2 specify minimum requirements with respect to end-to-end connection. The number of nodes employed in a network depends on the size of the mission area under consideration, the transceiver characteristics, as well as the types of nodes employed in the network. It is intuitive to think that a larger network area may require larger node density for coverage. However, certain transceivers offer a longer range of communication and may enable the use of fewer devices than other low range transceivers. Nonetheless, usually, such transceivers are specially designed, expensive and heavy devices, and it may be more beneficial to use cheaper, off-the-shelf counterparts with the commercially available, payload constrained UAVs (as in [460, 31, 40, 459]).

## 3.5 Aerial Wi-Fi Networks: Results from Existing Real-World Measurements

Similar to the diversity of aerial network applications, there is a myriad of wireless technologies that can be used in UAV networks. The most common wireless interface deployed on commercial small-scale UAVs is IEEE 802.11 for data transfer and IEEE 802.15.4 (XBee) for telemetry and control data traffic. While the motivation behind this choice is their high availability, size, cost, etc., we can also observe from Table 3.2 that most IEEE 802.11 interfaces can support the quality of service requirements of a wide range of applications. In this section, we will first provide potential network architectures for different UAV applications and then report on real-world measurements for multi-hop aerial Wi-Fi networks.

### 3.5.1 Network Architecture

In the following, we look at the types of network employed for UAV communication systems in some real-world scenarios. We classify the network implementation into infrastructure and ad hoc mode. The performances of infrastructure networks and

**Table 3.2** Quantitative communication requirements for UAV applications

| Application | Data type | Frequency | Throughput | Traffic type | QoS [Delay] |
|---|---|---|---|---|---|
| Search and rescue | Coordination | depends on plan-ahead timesteps [184], >1.7Hz with 2 nodes for required network throughput of 1Mbps [354] | 4.8kbaud [377] | real-time | 50–100ms [60] |
| | Sensed | image sensor dependent: >20fps [251], 30fps [138], laser sensor: 10Hz [412] | 2Mbps [138] in case of video streaming, >=64kbaud [377] | real-time | 50–100ms [60] |
| Monitoring and mapping | Coordination | 0.1Hz [458], 1Hz [394], 4-5Hz [153] | 4.8kbaud [377] | periodic or DTN | – |
| | Sensed | 1Hz [179], 12Hz [98] | 9.6–64kbaud [377] | DTN | can be offline |
| Visual tracking and surveillance | Coordination | Similar to SAR | 4.8kbaud [377] | real-time | <3sec [180] |
| | Sensed | >10Hz to incorporate mobility and noise in urban area [210], >15Hz [353], 300Hz [200] | 1Mbps [251] for images, 2Mbps [138] for video streaming, >=64kbaud [377] | real-time | 50–100ms |
| Network coverage provisioning | Coordination | depends on plan-ahead timesteps | – | periodic | – |
| | Sensed | 50Hz for voice traffic and 30Hz for video (H.264) [352] | 12.2kbps for voice, 384kbps for video [352], 9.6kbaud [377] | real-time | 50–100ms |
| Construction | Coordination | Control Command: 100Hz–50Hz, Motion Tracking: 150Hz | <250Kbps | real-time | – |
| Delivery of goods | Coordination | Control Command: 100Hz–20Hz, Motion Tracking: 100Hz | <250Kbps | periodic | – |

ad hoc networks are compared in [459], which reports the network characteristics such as throughput, delay, and jitter. While the infrastructure mode offers a central access point (AP) for connectivity amongst the devices, the ad hoc mode promises peer-to-peer (P2P) connectivity. Thus, intuitively, infrastructure networks may be more feasible if centralized decision making is implemented, and each device obtains its tasks from and provides information back to a central entity. On the other hand, distributed decision making may require the devices to form P2P connections, with each device contributing to the decision-making process either on an individual or a consensus basis. As mentioned in Section 3.3, applications employing UAV networks can be implemented using more than one methodology based on the decision matrix introduced in Table 3.1. For instance, in SAR projects, such as [404], focusing on small-sized rural areas, the infrastructure mode has been the choice for network implementation. However, for SAR operations performed in medium-sized urban and rural environments, the choice of network implementation is methodology dependent, with some projects favoring infrastructure implementation [184, 77], and others implementing the ad hoc mode [412, 435]. We can find examples of and implementation of infrastructure and ad hoc networks for any combination of area size, terrain, and traffic type for the area and network coverage applications. Infrastructure networking is implemented in [429] to establish an airborne base station to provide network coverage in emergency scenarios for large suburban or rural areas. Ad hoc networking is used in [352] to provide emergency network coverage in large urban areas. For both scenarios, the traffic constitutes real-time, periodic, and delay tolerant data. For real-time traffic, ad hoc networking has been used to form relay-chains in an urban environment considering medium-sized areas in [75]. On the other hand, infrastructure network has been employed for the same traffic, area size, and terrain in [353]. Most area coverage applications like vegetation monitoring, agricultural monitoring, archaeological site monitoring, and photographic mapping of large areas [129] do not require real-time data transfers, and thus use the infrastructure mode with offline data transfers after mission completion. However, in other cases where, for example, coverage of large crowds for target detection, large groups of animals, or multiple wildfire spots is required [301], distributed swarms of UAVs, which are able to cover dynamic targets over large areas in an efficient manner [97] may be more useful than single UAVs. In such cases, P2P connectivity may be required for cooperative coverage, ensuring collision avoidance [48]. For coverage in small-sized areas, where a single UAV is employed for mission completion, infrastructure network formation has been the preferred choice in [152].

### 3.5.2 Experimental Results

In an aerial network, the devices carry different types of traffic, each with their own network requirements [35]. The network interface modules on-board the UAVs need to be capable of satisfying these specific requirements. Moreover, as mentioned above, ad hoc multi-hop networking may be desirable in scenarios requiring deployment of multiple UAVs over large areas. Some experimental work has been carried out to test the feasibility of different wireless interfaces to facilitate networking amongst multiple devices. The focus of these tests has generally been on sensed data delivery, which likely demands more capacity than coordination and telemetry data.

Aerial network characterization for sensed traffic has been evaluated using various 802.11 technologies. Throughput and range analysis for aerial nodes and connectivity analysis for ground clients, with an ad hoc network of UAVs, have been the focus of [70], employing 802.11b radios. Similarly, 802.11g radios have been installed onboard the UAVs to form a mesh network, using the standard 802.11s mesh implementation [302]. This work compares aerial relaying versus ground relaying between two disconnected ground clients. The throughput performance of 802.11n has been analyzed in [39], where fixed physical layer data rates are compared to adaptive rate control. The experimental work shows that the performance of 802.11n radios was much lower than expected using adaptive rate control of commercially available off-the-shelf network interface modules. The authors employ internal planar, and external circular antennas with fixed-wing as well as quadcopter UAVs. They conclude that the degraded performance may be caused by the chassis of the quadcopters blocking the communication link and causing packet loss. On the other hand, [460] illustrates the impact of antenna orientation on the networking performance of UAVs and proposes a three-antenna structure to provide omnidirectional coverage. The performance of the proposed setup is tested with 802.11a, 802.11n, and 802.11ac compliant radios [460, 207]. It is shown in [207] that higher throughput over longer distances can be achieved using commercially available 802.11n modules employing the three-antenna structure on the quadcopter platforms. Figure 3.2 provides the throughput comparison of 802.11a and 802.11n at $P_{\text{TX}} = 12\text{dBm}$ in infrastructure mode for single hop with TCP traffic when the quadrotor is equipped with three external antennas [207]. The results show mean throughput and standard deviation ($\sigma$). Observe that the throughput of 802.11n is five times higher than that of 802.11a for distances up to 100m. However, as compared to 802.11a links, the link quality in the 802.11n network drops more steeply, and it can be witnessed that in the range of 150–350m from the BS, the throughput improvement is

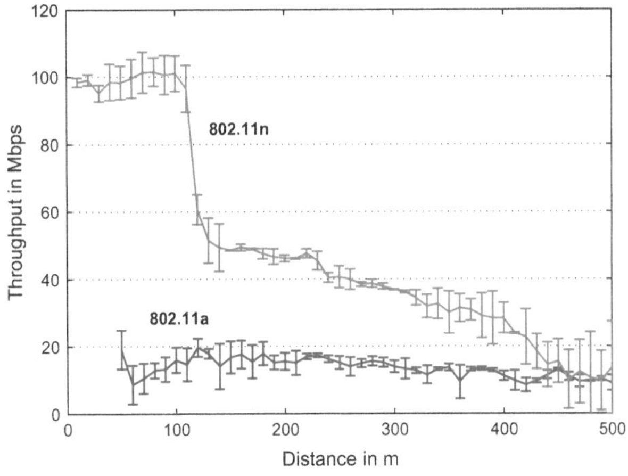


**Figure 3.2** TCP throughput over single hop links with infrastructure mode for $P_{tx} = 12\text{dBm}$

only twofold. At a distance of 500m, where a hovering UAV is still capable of achieving a throughput of 30Mbps, for a mobile UAV, the average throughput is only slightly better than that in the case of 802.11a links. It can also be seen that the deviation in the throughput at higher distance is much higher in 802.11n links than at close range. The performance difference for 802.11n compared to [39] shows that when deploying a communication system on-board the UAVs, care must be given to the 3D nature of the network, especially the on-board antennas need to be tuned for the application at hand to optimize performance. A newer technology 802.11ac has also been tested in [207]. While the laboratory measurements show significant improvement over 802.11n, the outdoor tests result in similar throughput for 802.11n and 802.11ac for distances over 100m. Furthermore, the performance of a two-hop UAV network has been analyzed in [459], where the performance of different network architectures, i.e., infrastructure vs. mesh mode using standard 802.11s implementation over 802.11a radios, has been compared. In addition to [70, 459, 302], [241] also analyzes a multi-hop network of fixed-wing small-scale UAVs equipped with the IEEE 802.11n wireless interface in ad hoc mode, with an implementation of optimized link-state routing (OLSR).

Throughput measurements for aerial Wi-Fi networks for several IEEE 802.11 technology standards are summarized in Table 3.3 [35, 460, 40, 70, 241, 302, 459, 207]. Multi-hop network tests measure TCP throughput, single-hop tests measure UDP throughput, and the maximum transmit power ($P_{tx}$) is used if not stated otherwise.

## 3.6 Conclusions and Outlook

Teams of small-scale UAVs are envisioned to be part of future air space traffic. Reliable communication and networking is essential in enabling successful coordination of aerial vehicles. Due to their wide availability in current networking devices, high-performance links, and their suitability for small-scale UAVs, IEEE 802.11 technologies are commonly used on-board the commercial small-scale UAVs to enable connectivity. However, the communication demands, in terms of quality-of-service, depend on the application the UAVs are deployed for and further research is necessary to determine whether Wi-Fi is the right communication technology for the application at hand. Without paying special attention to an application, several real-world measurements are conducted using UAVs equipped with different IEEE 802.11 standards for line-of-sight scenarios. Reported results show that in terms of required average throughput and delay, Wi-Fi technology can support many applications that require a small number of hops between the communication nodes. It is not clear, however, if the findings would scale to larger networks of UAVs and the existing ad hoc networking protocol implementations need to be adapted for multi-hop aerial networks. Furthermore, as has been the practice with VANETs, it is not yet clear if a new IEEE 802.11 standard needs to be developed for aerial networks. A deployable aerial communications architecture is nevertheless under discussion. Due to the need to support communications over air-to-air, ground-to-air, air-to-ground links in line-of-sight and obstacle-ridden environments, regardless of height or orientation differences, antenna structures tailored for small-scale UAVs

**Table 3.3** Throughput measurements of aerial Wi-Fi networks for line-of-sight links including A2A, A2G, G2A and, for comparison, ground-to-ground (G2G)

| Technology | Link | Topology | Throughput |
|---|---|---|---|
| 802.11b | A2G | single-hop | 1.4Mbit/s (2km) [70] |
| 802.11a (three-antenna) | A2G, G2A, | single-hop | UDP: 14Mbit/s (350m), 29Mbit/s (50m) [460] |
| | A2A | single-hop | TCP: 10Mbit/s (500m), 17Mbit/s (100m) [459] |
| 802.11n (three-antenna, $P_{tx} = 12$dBm) | A2G, G2A, | single-hop | TCP: 10Mbit/s (500m), 100Mbit/s (100m) [207] |
| 802.11ac (three-antenna, $P_{tx} = 10$dBm) | A2G, G2A, | single-hop | TCP: 5Mbit/s (300m), 220Mbit/s (50m) [207] |
| 802.11n (internal antenna) | A2A | single-hop | plane: 1–2Mbit/s (300m), 22Mbit/s (20m) [38] |
| | | single-hop | copter: 20Mbit/s (80m), 60Mbit/s (20m) [40] |
| 802.11g+802.11s [302] | G2G | multi-hop | no UAV: 0.064Mbit/s (75m) |
| | G2A–A2G | | 2-hop: 8Mbit/s (75m) |
| 802.11n+OLSR [241] | A2G | multi-hop | 1-hop: 1Mbit/s (600m) |
| (fixed rate: 13Mbit/s) | A2A–A2G | multi-hop | 2-hop: 2Mbit/s (600m) |
| 802.11a+802.11s ($P_{tx} = 12$dBm) [459] | A2G | multi-hop | 1-hop: 5Mbit/s (300m) |
| (fixed PHY rate: 36Mbit/s) | A2A–A2G | multi-hop | 2-hop: 8Mbit/s (300m, AP mode) |
| | | | 2-hop: 5 Mbit/s (300m, mesh mode) |

are necessary. The impact of antenna structures has been illustrated in the literature and some solutions are proposed; however, these antenna-enhanced aerial Wi-Fi networks are not tested for different application scenarios, e.g., that require real-time traffic support or reliable networking. Especially, the coordination traffic among UAVs needs to be reliable to ensure safety and avoid collisions. While some UAV swarms have been deployed, it is still not clear whether the IEEE 802.11 standard is suitable for distributed coordination of the vehicles, when strict latency deadlines must be adhered to. Therefore, while the reported results so far are encouraging, there is still a need for more efficient routing and medium-access control protocol solutions for multi-hop communications beyond two-hops using the IEEE 802.11 standard.

# 4 Disruption-Tolerant Airborne Networks and Protocols

James P.G. Sterbenz, Justin P. Rohrer, Mohammed J.F. Alenazi,
Truc Anh N. Nguyen, Egemen K. Çetinkaya, Hemanth Narra,
Kamakshi S. Pathapati, and Kevin Peters

## 4.1 Introduction

Traditional Internet protocols are not suitable for the aeronautical environment. The TCP/IP protocol stack has evolved for an Internet that is mostly wired (although that is changing at the edges) with stable topologies and connectivity to which disruptions require terminating of TCP connections, and reconvergence of routing algorithms.

Civilian and commercial airborne networks require some modifications due to wireless links and moderate but often predicable mobility. However, these scenarios are not necessarily multi-hop, and traditionally rely on point-to-point links to ground stations for ATC (air traffic control) and satellites (for Internet access and entertainment distribution). Drone networks may be more challenging if multi-hop communication is desired with unpredictable movements.

The most challenging scenario is for highly dynamic high-velocity multi-hop airborne networks, currently the domain of military communication. However, it is reasonable to expect future civilian, commercial, and government airborne networks to become more challenging as is becoming the case for ground-based vehicular networks, motivating VANETs (vehicular ad hoc networks) and VDTNs (vehicular disruption-tolerant networks).

This chapter is organized as follows: Section 4.2 introduces the communication environment for aeronautical networks, which can vary significantly depending on the scenario (e.g., civilian vs. military networks). Section 4.3 relates airborne networks to the traditional Internet, as well as other mobile wireless environments, including WMNs (wireless mesh networks), MANETs (mobile ad hoc networks), and DTNs (disruption-tolerant networks). Section 4.4 then presents an architecture and protocol suite suitable for the most demanding aeronautical environment: highly dynamic, high-velocity multi-hop networks that require the greatest change to past networking architectures, with comparisons to traditional end-to-end transport and routing protocols. Section 4.5 presents selected performance evaluation of this aeronautical protocol suite, with references to further analysis. Finally, Section 4.6 summarizes this chapter.

**Table 4.1** Aeronautical networking scenarios

| Scenario | Velocity | Mobility | Link | Net type |
|---|---|---|---|---|
| commercial | moderate | scheduled | gnd sta satellite | point-to-point *multi-hop DTN* |
| civilian | moderate | predictable | ground station | point-to-point *MANET* |
| drones | low–moderate | unpredictable | variable | *MANET DTN* |
| tactical military | high | unpredictable | multi-hop | MANET DTN |

## 4.2 Airborne Network Environment

Airborne networks are a class of mobile wireless networks: *wireless* due to the untethered communication links between aircraft, and *mobile* due to the movement of aircraft. Depending on the communication paradigm, they may or may not need to be ad hoc networks that self-organize.

Table 4.1 shows the key characteristics of airborne network scenarios, ordered in increasing challenge to network protocols. The evolution from current to future network types, described for each scenario below, is depicted in italic font in the last column.

*Commercial aviation* is the least challenging due to scheduled predictable flight paths and moderate airspeed (below Mach 1). Connectivity is traditionally served by point-to-point links to ground stations for ATC [151] and satellites for Internet access and entertainment distribution. Entertainment media content is either pre-loaded on the ground or broadcast from satellites. Internet access is most commonly provided by satellite links [237], but these are relatively low rate compared to ground-based Internet access, restricting users from bandwidth-intensive applications such as media streaming. Future higher-bandwidth services might benefit from multi-hop networking either as MANETs in dense airspace, or more likely DTNs on long-haul transoceanic, sparsely populated, or polar routes, the latter without good satellite coverage even for current Internet access and entertainment distribution. The predictable and shared flight paths of commercial aircraft will make these multi-hop networks an attractive solution.

*Civilian aviation* is similar to the commercial case, but with non-scheduled and less predictable flight paths, which may be subject to a flight plan filed near the beginning of each trip. Therefore, future multi-hop civilian aeronautical networks will have to be *self-organizing* MANETs or DTNs, with the practical consideration that small aircraft are more sensitive than large commercial airliners to the weight of on-board communications, processing, and storage.

*Drones* or unpiloted aerial vehicles (UAVs) are increasing dramatically in use, with regulation of them still in its infancy [385, 83]. Uses include situational awareness in hazardous areas, recreation, and possibly the delivery of small parcels [307]. Unlike commercial and civilian aircraft, drones generally have unpredictable flight paths at a lower air speed. The wide variety of current and emerging scenarios makes it difficult to generalize drone networks, but in some cases multi-hop MANETs and DTNs will be useful; this is an intensive area for the research community. It is also likely that the velocity of civilian drones will increase in the future, expanding beyond the current small quadcopters to include fixed-wing aircraft. In densely populated drone networks that are generally connected, MANET protocols will be appropriate; in sparsely populated networks, DTNs will be required.

Furthermore, drones and relatively stationary balloons can have an important role in a variety of airborne networks to enhance connectivity in sparse air space. These *relay nodes* can be deployed with the explicit purpose to increase the probability that an airborne MANET remains connected, or as nodes to increase the performance in airborne DTNs. In the latter case, a stationary relay node can provide for store-and-forward operation, or a mobile relay node can transport data in *store-and-haul* [399] (or ferry) mode.

Finally, drones can be an important part of *non-airborne* networks, by providing connectivity where other network infrastructure is unavailable. Examples of this include relatively stationary drones such as balloons to provide Internet relay connectivity in rural areas [56] and for rapid deployment of temporary network infrastructures in disaster scenarios – for situational awareness and disaster assessment, first responder connectivity, and temporary Internet access to an area before permanent infrastructure is restored [398].

*Tactical* military networks provide the most challenging scenarios, characterized by extreme and unpredictable mobility, for which even traditional MANET and DTN protocols and architectures are insufficient. Highly dynamic airborne tactical networks pose unique challenges to end-to-end data transmission.

Table 4.2 shows numerical baseline values from an example military network scenario [9, 356] to estimate the expected stability of the links in such a network between airborne nodes (AN) and ground stations (GS). In this scenario, the contact duration between two neighboring nodes can be as low as 15s in the case that they are closing and passing. On the other hand, there is no maximum contact duration (for example, when flying in formation or flying in a confined area), so a protocol used in this environment will need to be able to make efficient use of available spectrum and manage multiple traffic priorities for long-lived flows as well as short-lived ones.

**Table 4.2** Link stability analysis

| Scenario | Tx range [km] | Relative velocity | Contact duration [s] |
|---|---|---|---|
| *Single hop best case* | | | |
| AN – GS | 260 | 206m/s | 2520 |
| AN – AN | 28 | 412m/s | 135 |
| *Single hop worst case* | | | |
| AN – GS | 185 | 1191m/s<br>Mach 3.5 | 300 |
| AN – AN | 18 | 2382m/s<br>Mach 7.0 | 15 |

The current TCP/IP-based Internet architecture is not designed to function in this environment; however, this architecture is almost exclusively used within the embedded components that make up modern tactical communications systems, as well as across the Global Information Grid (GIG) [8]. This necessitates that any domain-specific solution designed to optimize performance in a tactical environment must at the same time maintain some compatibility with the TCP/IP stack. Therefore, any new domain-specific protocol suite must be fully interoperable with TCP/UDP/IP via gateways at the airborne network edges.

In particular, given the constraints and requirements of the aeronautical environment, the current Internet protocols are not suitable in a number of respects. These constraints include the physical network characteristics, such as intermittent connectivity, and mobility-induced dynamic topology that present severe challenges to reliable end-to-end communication. In order to build a resilient [395] and survivable mobile wireless [399] network infrastructure, cross-layer enabled protocols that are particularly suited for airborne networks are needed at the transport, network, and MAC layers. At the same time, there is a need to be compatible with both TCP/IP-based devices located on the airborne nodes as well as with ground-based control applications.

A typical airborne tactical network as depicted in Figure 4.1 consists of three types of nodes: airborne nodes (AN), ground stations (GS), and relay nodes (RN). The airborne nodes span the variety of classes listed in Table 4.1: commercial, civilian, drones (shown with a quadcopter icon), and military. At the top, satellite links connect commercial aircraft, some of which are also connected to ground stations. These planes are also connected by multi-hop links, with the one on the right being out of range of both the satellite and ground station. At the left is a civilian aircraft with only a point-to-point link to a ground station. Drones are shown in the middle, with a ground station link to a MANET-connected swarm and a more distant DTN-associated drone on the left. The RN drone is providing relay access to the Internet to a rural or post-disaster area out of range of network infrastructure. At the right is a military tactical network consisting of an individual fighter jet with a DTN-association to a swarm, with connectivity to a UAV relay node.

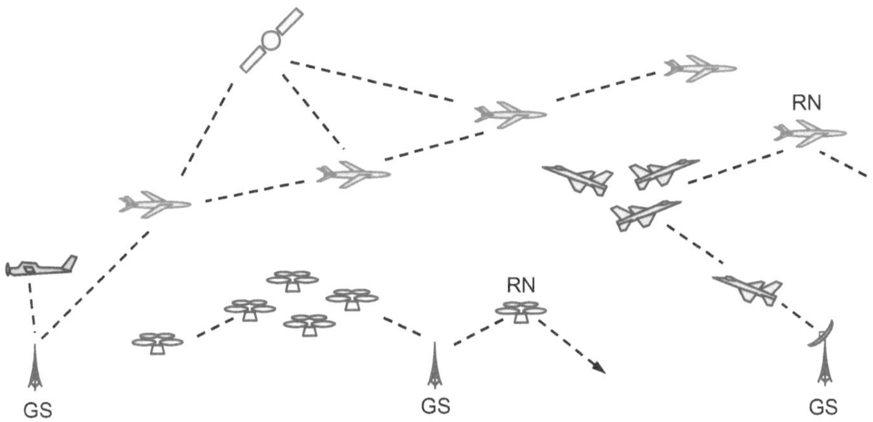


**Figure 4.1** Airborne network environment

## 4.3 Related Work

Given that the airborne network community relies in large part on existing TCP/IP-based embedded devices, communicating the data to existing IP-based systems and applications, it is important to understand the implications of using the traditional Internet protocols (UDP, RTP, TCP, and IP) in the airborne environment. There has been substantial research in transport protocols specific to wireless links and satellite networks, as well as routing protocols for MANETs, which share some characteristics with airborne networks, as shown in Table 4.1. This section considers traditional Internet protocols, as well as some domain-specific protocols and examines their suitability for these airborne networks. The particular issues of interest are mobility, spectrum availability, transmission range, and intermittent connectivity:

- *Mobility*: Airborne nodes can travel at speeds as high as Mach 3.5 (1191m/s), possibly faster in the future; the extreme is then two airborne nodes closing with a relative velocity of Mach 7 (2382m/s). Because of potentially high speeds, the network may be highly dynamic with constantly changing topology.
- *Spectrum availability*: Due to limited spectrum allocations available to airborne networks and the high volume of data to be transferred, particularly for situational awareness, airborne networks may be severely bandwidth-constrained.
- *Transmission range*: The energy available for data transmission on some airborne nodes is limited due to power and weight constraints, particularly with smaller vehicles, such as drones, requiring multi-hop transmission to reach a ground station.
- *Intermittent connectivity*: Given the transmission range of an airborne node and high mobility, the contact duration between any two nodes may be extremely short leading to network partitioning. Furthermore, wireless channels are subject to interference and jamming.

### 4.3.1 Traditional Internet Protocols

#### Transmission Control and User Datagram Protocols

The most widely used end-to-end transport protocol in the Internet is the transmission control protocol (TCP) [341, 292, 30, 345], which is optimized for terrestrial wired networks. TCP provides a connection-oriented reliable data-transfer service with congestion control, and uses constant end-to-end signaling to maintain a consistent state at the source and destination. This introduces overheads, prevents utilizing all available bandwidth, and prevents operation in partitioned network scenarios. Each new TCP session requires connection establishment or three-way handshake before data are transmitted. This wastes one RTT (round-trip time) of valuable transmission time on short-lived connections, such as those in the airborne network environment, and prevents the sending of any data before a stable end-to-end path exists. T/TCP (TCP for transactions) is a modification to the handshake process that bypasses the three-way handshake for *subsequent* connections between the same two hosts, but does not improve the process for the initial connection [67]. Even after the handshake is completed, TCP's slow-start algorithm prevents full utilization of the available bandwidth for many RTTs. This is a well-known problem in long delay scenarios such as satellite networks [242, 29], but also in highly disconnected scenarios in which splitting application data units into many TCP segments may prevent communication. TCP also assumes that all loss is due to congestion, and its standard congestion control algorithm operates by halving the transmission rate every time there is a packet loss. This is the wrong approach for wireless networks and satellite networks [109, 149] in which noisy channel conditions are expected to be the dominant cause of packet loss [267]. TCP's flow control requires a reliable ACK stream, which limits its ability to handle highly asymmetric links even when the data are flowing in the high-bandwidth direction [44]. The practical limit to asymmetry for TCP flows is about 75:1 [28]. There is also substantial overhead with the 20 byte TCP header per packet, especially when using small segments for ACKs or to decrease the probability of suffering an errored packet. TCP was not designed with intermittent connectivity in mind; short-term link outages invoke congestion control and repeated retransmission timer back-offs, which results in an inability to detect link restoration and to utilize the link in a timely manner [399]. A longer link outage results in TCP dropping the connection. Varying RTT can also pose a problem for TCP, because if the actual RTT becomes much larger than the current estimate, TCP will incorrectly assume a packet loss and retransmit unnecessarily as well as reducing the congestion window. Hence, many standard TCP mechanisms are unsuitable for wireless networks in general and the dynamic airborne environment in particular.

The other commonly used Internet transport protocol is the user datagram protocol (UDP) [340]. UDP is far simpler than TCP, but does not offer any assurance or notification of correct delivery, which does not meet the reliability requirements of many airborne network applications. UDP does not undertake any connection setup, congestion control, or data retransmission, and therefore does not need to maintain a consistent state at both ends of the connection. UDP does not do flow control, so the need for ACKs to self-clock is eliminated completely. An extension to UDP is the real-time transport protocol (RTP) [379], which adds timing information to support real-time media, but does not add reliability or delivery assurance.

In airborne networks, we expect multiple classes of traffic with different characteristics, loss tolerance, and priorities. Neither TCP nor UDP can express differentiated levels of precedence or QoS that permit the network to meet these requirements. A number of these shortcomings have been researched, and a few alternative protocols exist, such as SCPS-TP (Space Communications Protocol Standards – Transport Protocol) [146], from which we can draw some mechanisms, but they are only a partial solution.

## SCPS-TP

SCPS-TP [146] is a set of extensions and modifications to TCP that improve operation in the space environment, particularly for satellite communications [439]. It adds mechanisms to deal with specific environmentally induced problems, and modifies existing mechanisms to reduce undesirable behaviors. The use of SCPS-TP options is negotiated at the time of connection establishment, which allows the SCPS-TP agent to emulate TCP when communicating with a non-SCPS peer.

In SCPS-TP, the default loss assumption is a user-selectable parameter on a per-path basis, so it will not assume link congestion, which is unlikely. It also allows for signaling of congestion, corruption, and link outage from both the destination host and intermediate routers to explicitly determine the source of packet loss. SCPS-TP implements the TCP Vegas [127] slow-start algorithm and congestion control based on RTT estimates. Additionally, SCPS-TP asks the user for the path bandwidth-×-delay product and enters congestion avoidance once the congestion window size reaches this value, similar to the congestion avoidance algorithm described in [212]. This is beneficial for paths with high RTT; however, given the rapidly changing topology of an airborne network, it is practically impossible to maintain consistent RTT and bandwidth-×-delay estimates. To attempt to do so would require the use of extremely conservative estimates, resulting in low utilization of the already limited bandwidth. SCPS provides explicit congestion notifications (ECN) using source quench SCMP (SCPS specific version of ICMP) messages [170]. It also uses an open-loop token-bucket rate-control mechanism [333] for each space link to avoid congestion, with the available capacity shared in the global routing structure. The highly dynamic nature of the aeronautical environment makes it difficult to maintain globally consistent routing information, and requires flow control to be handled locally.

For loss due to corruption, SCPS-TP relies on the receiving ground station of each space link to maintain a moving average of the ratio of corrupted frames received and uses explicit cross-layer messages to inform the SCPS-TP destinations when that ratio exceeds a threshold. The destinations are then responsible for continuously notifying their respective sources of the corruption, during which the sources will not reduce the congestion window or back-off the retransmission timer in response to packet loss. In the case of a link outage, SCPS-TP assumes that the outage is bi-directional, so the endpoints of the space link are responsible for notifying the SCPS-TP source and destination nodes on their side of the link. SCPS-TP then enters a *persist state* in which it periodically probes for link restoration so it can resume transmission where it left off without multiple timeouts, retransmissions, or going through slow-start again.

To deal with the problem of highly asymmetric channels, SCPS-TP reduces the number of ACKs required by the TCP [66] from every other segment to only a few per RTT. This requires other TCP mechanisms, such as fast retransmit [401], to be disabled. To deal with constrained bandwidth in general, SCPS-TP employs header compression and selective negative acknowledgments (SNACKs) [236, 292]. The header compression is end-to-end, as opposed to the TCP/IP header compression that is done hop-by-hop [234]. This is because hop-by-hop header compression requires a costly resynchronization process and loses all segments in flight every time a packet is lost or arrives out of order. The end-to-end compression achieves a reduction of about 50% in header size by summarizing information that does not change during the course of the transport session. It also avoids the problems incurred by changing connectivity because the compression takes place at the endpoints that remain constant. The SNACK option allows a single NAK [173] to identify multiple holes in the received data out-of-sequence queue.

SCPS-TP also uses TCP timestamps [235] to keep track of RTTs even with lossy channel conditions, and uses the TCP window scaling option [235] so that the channel can be kept full even while recovering from losses. Many of these techniques for handling highly asymmetric channels are applicable to the airborne network environment.

While SCPS-TP solves a number of the problems associated with tactical networks, SCPS-TP is not ideal for *highly dynamic* airborne networks because it relies too heavily on channel condition information that is either pre-configured or learned gradually over multiple end-to-end connections. This process cannot adapt adequately to a rapidly changing airborne environment, or opportunistically make use of the available bandwidth on a hop-by-hop basis.

#### Internet Protocol (IP)

The global Internet uses IPv4 at the network layer, with various routing protocols such as OSPF [305] and BGP [350]. TCP over IPv4 adds a header of 40 bytes per packet, increasing to 60 bytes for TCP over IPv6 [131]. This overhead becomes significant if there are many small packets (e.g., control traffic), which is the case with the per-segment acknowledgements of TCP. The current Internet architecture is based on the fundamental assumption of long-lasting stable links that do not hold true for mobile airborne networks, which not only challenges the TCP as described above, but also network routing. Internet protocols require convergence of the routes and do not naturally support dynamic topologies inherent in the airborne environment. Furthermore, the current architecture was not systematically designed to be a solution to a global optimization problem [95] and it does not support explicit cross-layer information exchange to leverage unique information such as position and trajectory to be available on the network.

### 4.3.2 Mobile Wireless Network Protocols

There are several categories of protocols designed for the mobile wireless environment: wireless mesh, ad hoc, and disruption-tolerant networks.

#### Wireless Mesh Networks

Wireless Mesh Networks (WMNs) are fixed wireless networks in which the nodes do not move [24]. WMNs may have the problem of intermittently connected links due to interference, jamming, and weather-based attenuation. A number of network-layer techniques can be used to ameliorate these factors, such as weather disruption-tolerant cross-layered and predictive routing [233, 232]. WMN techniques may be applicable to airborne networks when stationary or slowly moving nodes are deployed, such as balloon drones for relay connectivity.

#### Ad Hoc Routing Protocols

MANETs (mobile ad hoc wireless networks) [392, 347, 338] are more challenging than WMNs due to the dynamic topology in which node mobility requires frequent changes to the link association, but in which the network generally remains connected. Several routing protocols have been developed that adapt to changes in topology.

Reactive routing protocols such as AODV [335] and DSR [244] attempt to construct source-to-destination paths on demand, but therefore suffer from the delay involved in finding paths on-demand and such delay may exceed the duration nodes are within range in highly dynamic airborne networks. On the other hand, proactive routing protocols such as DSDV [337] and OLSR [108] attempt to maintain complete network topology in forwarding tables that depend on route convergence. This generates increasing overhead due to frequent route updates (assuming convergence is even possible) and is also not suitable for highly dynamic airborne networks, particularly when bandwidth is constrained.

There are several other protocols that adapt to mobility by forwarding packets one hop at a time without attempting to construct the entire path. These include simplistic approaches such as flooding, optimizations such as epidemic routing [427], and other greedy algorithms that send multiple copies in the network [409]. More complex routing schemes leverage specific information from the network [130, 293]. Most notable are the location-based routing protocols such as DREAM [47], GPSR [253], GRID [278], LAR [259], MACRO [177], SiFT [80], and TBR [464], which use GPS coordinates of the nodes to determine the next hop. LAR uses the geolocation information to limit the region in which potential routes are searched in order to reduce the overhead associated with the route discovery phase. On the other hand, DREAM uses the stored location information of the nodes to forward data in the direction of the destination. The same is true of APRAM [228], which is a hybrid protocol for commercial aviation networks that utilizes geographic location to discover the shortest but complete end-to-end path between source and destination, but this is challenged by highly dynamic networks.

MANET protocols specifically targeted to highly dynamic airborne networks include AeroRP [339], which includes position and velocity vectors in packet headers (described in more detail in Section 4.4.3). Anticipatory routing [408] tracks highly mobile end-points that reach the reactive limit in which the speed of the nodes is comparable to the time it takes for the location tracking to converge upon the position of the node. Spray routing [407] unicasts towards the destination, and additionally multicast sprays over an area to reach highly mobile airborne nodes.

There have also been mechanisms proposed to improve data delivery in ad hoc networks. One such mechanism is the use of network coding [21] that combines multiple packets so that they are carried together in the network, resulting in an increase of throughput and a reduction in energy cost. While the objective of network coding is to make efficient use of network resources, it requires accumulation of packets that causes increased packet delay [365]. However, this strategy is not particularly suitable for dynamic airborne networks in which delay becomes a significant problem due to the rapidly changing paths. Opportunistic routing [57, 281] has been proposed to enhance data delivery in ad hoc networks by exploiting the promiscuous nature of wireless channels. The basic idea is to forward a data packet to multiple neighbors who in turn collectively decide the most suitable nodes to forward the data packet. Another routing mechanism that exploits the broadcast nature of the wireless channels is beaconless geographic routing [368] in which the geographic locations of the source, destination, and neighbors are used to determine the best forwarding option. Several protocols have been proposed such as IGF [58], BOSS [369], and BLR [211] that vary in the algorithm used to select the forwarding node. However, both opportunistic routing and beaconless geographic routing, which are often implemented jointly with a MAC protocol, are more suitable for static or slowly moving nodes such that forwarding node election can be performed on a stable topology. When the node-to-node contact durations are extremely short, the performance of such mechanisms degrades significantly.

Furthermore, some airborne tactical networks require the routing protocol to be highly adaptive based on the particular mission and application requirements. Most existing routing mechanisms are *unimodal*, wherein the algorithm is optimized for a specific mode of operation. A varying set of operating conditions and service requirements justifies the need for a domain-specific *multi-modal* protocol that inherently supports multiple modes of operation.

#### Disruption and Delay Tolerant Networking

DTNs (delay-tolerant networks) [161], originally designed for the Interplanetary Internet [78, 84], are the most prominent approach to provide interoperable communications with and among challenged networks or intermittently connected networks that are characterized by high latency, limited bandwidth, high error probability, limited node longevity, and most importantly a lack of stable end-to-end paths [399]. While MANETs support dynamic topologies, they require that the network is connected and never (or rarely) partitions. DTNs assume that there are frequently cases in which a stable end-to-end path may not exist between communicating nodes, and network partitions are routine.

At the core of the architecture developed by the IRTF DTNRG [2] is the bundle protocol (BP), which is an application-layer protocol with the capability of constructing a store-and-forward overlay network to cope with intermittent connectivity and perform custody-based retransmission [380]. In order for the DTN architecture to accommodate various network environments, the BP operation requires the assistance of a convergence layer that adds missing features to the underlying protocol. The TCP convergence layer (TCPCL) [136] and the Licklider transmission protocol (LTP) [79] are two examples of this convergence layer.

The concept of delay-tolerant networking is generalized to *disruption*-tolerant networking (also using the acronym DTN) in which disruptions in connectivity are more general than interplanetary speed-of-light delays. The general technique is to *store-and-forward* information (as done by the BP) when a node is not able to immediately forward to its next hop, and to *store-and-haul* [399] (or ferry [467]) information by using a mobile node to physically transport information towards a next-hop node for transfer when in communication range. There have been numerous DTN routing research proposals and projects [466, 395, 257, 431].

### 4.3.3 Transportation Network Protocols

Mobile wireless networking protocols are being increasingly adapted to vehicular environments, in general, and the aeronautical environment, in particular.

### Vehicular Ad Hoc Networks

VANETs (vehicular ad hoc networks) [276, 204, 279, 273] are a type of MANET in which *vehicles* such as automobiles, motorcycles, buses, and trucks/lorries are the mobile nodes, with mobility constrained by roadways. There are three main types of VANET architectures: WLAN/cellular, ad hoc, and hybrid. In WLAN/cellular VANETs, there are fixed cellular gateways that provide Internet access, with WLAN (typically 802.11 wireless LAN) access points to provide traffic information, which can be used for rerouting purposes [277], and provide the mechanism for transferring data when vehicles are in range, for example [76, 329]. In ad hoc VANET architectures, all vehicles serve the role of traditional MANET mobile nodes that self-organize into an ad hoc network to communicate with each other. Combining WLAN/cellular and ad hoc architectures forms the hybrid network architecture, in which groups of vehicles form MANETs, but data are exchanged periodically as they drive past WLAN/cellular base stations for Internet access.

Note that it is increasingly common for vehicles such as automobiles and buses to provide 802.11 hotspots linked to a 3G UMTS or LTE cellular link for Internet access in the vehicle; these are not VANETs, but rather point-to-point links conceptually similar to commercial airline–satellite Internet service provision.

While VANETs have the advantage of mobility constrained by roadways and likely connectivity when vehicle density is high, the topology may be highly dynamic due to the short-range radio transceivers (typically 250m for 802.11), and high allowed speed limits (which may reach 85 meters per hour in some US states and 130km/hr in Europe or higher on German autobahns). In low-density rural traffic situations in which following vehicles are out of range, connectivity will be provided primarily by vehicles passing in opposite directions. At freeway/motorway/autobahn speeds, contact durations may be below 5s, providing an environment as challenging as the highly dynamic military tactical network case shown in Table 4.2. Furthermore, there are hard delay constraints for some VANET safety-related applications. Thus, some emerging VANET and airborne networking solutions may share common mechanisms.

In many MANET applications, energy is a scarce resource due to the small size of devices and their batteries, particularly for small drones such as quadcopters. However, VANET vehicles typically have sufficient energy to provide high transmission range communication and processing power at all times [277], in common with larger aircraft, and will benefit from new link technologies and features likely in 5G mobile networks [36, 195].

### Aeronautical Mobile Ad Hoc Networks

In recent years, there have been several attempts to introduce new aeronautical communication architecture [367]. For example, the Aircraft Communication and Reporting Systems (ACARS) was introduced to allow mobile nodes such as airplanes to only communicate to ground station nodes [37]. Furthermore, there have been several studies on airborne network routing protocols [42]. OSPF-MDR [322] is a protocol that extends OSPF for MANET routing. OLSR and OSPF-MDR were compared using several scenarios that vary node speed and density [162, 92, 269]. ARPAM [229] is a protocol that extends AODV to exploit geolocation. Compared to AODV, DSR, and TORA [230], ARPAM generates fewer overheards. Layernet is a self-organizing protocol for widely spread low-velocity nodes [216]. AANET is a multi-hop aeronautical MANET considering communication between long distance clusters of aircraft [366].

### 4.3.4 Cross-Layering

It is important to note that while airborne networks challenge some aspects of network operation, there are also aspects that can be *exploited* by domain-specific protocols, such as knowledge of the airborne node location and trajectory. Previous research has developed several intelligent network protocols in the context of MANETs and WSNs (wireless sensor networks) that attempt to exploit any additional information available [12, 217].

As generally recognized, strict layering in the network stack is not particularly suitable for wireless networks due to mobility, limited bandwidth, low energy, and QoS requirements. Therefore, it is commonly agreed upon that tighter, more explicit, yet careful integration amongst the layers will improve overall wireless network performance in general; and in the case of highly dynamic, bandwidth-constrained networks may provide the only feasible solution that meets the requirements of tactical applications. For example, location and trajectory information can be used to find better paths if there exists a mechanism, either implicit or explicit, for information exchange between the physical/link and network layers.

Thus, we need to facilitate cross-layering across the multiple layers, as shown in Figure 4.2. In cross-layering, *dials* $\mathbb{D}$ [397, 106] expose lower layer information (such as error rate, link stability, and position) to higher layers, while *knobs* $\mathbb{K}$ permit higher layers to influence the behavior of lower layers (for example, adjusting FEC forward error correction strength to meet application service needs) [233, 317, 358, 94, 396].

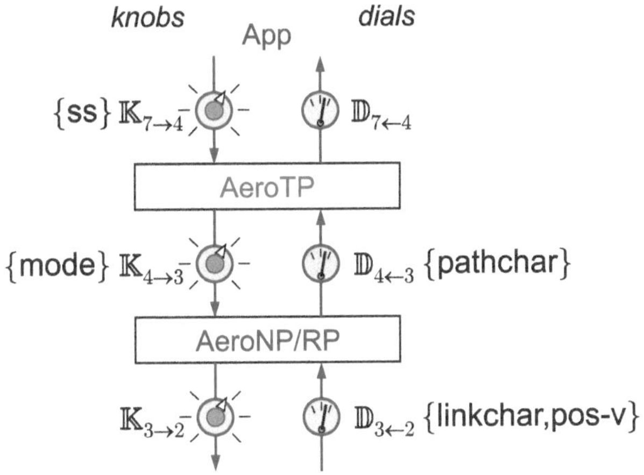


**Figure 4.2** Aeronautical protocol cross-layering

The application knobs pass a service specification ss to the aeronautical transport protocol AeroTP. AeroTP knobs pass mode information to influence the behavior of the aeronautical network and routing protocols AeroNP and AeroRP. Information on link characteristics linkchar and node position and velocity pos-v and is exposed by dials to AeroNP and AeroRP. These use dials to expose path characteristics pathchar such as error rate to AeroTP to adjust FEC strength. Significant benefits are achieved by employing cross-layer optimizations among transport, network, and MAC layers for this environment, as shown in Section 4.5.

## 4.4 Aeronautical Protocol Architecture

To provide a detailed example of a disruption-tolerant airborne networking architecture, this section describes a set of protocols designed for the extreme highly dynamic aeronautical environment [358, 357, 355, 359, 356]: AeroTP, a TCP-friendly[1] transport protocol; AeroNP, an IP-compatible network protocol; and the AeroRP routing protocol for highly dynamic airborne nodes. This protocol stack is shown in Figure 4.3 in which the airborne nodes (AN) or relay nodes (RN) communicate with one another, and to a ground station (GS) using this domain-specific protocol stack. Gateways (GW) in the ground station translate to the TCP/IP-based Global Internet; airborne nodes also have gateways for their own on-board TCP/IP networks.

The major functions of each of these protocols, as well as the control-plane relationships between them, are shown in Figure 4.4. The communications can include any type of packetized information and may be directed from GS to AN, from AN to GS, or from AN to AN, and may use intermediate RNs if available. Both the source and destination for data transmitted may be native Aero-protocol devices or TCP/IP-based systems; however, the IP protocol stack is not suitable for use within the highly dynamic airborne network itself, as described in Sections 4.2 and 4.3. To overcome this challenge

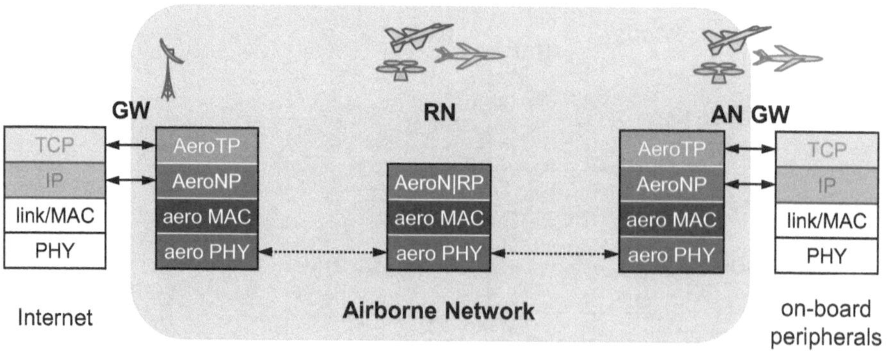


**Figure 4.3** Airborne network protocol architecture

[1] Note that this use of the term "TCP-friendly" is a more general sense than the established term "TCP-friendly rate control" (TFRC) [201]

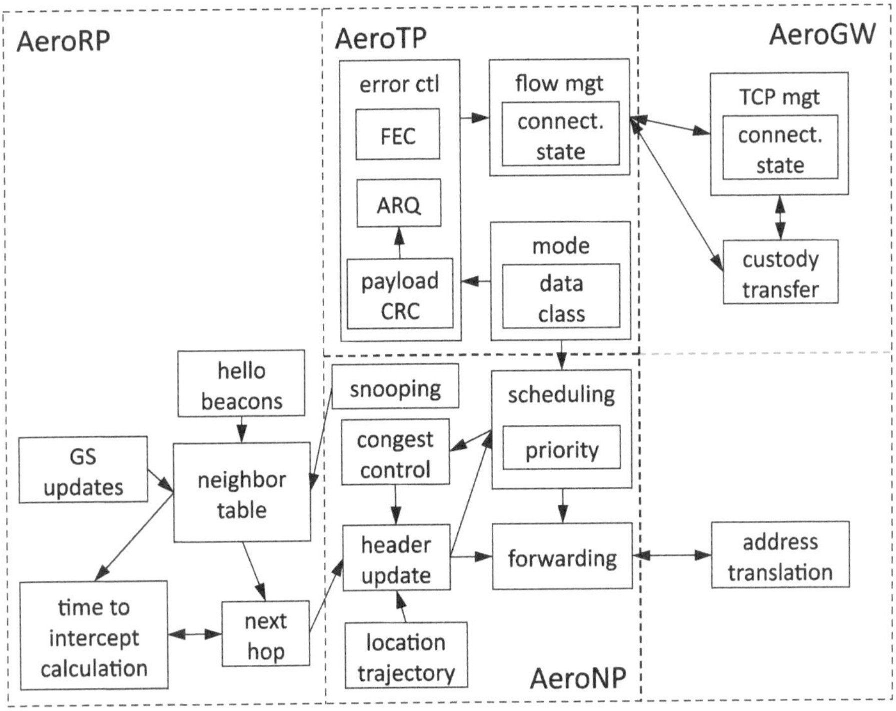


**Figure 4.4** Airborne network protocol functional block diagram

without requiring a total redesign of all on-board aircraft sensors, peripherals, applications, and workstations, the AeroGW gateway [85] translates between TCP/IP and Aero protocols. The gateway concept is well established in general [63] as a mechanism for bridging between disparate network environments. In this case, its operation is similar to TCP-Splice [288]; however, instead of splicing TCP with TCP, it translates TCP and UDP/RTP to AeroTP and IP to AeroNP.

### 4.4.1 AeroTP: TCP-Friendly Transport Protocol

*AeroTP* is a domain-specific transport protocol designed to meet the needs of the highly dynamic network environment while being *TCP friendly* to allow efficient splicing with conventional TCP at the AeroGWs in the GS and on the AN [359, 356]. Thus it transports TCP and UDP through the airborne network, but in an efficient manner that meets the needs of this environment: disruption tolerance, dynamic resource sharing, QoS support for fairness and precedence, real-time data service, and bidirectional communication. Table 4.3 identifies a number of key features of AeroTP and compares it to other traditional, modern, and research transport protocols of significance. AeroTP has several operational modes that support different service classes: reliable, nearly reliable, quasi-reliable, best-effort connections, and best-effort datagrams. The first of these is fully TCP compatible, the last fully UDP compatible, and the others TCP friendly with reliability semantics matching the needs of the application and capabilities of the

**Table 4.3** Feature comparison of AeroTP, TP++, UDP, and TCP variants

| Feature | AeroTP | TP++ | UDP | TCP NewReno | BIC/CUBIC-TCP | T/TCP | SCPS-TP |
|---|---|---|---|---|---|---|---|
| TCP Compatible | friendly | no | no | yes | yes | yes | interop |
| UDP Compatible | friendly | no | yes | no | no | no | no |
| 3-way handshake | no | no | no | per-flow | per-flow | per-endpoint | per-endpoint |
| partial-path support | yes | no | yes | no | no | no | no |
| header integrity check | CRC-16 | chksum | no | no | no | no | no |
| data integrity check | CRC-32 | chksum | 16-bit chksum | 16-bit chksum | 16-bit chksum | 16-bit chksum | 16-bit chksum |
| error correction | variable FEC | FEC | no | no | no | no | no |
| aggregated ACKs | yes | yes | no | optional | optional | no | yes |
| selective repeat | yes | yes | no | optional | optional | no | yes |
| negative ACKs | optional | no | no | no | no | no | optional |
| multi-path friendly | yes | yes | no | no | no | no | no |
| flow control | x-layer | out-of-band signals | no | windowed | windowed | windowed | windowed |
| congestion ctrl | x-layer AeroNP backpressure | none | none | slow-start, AIMD, fast rexmit | slow-start, (CU)BIC, fast rexmit | estimate, AIMD | estimate, Vegas, fast rexmit |
| error control | hybrid, modular, adaptive, | hybrid, modular | none | ARQ | ARQ | ARQ | ARQ |
| reliability modes | reliable<br>nearly-reliable<br>quasi-reliable<br>best-effort | reliable<br><br>quasi-reliable<br> | <br><br><br>best-effort | reliable | reliable | reliable | reliable |

airborne network. The AeroTP header is designed to permit efficient translation between TCP/UDP and AeroTP at the AeroGW.

AeroTP performs end-to-end data transfer between the edges of the airborne network and either terminates at native on-board aeronautical devices or splices to TCP/UDP flows at the AeroGWs. Transport-layer functions that must be performed by AeroTP include connection setup and management, transmission control, and error control, shown in Figure 4.4.

## Connection Management and Transmission Control

AeroTP uses connection management paradigms suited to the wireless network environment. An alternative to the overhead of the three-way handshake is an opportunistic connection establishment in which data can begin to flow with the ASYN (AeroSYN) setup message, shown in Figure 4.5. In this example, the flow of data is originated by an on-board peripheral (per) as a standard TCP session, translated into an AeroTP session by the gateway to traverse the airborne network, and then translated back into a standard TCP session by the gateway on the ground. The TPDU (transport protocol data unit) size may be discovered using the standard path MTU discovery mechanism [299]; however, given the specialized nature of these networks, it is expected that the best performance will be achieved by setting the peripherals to use an appropriate MTU as determined by the characteristics of the underlying aeronautical MAC. Closed-loop window-based flow and congestion control with slow start is not appropriate to the highly dynamic nature of this network, therefore AeroTP uses an open-loop *rate-based* transmission control with cross-layer instrumentation from the network layer to determine an initial rate, with backpressure to control congestion, as described in Section 4.4.2 for AeroNP. Error control is fully decoupled from rate control [107, 400], and is service specific as described below.

## Segment structure and gateway functionality

AeroTP is *TCP-friendly*, meaning it is designed to efficiently interoperate with TCP and UDP at the gateways. To support this, gateway functionality provides IP–AeroNP

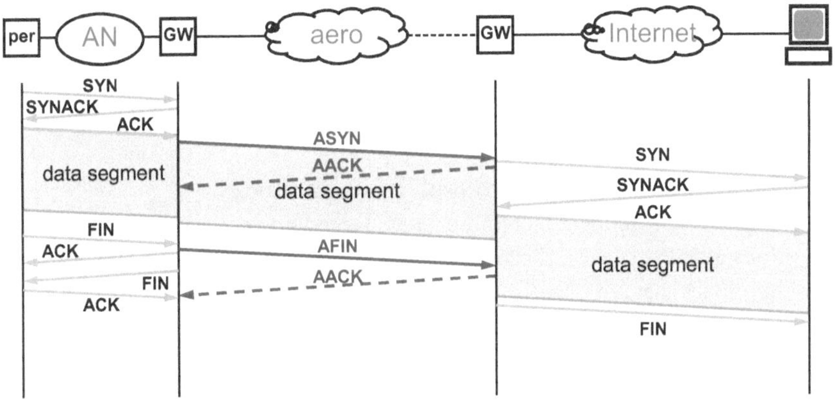


**Figure 4.5** AeroTP connection setup

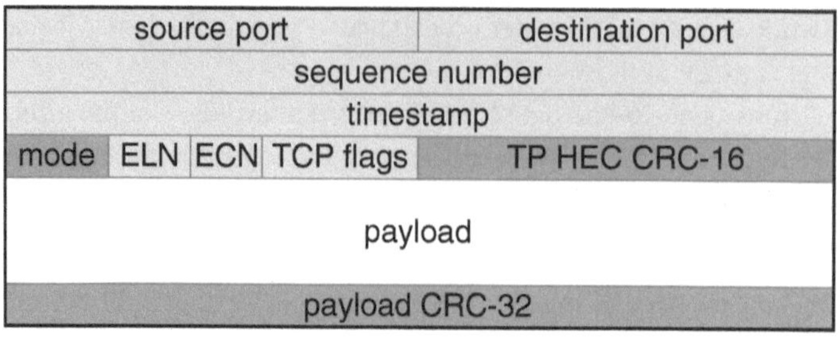


**Figure 4.6** AeroTP TPDU structure

translation [231] and TCP/UDP–AeroTP splicing. A packet may pass through two gateways on its path from source to destination. The ingress gateway will convert the TCP segments to AeroTPDUs, while the egress gateway will convert AeroTPDUs to TCP segments.

The AeroTPDU (transport protocol data unit) is shown in Figure 4.6. Since bandwidth efficiency is critical, AeroTP does not encapsulate the entire TCP/UDP and IP headers, but rather is capable of converting to the TCP/UDP format at the gateways. Some fields that are not needed for AeroTP operation but are needed for proper end-to-end semantics are passed through, such as the source and destination port number, TCP flags, and the timestamp. The sequence number allows reordering of packets due to erasure coding (as with TP++ [168]) over multiple paths or AN mobility, and is either the TCP byte-sequence number or a segment number, depending on the AeroTP transfer mode. The HEC (header error check) field is a strong CRC (cyclic redundancy check) on the integrity of the header to detect bit errors caused by the wireless channel. This allows the packet to be correctly delivered to the AeroTP at the destination where a corrupted payload can be corrected on an end-to-end basis using FEC (forward error correction), as long as the header itself is not corrupt. A payload CRC protects the integrity of the data edge-to-edge across the airborne network in the absence of a separate AeroNP or link layer frame CRC, and enables measurement of the bit-error-rate for error-correction code adaptation depending on the transfer mode. This method of error detection and correction implies that the AeroNP does not necessarily drop corrupted packets at intermediate hops, which is a key difference from the IP forwarding semantics [66, 267].

## Error Control and QoS-Based Transfer Modes

Based on the application requirements, there will be a number of classes of data being transmitted over the airborne network. For this reason, AeroTP supports multiple transfer modes that are mapped to different traffic classes: reliable connection, near-reliable connection, quasi-reliable connection, unreliable connection, and unreliable datagram.

All modes except unreliable datagram are connection-oriented for TCP-friendliness and use sequence numbers so that packets may follow varying or multiple paths and be reordered at the AeroTP receiver.

- ***Reliable connection*** mode (Figure 4.7) must preserve end-to-end acknowledgement semantics from source to destination as the only way to *guarantee* delivery. This is achieved using TCP ACK passthrough, which has the disadvantage of imposing TCP window and ACK timing onto the AeroTP realm, but will never falsely inform the source of successful delivery.

  A variant of this mode uses HARQ (hybrid ARQ) with the addition of FEC (as shown below for quasi-reliable mode) to reduce the probability of the need for retransmission. ARQ is only invoked if the CRC-32 check fails on the payload including the additional FEC overhead. This submode is particularly useful when reliable transfer is needed, the probability of bit errors is moderate-to-high, and the delay sensitivity is more important than the bandwidth overhead of FEC protection.
- ***Near-reliable connection*** mode (Figure 4.8) uses a custody transfer mechanism similar to that used in DTNs [78, 380] to provide high reliability, but cannot *guarantee* delivery since the gateway immediately returns TCP ACKs to the source on the assumption that AeroTPs reliable ARQ (automatic repeat request)-based delivery will succeed using SNACKs (selective negative acknowledgements)

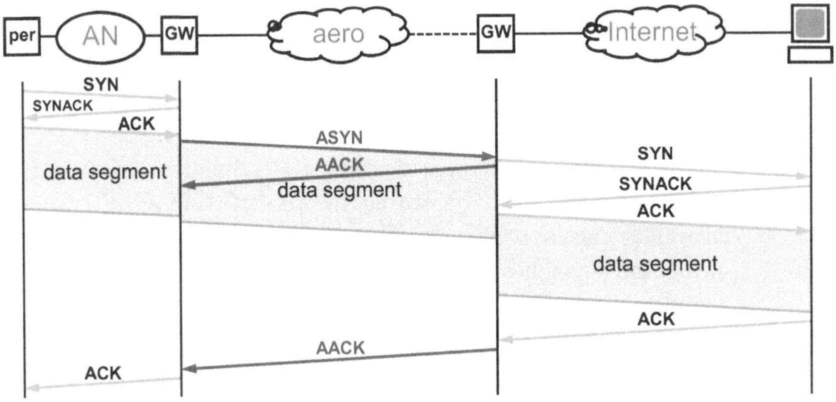


**Figure 4.7** AeroTP reliable connection transfer mode

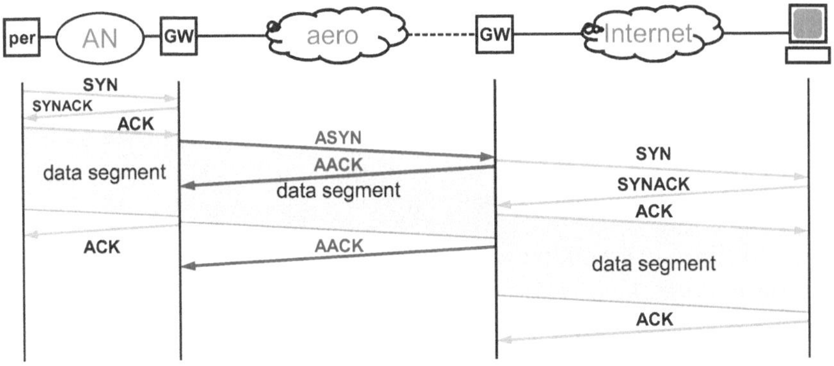


**Figure 4.8** AeroTP near-reliable transfer mode

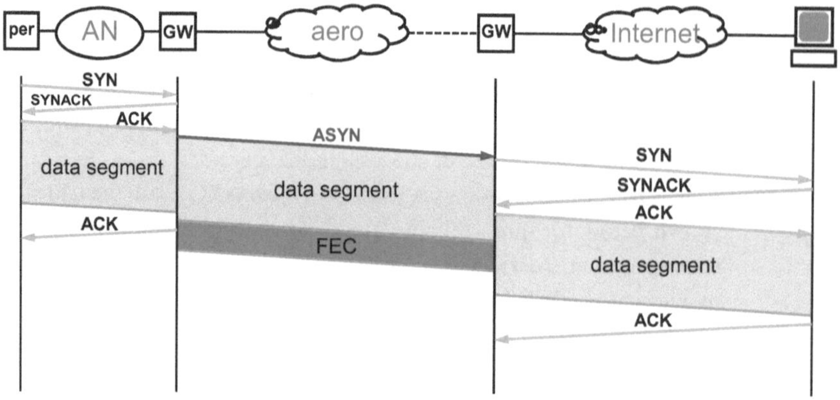


**Figure 4.9** AeroTP quasi-reliable transfer mode

[146] supplemented by a limited number of (positive) ACKs as well as ELN (explicit loss notification) [267]. This still requires that the gateway buffer segments until acknowledged across the airborne network by AeroTP, but is more bandwidth-efficient than full source–destination reliability because TCP's ACK-clocked behavior only operates over the well-connected on-board and terrestrial Internet links, while allowing AeroTP to efficiently utilize the airborne network. However, the possibility exists of confirming delivery of data that the gateway cannot actually deliver to its final destination.

- ***Quasi-reliable connection*** mode (Figure 4.9) eliminates ACKs and ARQ entirely, using only open-loop error recovery mechanisms such as FEC over unipath flows, as well as erasure coding spread across multiple paths if available [296]. In this mode, the strength of the coding can be tuned using cross-layer optimizations based on the quality of the wireless channel being traversed, available bandwidth, and the sensitivity of the application to packet corruption. This mode provides an arbitrary level of statistical reliability but without absolute delivery guarantees.
- ***Unreliable connection*** mode relies exclusively on the link layer (FEC or ARQ) to preserve data integrity and does not use any error correction mechanism at the transport layer. Cross-layering may be used to vary the strength of the link-layer FEC.
- ***Unreliable datagram*** mode is intended to transparently pass UDP traffic, and no AeroTP connection state is established at all.

### 4.4.2 AeroNP: IP-Compatible Network Protocol

AeroNP is a network protocol designed specifically for the highly dynamic airborne environment [356]; however, given the IP-based end devices on the ground for command and control, as well as TCP/IP peripherals on the AN, it is critical for the airborne network protocol to be compatible with the IP. The AeroGW converts IP packets to AeroNP packets and vice-versa [85]. The key features of AeroNP are to provide explicit support

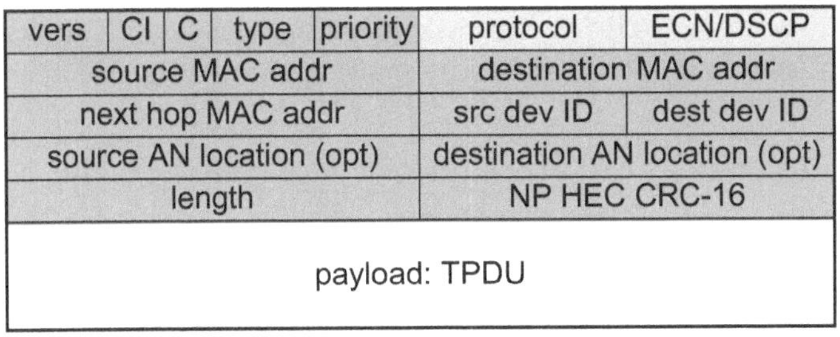


**Figure 4.10** AeroNP packet structure

for cross-layering messages discussed in Section 4.3.4, reduce the overhead by providing an efficient addressing mapping from the IP, and provide a strong header check to decode corrupted payloads that could be recovered using AeroTP error-correction mechanisms.

The AeroNP packet header format, shown in Figure 4.10, is 32-bits wide. The version is the AeroNP protocol version, the congestion indicator CI is set by each node to notify the neighboring nodes of its congestion level as discussed later. The type and priority fields specify the QoS level of a given packet. The number of QoS classes can be customized for a given scenario. Demultiplexing is governed by the protocol identifier to determine where packets are handed to the destination, passed through from the source. In order to provide IP transparency, the ECN/DSCP (explicit congestion notification/diffserv code point) nibble is carried over from the IP header. An AeroNP packet is handed to medium-access control, and thus contains the source, destination, and next hop MAC addr (which may be either a 16-bit address or TDMA slot id as shown for efficiency, or a 48-bit IEEE MAC address if necessary, governed by operational mode).

Furthermore, significant efficiency can be gained if the AeroNP header does not carry the 32-bit source and destination IP addresses (or the even worse 128-bit addresses for IPv6). By performing an ARP-like address resolution process, the IP address can be mapped to MAC addresses in the AeroGW. However, each AN can have multiple on-board peripherals, each of which has an IP address. Therefore, the destination MAC address, device id tuple is mapped to the peripheral IP address at the AeroGW. While dynamic mapping procedures are possible, it is more efficient to preload the translation table at the beginning of each mission. As with the MAC address case, full IPv4 or IPv6 addresses may be controlled by the operational mode, at the cost of significant protocol overhead.

Optionally, source and destination AN location are included, which are the GPS coordinates and velocity vector that are used in location-aware routing (and longer than implied by the header figure). These can be omitted when the velocities are low, but are included when highly dynamic for use by the AeroRP. The length indicates the actual length of the header in bytes. A strong check on the integrity of the header, HEC, is included to protect against bit errors. Unlike Internet protocols [66], the default behavior of AeroNP is to repair the corrupted bit and forward the errored packets to the transport layer instead of dropping them at the network layer. The corruption indicator C bit is set by AeroNP to notify AeroTP that corruption has been experienced. This permits FEC at the transport layer to correct errors in the AeroTP quasi-reliable mode, as described in Section 4.5.1.

### 4.4.3 AeroRP: Location-Aware Routing Algorithm

The small contact duration among ANs at high velocity results in frequent routing changes and is indicative of the need for an intelligent multi-hop routing protocol, supporting reliable communication over the highly dynamic physical topology. As discussed previously, existing routing mechanisms generate significant overheads and do not converge quickly (if ever) in the presence of frequent topology changes and hence are not suitable for highly dynamic networks. The AeroRP routing protocol is specifically designed to address the issues related to highly mobile aeronautical environments [339, 313, 356]. Important design aspects of AeroRP are:

- **Proactive behavior:** AeroRP is a fundamentally proactive routing protocol, but with limited updates thereby lowering protocol overheard.
- **Exploits cross-layer controls:** AeroRP is designed to exploit the explicit cross-layering support provided by AeroNP and the geographic node location and trajectory information available at nodes.
- **Per-hop behavior:** Unlike existing protocols, AeroRP forwards data per-hop based on partial local information and routes, thereby avoiding the necessity for global convergence, making it especially suitable for highly dynamic environments.
- **Multi-modal:** Airborne networking scenarios and applications present a high level of variation in their operational parameters. For example, based on the security requirements of the application, the geolocation of the nodes may or may not be available. In order to support these dynamics in operation, policies, and constraints, AeroRP provides multiple modes of operation.

#### Protocol Operation

The basic operation of AeroRP consists of two phases. In the first phase, each AN learns and makes a list of available neighbors at any given point in time. It utilizes a number of different mechanisms to facilitate neighbor discovery, discussed later in this section. The second phase of the algorithm is to find the appropriate next hop to forward the data packets. In order to forward the packets toward a specific destination, additional information such as location data or route updates are required. For each of these two phases, the protocol defines a number of different mechanisms. The particular choice of mechanism to be used is dependent upon the mode of operation. The protocol does not specify a predefined set of discrete operational modes; the total number of supported modes is merely the combination of all the different mechanisms available. The two phases of operation are *neighbor discovery* and *data forwarding*.

**Neighbor discovery:** The first objective of an airborne node is to determine its neighboring nodes. In order to achieve this, several different mechanisms are employed with the objective to minimize overhead and increase adaptability:

**Table 4.4** Feature comparison of AeroRP and other routing protocol categories

| Feature | AeroRP | Traditional MANET (AODV, OLSR, DSDV, DSR) | Opportunistic Routing (OR, EOR) | Geographic Routing (LAR, DREAM) | Beaconless Routing (IGF, BOSS) |
|---|---|---|---|---|---|
| partial-path support | yes | no | yes | yes | yes |
| store & haul | yes | no | no | no | no |
| cross-layering | yes | no | no | yes | yes |
| snooping | yes | no | no | no | yes |
| location aware | yes | no | no | yes | yes |
| beaconless | optional | no | no | yes | yes |
| update frequency | aperiodic topology dependent | periodic or on-demand | no updates | periodic | no updates |
| route reconfig | hop-by-hop | source init or update based | hop-by-hop | update based | hop-by-hop |
| multiple modes | yes | no | no | no | no |

- ***Active snooping*** is the primary mechanism used by the node to locate and identify its neighbors. In the wireless network, a node that is not transmitting listens to all transmissions on the wireless channel. AeroRP adds the transmitting MAC address of each overheard packet to its neighbor table. The protocol assumes cooperative nodes and symmetric transmission ranges among ANs. This implies that if a node can hear transmissions from a node, it can also communicate with that node. Stale entries are removed from the neighbor table if no transmissions from a node are heard for a predetermined time interval related to the anticipated contact duration.
- ***Hello beacons*** are used by idle nodes to advertise their presence. When neighboring nodes hear a hello beacon, they update their neighbor table appropriately. The frequency of the hello beacon is inversely proportional to the minimum calculated contact duration.
- ***Ground station updates*** may be used to augment or replace *active snooping* in some scenarios, in which the ground station has a partial or even complete flight plan. The ground station sends periodic updates containing the location and trajectory vectors predicted by the flight plan to all nodes.

Security requirements may impose restrictions on airborne networks, particularly in military tactical networks. In certain cases in which the node location or trajectory is considered sensitive, individual nodes may not include this information in the header of the data packets or hello updates. In this case, the ground station may send location updates of all the nodes on an encrypted channel. Finally, in the most secure mode, no geographic node information is available and the routes have to be discovered using traditional MANET methods, such as explicit routing updates and the exchange of node contacts between neighbors.

Given the dynamic nature of the aeronautical network, neighbor discovery consists not only of finding nodes within the transmission range, but also in determining the duration for which a discovered node will remain within range. Depending upon operational constraints, this information is obtained via different mechanisms: location and trajectory information is included in the AeroNP header, or in updates sent by the ground station.

**Data forwarding:** After neighbor discovery, the second phase of AeroRP is for individual nodes to determine the next hop for a particular transmission. Unlike conventional protocols, AeroRP performs hop-by-hop forwarding based on partial paths without full knowledge of the end-to-end paths [209]. Each node forwards packets such that they end up geographically closer to the destination, which will frequently be a GS in many scenarios.

When a node needs to transmit data, and assuming that one or more neighbors are discovered, packets are forwarded to the node that is nearest to the destination as calculated from its current coordinates and trajectory. The destination location is obtained in a manner similar to that of discovering neighbors. Furthermore, in many cases the destination is the stationary GS whose coordinates are known to many ANs.

## Data Forwarding Algorithm

The position of the $i$th airborne node, $n_i$, is represented by the vector $\mathbf{P}_i = (x_i, y_i, z_i)$ and the trajectory is defined by the vector $\mathbf{T}_i = (s_i, \theta_i, \phi_i)$, where $x$, $y$, and $z$ are the absolute node coordinates, $\mathbf{T}$ is the spherical direction vector (speed, inclination, and azimuth). Since the network is highly dynamic, both the position and trajectory of nodes are time dependent. For a given source–destination pair at time $t$, the source node $n_\mathrm{s}$ has the position $\mathbf{P}^t_\mathrm{s}$ and trajectory $\mathbf{T}^t_\mathrm{s}$. Similarly the destination node $n_\mathrm{d}$ is represented by $\mathbf{P}^t_\mathrm{d}$ and $\mathbf{T}^t_\mathrm{d}$. The congestion status of the node is given by the vector $\mathbf{C}_i = \{CI, \text{priority}\}$, where *CI* and *priority* are the congestion indicator and priority fields from the AeroNP header.

**Step 1:** Each node maintains two tables: a *neighbor table* that stores the information about the nodes that are currently in the transmission range, and a *destination data table* that stores the information of all destinations, which may or may not be currently in the transmission range.

**Step 2:** When the node receives any packet, it updates the neighbor and destination data tables. If the captured packet is an overheard transmission or hello advertisement from node $n_i$, the node $i$ is assumed to be in the transmission range of the current node. Hence the neighbor list is updated as $N = N \cup \{n_i\}$. Furthermore, the macID, position, trajectory, and congestion status of the node are derived from its header and stored in the neighbor table as the tuple $\{\text{macID}_i, \mathbf{P}^t_i, \mathbf{T}^t_i, \mathbf{C}^t_i\}$. If the received transmission is a *ground station update*, each entry in the update is stored in the destination data table as the tuple $\{\text{time}, \text{macID}_i, \mathbf{P}^t_i, \mathbf{T}^t_i, \mathbf{C}^t_i\}$. Since the GS update may contain information on node positions in the future, the entries in the destination data table are time stamped. Lastly, when a ground station update is received, the location and trajectory fields of neighbor table entries are updated with the latest values.

**Step 3:** At the completion of step 2, assume that a given node $n_0$ has $k$ discovered neighbors. From this set, all congested neighbors (CI bit set for priority equal to or greater than the priority of the data to be sent) are removed. Furthermore, each node adds itself as the first neighbor in the list: $N_0 = \{n_0, n_1, \ldots, n_i, \ldots\}$. Assume that node $n_0$ wants to send a data packet to the ground station $n_\mathrm{d}$ with position $\mathbf{P}_\mathrm{d}$ and that the transmission range of all nodes is $R$. The *time to intercept*, TTI is then calculated for all neighbors. The $\mathrm{TTI}_i$ represents the time it will take for node $n_i$ to get reach within the transmission range of the destination if it continues on its current trajectory, calculated as

$$\mathrm{TTI}_i = \frac{|\mathbf{P}_\mathrm{d}^t - \mathbf{P}_i^t| - R}{s_\mathrm{d}}$$

where $|\mathbf{P}_\mathrm{d}^t - \mathbf{P}_i^t|$ gives the euclidian distance between the current location of node $n_i$ and the destination node $n_\mathrm{d}$ and $s_\mathrm{d}$ is the component of the actual speed $s_i$ of node $n_i$ in the direction of the destination, calculated as: $s_\mathrm{d} = s_i \times \cos(\theta_i - \theta_\mathrm{d})$, where $\theta_\mathrm{d}$ is the angle of the destination with respect to the current node position.

**Step 4:** Finally, the data are forwarded to the $j$th node, $n_j$, such that

$$\mathrm{TTI}_j = \min\{\mathrm{TTI}_i\} \ \forall i : n_i \in N_0$$

The process is repeated at every node, until the data reach the destination.

Ground stations are special nodes in the aeronautical network, which listen to transmissions and forward packets that are destined to other GSs. For uplink data, a GS forwards packets to the node that is closest to the destination node. The GS is aware of the location of airborne nodes either from the flight plan or by learning from header information in received packets.

RNs (relay nodes), if explicitly designated, are always the default next-hop. They accept packets from all the ANs and forward them directly to the GS or another AN.

### Quality of Service

The wireless links in airborne networks may be bandwidth-constrained and under-provisioned for the traffic generated. Hence, it is essential to implement a quality of service mechanism in this network to ensure that high priority packets can be reliably delivered. The AeroNP protocol uses two fields in the header to specify the quality of service of packets in the network: data type (e.g., network control traffic, emergency response (e.g., for first responders in a disaster scenario), airborne operations, general Internet access, and civilian entertainment) and priority within a given type. The scenario and application requirements determine the *type* and *priority* for a given data flow, which are passed to AeroNP through AeroTP via out-of-band signaling. The scheduling algorithm used at the nodes is "weighted fair queuing" based on type and priority.

### Congestion Control

In a heavily loaded network, multi-hop routing can induce severe congestion at nodes involved in forwarding as well as transmitting their own data. To overcome this, AeroNP uses a simple congestion control mechanism at the network layer using *congestion indicators* and *back pressure*, chosen for their simplicity in operation, based on little feedback. The objective is to avoid local congestion, but does not guarantee global optimization or fairness. A more rigorous rate control mechanism (such as [255]) is not suitable due to the highly dynamic nature of the network, in which an optimal solution would become stale by the time it is achieved.

In the first mechanism, the node uses a congestion indicator [346, 289] field to indicate its own congestion level. All packet transmissions from a node carry the CI field along with the type and priority of the data. When the transmit queue of a node exceeds a predetermined threshold, the node sets its CI field. Neighboring nodes eavesdrop on the transmission and are made aware of the congestion at a given node. If a node is congested, the neighbors back off if the data that they have are of equal or less priority, however higher priority data are still forwarded to a congested node because the priority queue at that node will service this traffic first.

The second mechanism through which congestion control is achieved in the airborne network is *back pressure* [330, 437]. As a source sends packets to an intermediate node, it simultaneously eavesdrops on that node to see if the packets are being forwarded at the same rate they are being sent. If not, and other packets are being forwarded instead, then the source can infer that the next hop it has chosen is queuing its packets due to congestion. The source node then backs off and if possible chooses an alternate next-hop. Similarly, in a multi-hop scenario, if a bottleneck is encountered, each intermediate hop either stops or slows down its transmissions on the congested path successively until the source of the traffic is reached.

## 4.5 Performance Evaluation

This section presents selected results from simulations of the AeroTP and AeroRP protocols performed using the ns-3 [10] (and in one early case predecessor ns-2) open-source network simulators. The performance of AeroTP is compared to TCP and UDP, and the performance of AeroRP is compared to that of DSDV, AODV, and OLSR. Additional analysis and a more comprehensive explanation is provided in the references for each section; testbed prototype implementation is described in [27].

### 4.5.1 AeroTP Simulation Results

Simulation of AeroTP connection establishment, as well as data transfer in fully reliable (ARQ selective repeat) and quasi-reliable (FEC) modes is evaluated against TCP and UDP using the ns-3 (and in one case ns-2) network simulator [359].

#### AeroTP Connection Establishment

As mentioned previously, one of the drawbacks of TCP for highly dynamic airborne environments is the delay of the three-way-handshake used for connection establishment. For this reason, AeroTP is designed to establish a connection when the first data

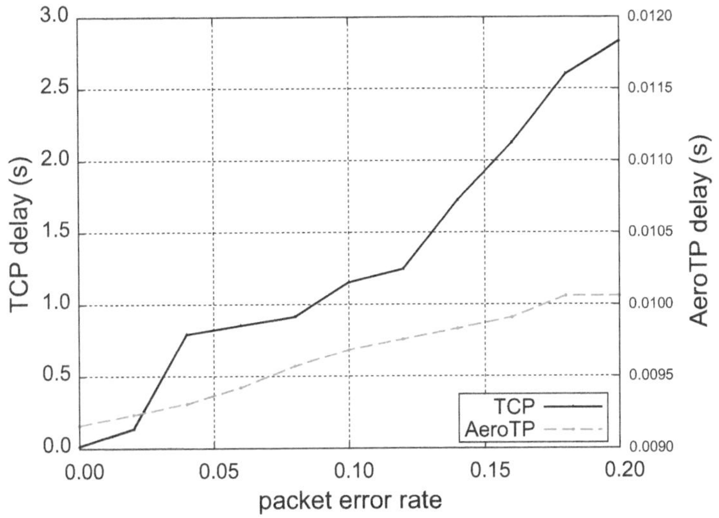


**Figure 4.11** TCP and AeroTP connection establishment delay

TPDU (transport protocol data unit) in a flow is received with the ASYN bit set. If the first packet is lost, the connection can still be established using header information from the second or subsequent data packet, and the first packet can be retransmitted later if required by the specified reliability mode.

Figure 4.11 shows the time required to establish an AeroTP connection compared to a standard TCP Reno [360].[2] Two airborne nodes are connected by a 10Mb/s link with 5ms latency and a fixed probability of packet loss, which varies between 0% and 20%. Each data point consists of a single connection attempt by either an AeroTP or a TCP, averaged over 100 runs. The delay is from when the connection setup is issued to the transport protocol at the transmitting node to when the first *data* packet is received by the destination node.

Both the TCP and AeroTP results are presented in a single plot; however, note that they are plotted against two different *y*-axes: TCP on the left and AeroTP on the right. The TCP delay starts at about 20ms when no losses occur, and increases linearly until it approaches 3s when the packet-loss rate is 20%. AeroTP on the other hand, has a delay of 9.2ms when no losses occur, and increases linearly to 10.1ms when the packet-loss rate is 20%. This shows an improvement of two orders-of-magnitude, which contributes significantly in enabling AeroTP to successfully send data over paths that only exist for a few seconds, while TCP would still be trying to establish the connection.

## AeroTP Data Transfer

The data-transfer simulations that characterize AeroTP performance consist of two airborne nodes transmitting 1MB at a constant bit rate (CBR) of 4.416Mb/s: 3000 packets/s with an MTU of 1500B. The path is three hops consisting of a 10Gb/s link

[2] This early simulation was performed in ns-2 rather than ns-3.

representing the LAN on the AN, a 5Mb/s link with a latency of 10s representing the airborne network path, and a second 10Gb/s link representing the LAN at the GS (ground station). The airborne network link is lossy, with the BER (bit-error rate) ranging from 0–$10^{-4}$. Each simulation case is averaged over 20 runs using a different random seed.

**Quasi-reliable mode characterization:** In *quasi-reliable* mode, AeroTP uses end-to-end FEC as its reliability mechanism, trading overhead on each packet for reliability. TPDUs for which the FEC cannot correct errors (as checked by the CRC-32) are *not* retransmitted on the assumption that the applications using this mode have some loss tolerance. The advantage to this mechanism is low delay, because no retransmissions are required to correct errors covered by the FEC code. The first set of plots compares varying FEC strengths, from zero FEC 32-bit words per TPDU, to 256 FEC words. In all cases, 1500-byte packets are used; thus, as the number of FEC words in each packet is increased, the number of bytes of data in each packet decreases, meaning that more packets are required to transfer the same amount of data.

Figure 4.12 shows that the goodput (throughput successfully received) decreases due to uncorrectable packets as the error-rate increases; however, this effect can be mitigated by increasing the FEC strength. For very high FEC strengths (128 and 256), there is virtually no decrease in performance across the range of error-rates tested, however the performance is decreased at low error-rates due to the high level of overhead. Due to the fact that retransmissions are not involved, the latency of data transmission is not affected by packet errors.

However, as very high levels of FEC result in link saturation this translates into added latency due to queuing delay. Similar to the goodput plot, Figure 4.13 shows the total amount of data received. Depending on the FEC strength, this quantity decreases as the

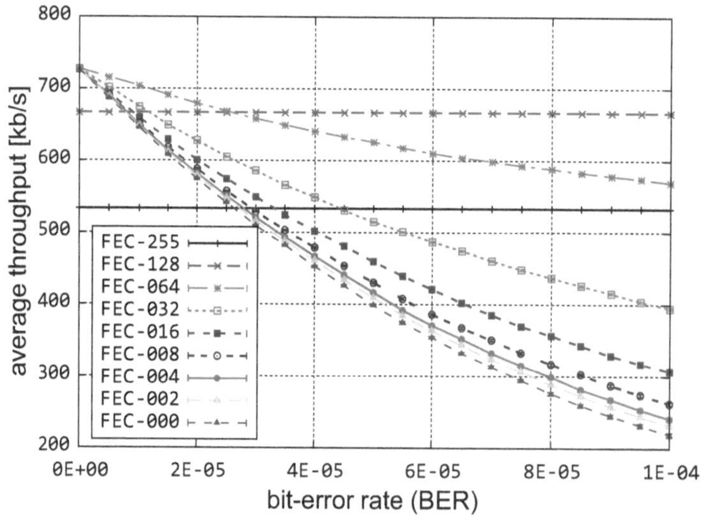


**Figure 4.12** AeroTP quasi-reliable average goodput

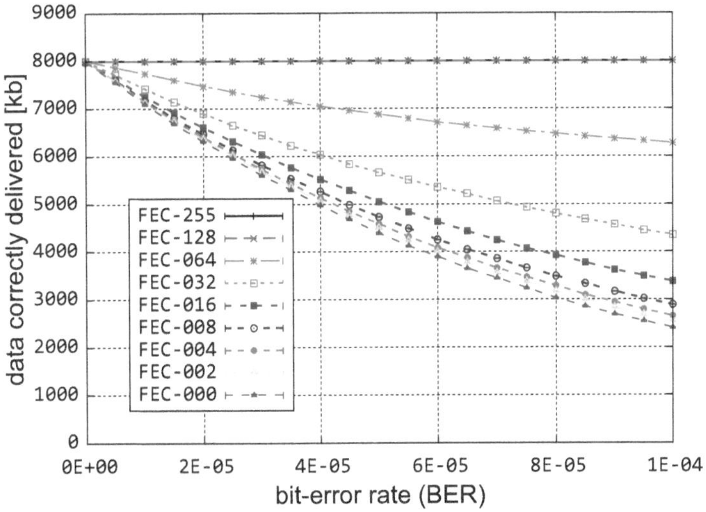


**Figure 4.13** AeroTP quasi-reliable cumulative goodput

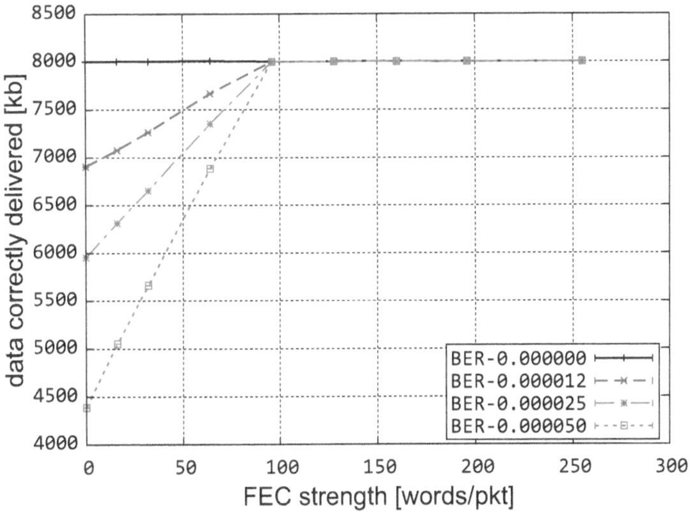


**Figure 4.14** AeroTP quasi-reliable cumulative goodput

error-rate increases, except for very high FEC strengths (128 and 256) at which all errors are able to be corrected at the rates tested.

Figure 4.14 shows that an FEC strength of 96 words/packet or greater is able to correct all errors at the error rates tested.

**Fully reliable mode characterization over lossy links:** In reliable mode, ARQ retransmissions are used when the CRC-32 check indicates that a TPDU has been corrupted. Figure 4.15 shows that AeroTP is able to achieve significantly better performance than TCP, which backs off substantially as the BER (bit-error rate)

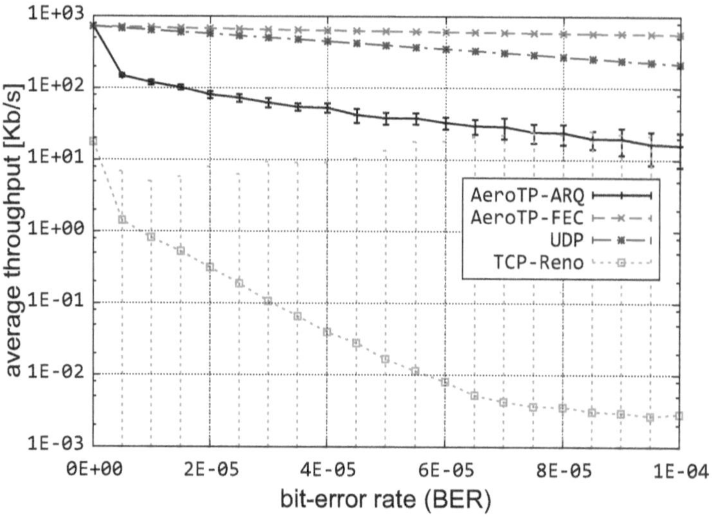


**Figure 4.15** AeroTP and TCP reliable average goodput

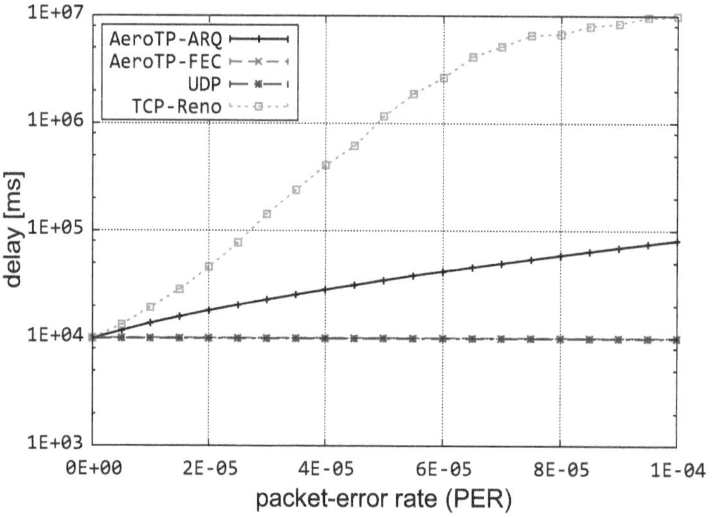


**Figure 4.16** AeroTP and TCP reliable average delay

increases. TCP also becomes highly unpredictable in its performance, as shown by the large error bars. At the same time, TCP's end-to-end delay increases by three orders-of-magnitude with a BER of $1 \times 10^{-4}$, while AeroTP reliable (ARQ) increases by less than one order-of-magnitude as shown in Figure 4.16; since neither AeroTP quasi-reliable (FEC) nor UDP retransmit corrupted packets, the delay is constant at the cost of unrecoverable errors. Over the course of the simulation, both TCP and AeroTP are able to deliver the full 1MB of data transmitted for low error rates <0.000035, but above that TCP performance drops rapidly due to retransmissions, while AeroTP is still

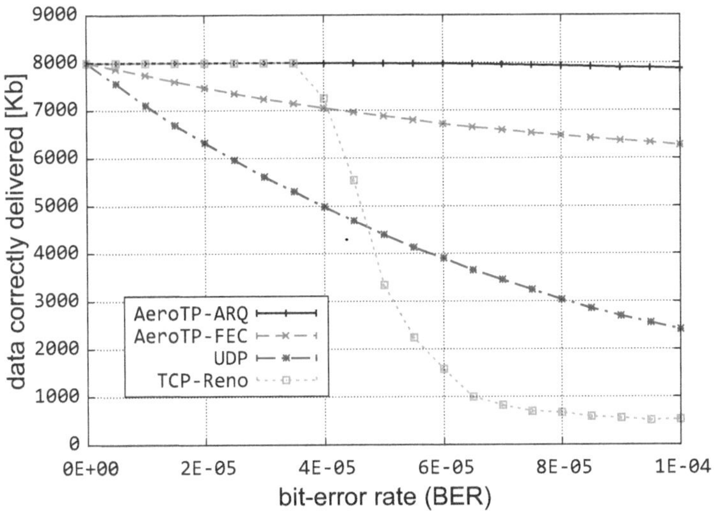


**Figure 4.17** AeroTP and TCP cumulative goodput

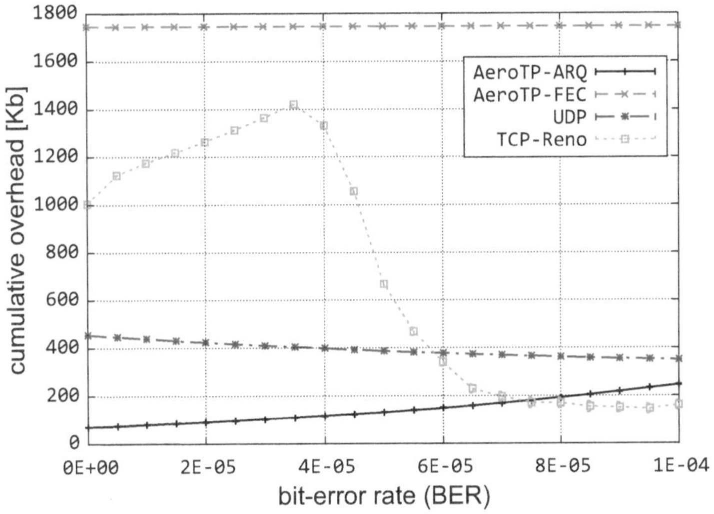


**Figure 4.18** AeroTP and TCP cumulative overhead

able to deliver nearly all the data at the highest error rates as shown in Figure 4.17. In the same plot, we see that UDP loses a percentage of the data due to corruption as the BER increases, but that the AeroTP quasi-reliable mode loses a much smaller percentage. Lastly, in Figure 4.18, we see that the performance improvement of the AeroTP reliable-mode is achieved with a much lower overhead than TCP, while the quasi-reliable mode does incur significant overhead, but does not cause any increased delay as the BER increases.

## 4.5.2 AeroRP and AeroNP Simulation Results

Simulation of various modes and location information of AeroRP and AeroNP is evaluated against the traditional MANET (mobile ad hoc networking) protocols DSDV, AODV, and OLSR using the ns-3 network simulator [313, 355, 339]. Simulations are performed with 5–60 or 10–100 randomly placed nodes in an area of 150km$^2$, over 10 runs with each simulation running for 1000s, plotted with 95% confidence interval bars. The communication model is multi-hop peer-to-peer with as many flows as the number of nodes in the network to a GS (ground-station sink). All the ANs are configured to send 1 packet/s. Using this lower packet rate, we can correctly evaluate the performance of the protocol by minimizing the probability of congestion at the MAC layer.

The On-Off application in ns-3 is used to generate CBR (constant-bit rate) traffic. The 802.11b MAC is used over the Friis propagation loss model to limit the transmission ranges of nodes. The transmit power is set to 50dBm to achieve a 27800m (15 nautical mi) transmission range. The mobility model is 3D Gauss–Markov (G-M) [72] with node velocities ranging from 50–1200m/s. Setting the G-M parameter $\alpha$ between zero and one allows us to tune the model with degrees of memory and variation, set here to 0.85 to have some predictability in the mobility of the nodes, while avoiding abrupt AN direction changes. Ferrying of packets is enabled by default in AeroRP using a drop-tail queue implementation. The maximum buffer size is set to 400 and the maximum buffering time is set to 30s. Any packet that reaches its buffer expiry time is purged. Active snooping is enabled by default.

The performance metrics for the evaluation of AeroRP are packet delivery ratio (PDR), routing overhead, and delay: PDR is the number of packets received divided by the number of packets sent by the application; routing overhead is the fraction of bytes used by the protocol for AeroRP control messages; delay is the time taken by the packet to reach the destination node MAC protocol from the source node MAC protocol. *Accuracy* is the number of packets received divided by the number of packets sent at the MAC layer. This allows us to measure how *accurate* a route is for a given routing protocol based on whether or not the route that was chosen for the packet results in a successful reception at the destination. This is a good metric to gauge the quality of a route in a highly dynamic topology in which the validity of a route can rapidly change.

### Performance Comparison by Mode

A single stationary sink node is located in the center of the simulation area representing the ground station. AeroRP is tested in the ferrying (store-and-haul), buffer, and drop modes [339]. In buffer mode, the packet queue size and timeout are configured to be in line with what AODV and DSDV implement. AeroRP is tested with these three modes as well as in both beacon and beaconless promiscuous mode.

Figure 4.19 shows the average PDR as the number of nodes is increased. The node density of the network affects all of the routing protocols with AeroRP ferrying packets in beaconless promiscuous mode performing the best (no hello beacons, but neighbors snoop on transmission). The PDR for all AeroRP modes increases as the number of nodes increases with the exception of a slight performance degradation as the number of nodes approaches 90 and higher. The PDR for both DSDV and AODV immediately

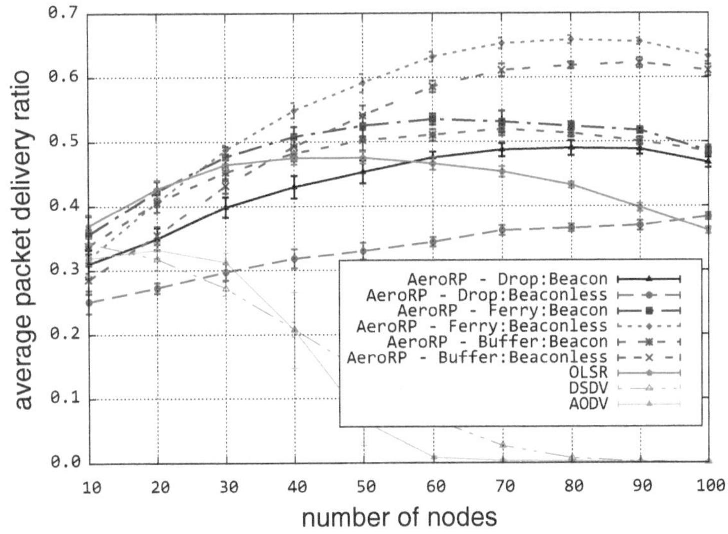


**Figure 4.19** Effect of node density on PDR

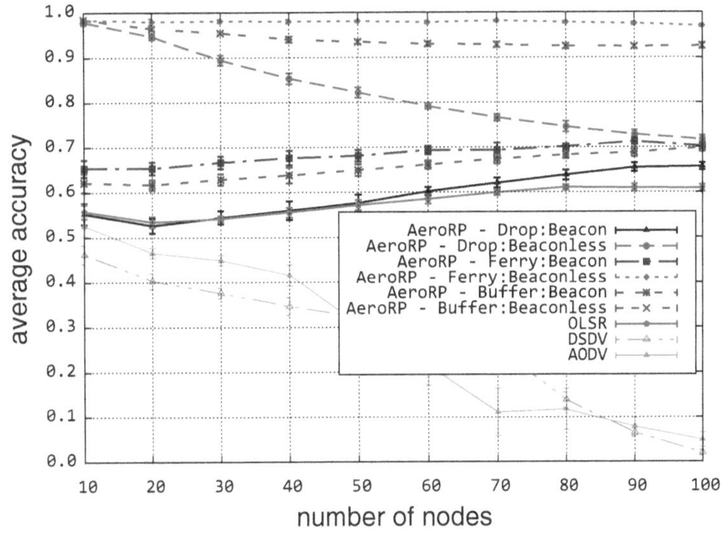


**Figure 4.20** Effect of node density on accuracy

degrades as the number of nodes increases. This is most likely due to the increase in overhead observed as the number of nodes increases. The performance of OLSR starts to degrade around 50 nodes. This suggests that as the number of nodes increases, AeroRP is able to make more intelligent decisions on how to move the data packets towards the destination whereas the non-AeroRP routing protocols are relying on non-geographic-based links to move the packet to the destination.

Figure 4.20 shows how accurate the various routing protocols are as the number of nodes is increased. OLSR, DSDV, and AODV yield an accuracy of less than 65%. All of the various modes of AeroRP have an accuracy of 50% or higher at all node densities. The accuracy of ferrying and buffering packets with AeroRP stays constant at almost 100% as the number of nodes increases. This illustrates AeroRP's ability to accurately deliver packets, largely facilitated by predicting whether fast moving neighbors are in or out of transmission range by using the last known distance between the source and neighbor as well as the neighbor's velocity. Of the AeroRP modes, the beaconless promiscuous mode is more accurate than the beacon modes for two reasons. First, the surrounding nodes overhear *all* data packets in range and thus get trajectory data every time a packet is transmitted. This results in sharing trajectory information more often than sending out periodic hello beacons. Second, putting the control data in the actual data packets makes the communication more symmetric than sending separate control packets. A control packet that is 44B may be transmitted successfully to a neighbor. However, this does not necessarily mean that a 1000B payload plus the control overhead can be successfully transmitted to that same neighbor, especially if that neighbor is on the edge of the transmission range.

For AeroRP, the accuracy increases with the node density with the exception of AeroRP running in beaconless promiscuous mode but with no ferrying or buffering of packets. This slight decrease in accuracy can be attributed to the nodes having to rely on data transmissions to communicate their trajectories to nearby nodes. The buffering and ferrying allows data to be delivered at different times in the simulation whereas AeroRP that is not ferrying or buffering packets is not sharing this information as often. This does not occur when AeroRP is in beacon mode but not ferrying or buffering packets because it still regularly shares its trajectory information with its neighbors in the form of periodic hello beacons. OLSR yields higher accuracy as the number of nodes increases but still not as high as the AeroRP modes.

As illustrated in Figure 4.21, the average overhead of the network increases with the node density. However, the proactive discovery and maintenance of routes in AODV

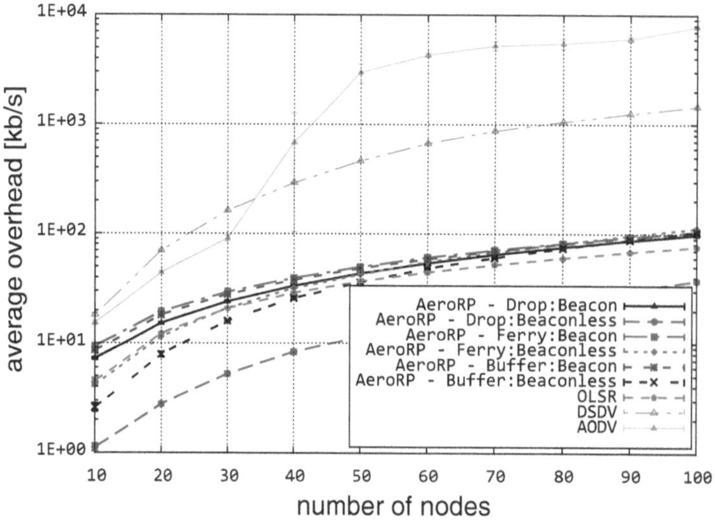


**Figure 4.21** Effect of node density on overhead

results in exponentially increasing overhead as the number of nodes increases from 30 to 50 nodes. The overhead of AODV would probably continue to increase exponentially and monopolize the bandwidth if it did not hit the network saturation point. The overhead of DSDV increases linearly from 18kb/s at 10 nodes to 1.5Mb/s at 100 nodes. The overhead of OLSR and the AeroRP routing protocols also increase linearly, but not as drastically as DSDV, from around 15kb/s at 10 nodes to around 100kb/s at 100 nodes. Note that AeroRP in beaconless promiscuous mode with no ferrying has the lowest overhead. This makes sense because the beaconless promiscuous mode removes the need for separate control packets, thus cutting down on overhead. Also, this mode of AeroRP transmits fewer packets because it drops them immediately if there is no route, which also cuts down on overhead.

The effect that node density has on the delay of data packet transmissions is shown in Figure 4.22. The AeroRP modes that ferry or buffer packets have the highest delay because the packet will be held as long as is required for delivery or until the packet expires in the queue in buffer mode. Although the delay is higher, these modes usually deliver more packets than the other routing protocols as previously shown in Figure 4.19. The two AeroRP modes that buffer packets have delay in the middle compared to the other routing protocols because they will only hold packets for a finite amount of time. The two AeroRP modes that do not do any packet ferrying or buffering have the lowest delay because they will drop the packet immediately if there is no route.

It is interesting to note that the ferrying and buffering modes of AeroRP actually decrease in delay as the node density increases while the other AeroRP modes, AODV, DSDV, and OLSR actually increase in delay as the node density increases. Perhaps as the node density increases, the buffering and ferrying modes are able to get their buffered and ferried packets to their destinations more quickly while the increase in nodes results in more hops and thus longer delay to the destination. It is suspected that calculating

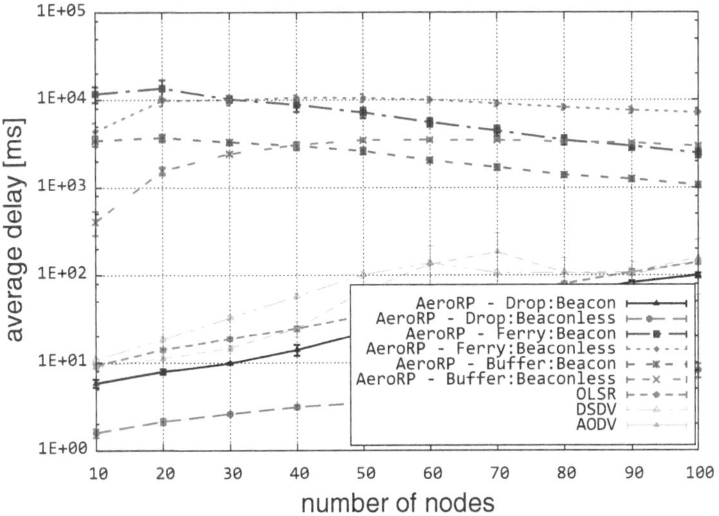


**Figure 4.22** Effect of node density on delay

just the average delay for data packets that were not ferried or buffered for the AeroRP modes that do this would result in the same increasing trend of delay as the number of nodes increases.

### Effect of Location Information

The previous analysis used location and trajectory in each packet to make routing decisions (either from only hello beacons or from all data packets). There are four possible combinations of information based on whether AN packets contain geolocation and velocity vectors (NoGS-TopologyInformation) or do not (NoGS-LinkInformation), and whether or not a GS (ground station) is *additionally* broadcasting to all in-range ANs information about complete airborne network topology (GS-TopologyInformation – most relevant to scheduled commercial, civilian with flightplans, and planned military missions), or only information discovered from other ANs (GS-LinkInformation – most relevant to other unplanned scenarios or confidential military missions) [313].

Figure 4.23 shows the performance of these four levels of location information of AeroRP in terms of PDR as the airborne node density increases from 5 to 60, moving at a velocity of 1200m/s. The GS update mode with LinkInformation performs better until the node density reaches 35 nodes and later starts to degrade as the density further increases. The increase in node density leads to more link breakages, which may result in disruptions to the end-to-end paths determined by the AN. Though the GS tries to update the changes in link information by sending out trigger updates, there is a chance that some nodes might not receive all the updates leading to inconsistent routing of packets. As the node density increases, AeroRP update modes without GS updates and using geolocation information begin to perform comparatively better as the nodes in this mode exchange geolocation information.

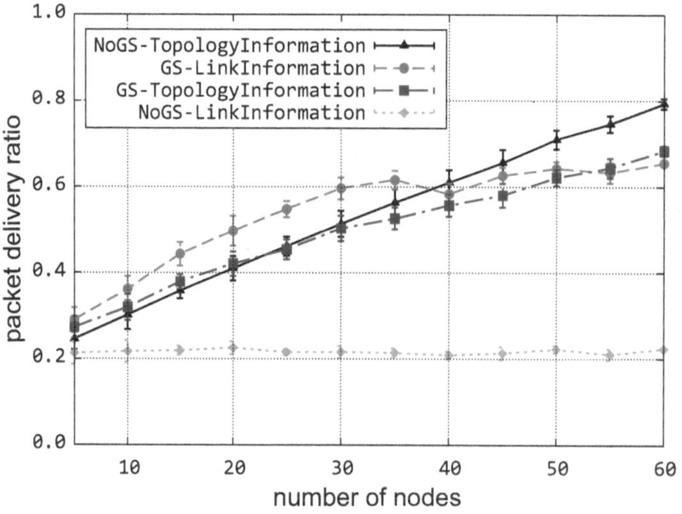


**Figure 4.23** Effect of node density on PDR

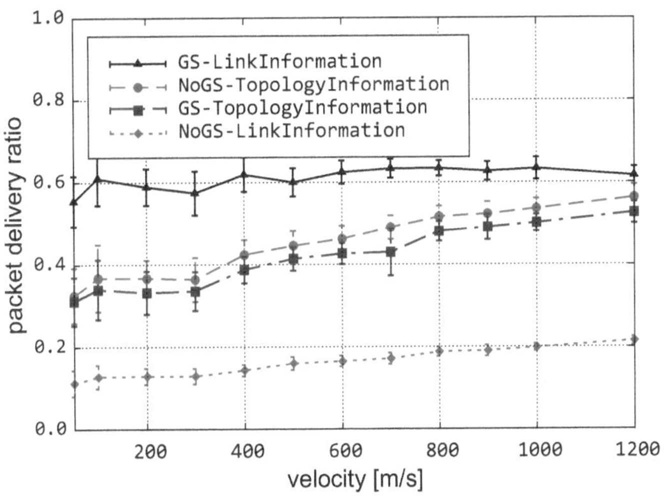


**Figure 4.24** Effect of node velocity on PDR

Figure 4.24 shows the variation of PDR as the velocity increases from 50m/s to 1200m/s for 35 nodes. At 35 nodes the average network connectivity is 75%, determined from the geolocation information of nodes and their transmission ranges. Average network connectivity of 75% is chosen to permit the analysis of PDR in a not fully connected network. AeroRP in the presence of GS and using LinkInformation performs better. This is because with the 3D Gauss–Markov mobility model, ANs move without any abrupt change in their direction. Thus the links predicted by the GS remain stable most of the time. However, as the velocities increase there are many link breakages leading to the GS sending an increased number of updates. Since the ANs run the Dijkstra shortest-path algorithm, they need to have consistent information from the GS, otherwise this may lead to inconsistent routing of packets. The other modes of AeroRP using TopologyInformation, with and without the presence of GS updates, perform better as the velocities increase. This is because at high velocities the nodes have a better chance of identifying a path to the destination as they come across different neighbors, which increases the probability of identifying a good route to the destination. Furthermore, they are aware of the entire topology of the network as they exchange geolocation information in the AeroNP headers. The AeroRP mode using LinkInformation in the absence of GS updates does not perform well as ANs only know about their in-range neighbors.

The key points are that not only does increasing AN density improve performance, but so does (generally) increasing *velocity* since ANs are more likely to come in contact with more neighbors.

Figure 4.25 shows the end-to-end packet delay with varying node densities. The delay is greater while using TopologyInformation with and without the presence of GS updates. This is because the calculation of TTI (time to intercept in Section 4.4.2) is based on the AN current and predicted geolocation coordinates. By using

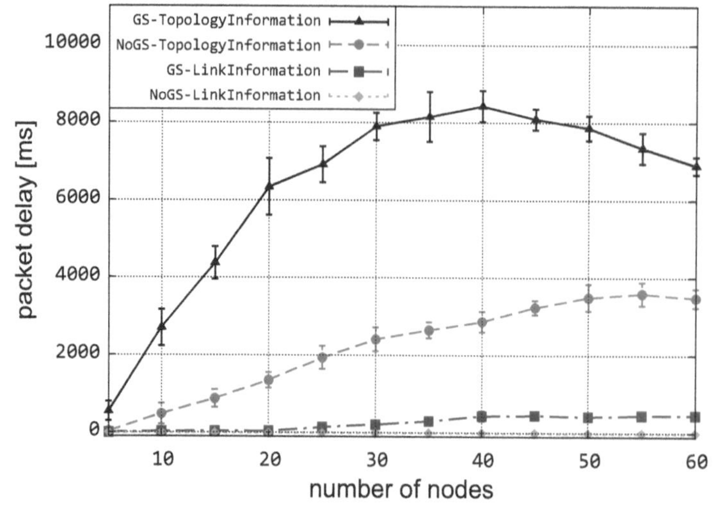


**Figure 4.25** Effect of node density on delay

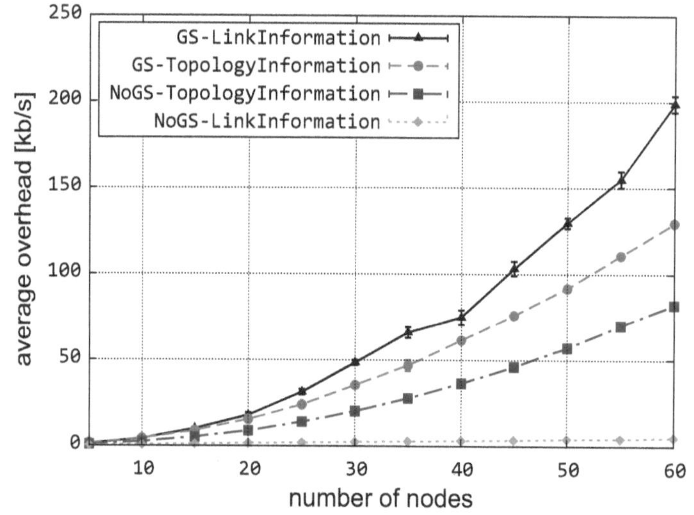


**Figure 4.26** Effect of node density on overhead

TopologyInformation from GS updates, the node trajectory information can be predicted for a longer duration as the expireTime provided by the GS is usually greater compared to the neighborholdTime. This results in higher TTI values while using TopologyInformation from the GS updates. TTI values specify how long a packet should be buffered before identifying a route to the destination. The other two modes using LinkInformation have very low delay as they do not use the TTI metric.

The variation of AeroRP control overhead with node density in its various modes of operation is analyzed in Figure 4.26. The control overhead in the absence of GS updates and using LinkInformation is significantly less as there are no AeroRP messages sent out except for the initial bootstrap beacons. Control overhead however increases exponentially with the increase of node density for AeroRP mode with GS updates enabled and sending out LinkInformation. This is because as the node density increases, so does the number of links between in-range nodes, leading to more updates from the GS whenever a new link is formed or an existing link is broken.

## 4.6 Summary

This chapter presents the characteristics of various airborne network scenarios. The existing TCP/IP-based Internet protocol architecture is not well suited to airborne networks. Traditional MANET routing protocols such as AODV, DSDV, and OLSR are not designed for highly dynamic topologies with high velocity and unpredictable mobility, as is the case for emerging drone and tactical military scenarios.

A new protocol architecture that addresses these issues with domain-specific transport and network layers is presented. It is observed that the exchange of information across layers provides significant benefit in the highly mobile environment. The domain-specific transport (AeroTP), network (AeroNP), and routing (AeroRP) protocols leverage cross-layer information in optimizing end-to-end performance.

AeroTP provides flexible and composable disruption-tolerant mechanisms that can deliver information end-to-end between airborne nodes in challenging scenarios. By predicting when links will be available based on trajectory information, as well as actively listening for nearby nodes, AeroRP can send data opportunistically towards its destination and make much more efficient use of available network capacity.

A simulation analysis shows significant improvement over traditional Internet and MANET protocols, as well as providing insight into the cost/benefit trade-offs of various operating modes.

# 5 UAV Systems and Networks: Emulation and Field Demonstration*

Jae H. Kim, Natasha Neogi, Claudiu Danilov, and Andres Ortiz

This chapter discusses the design, implementation and deployment of an unmanned aerial vehicle (UAV) network for the purpose of transmitting video surveillance data amongst nodes. It presents a heterogeneous network consisting of multiple stationary and mobile ground-based nodes, as well as multiple autonomous aerial vehicles. A design process to enable emergent system safety in an Unmanned Aerial Vehicle Platform System through the correct integration of critical UAV subsystems in a scalable fashion to enable collaboration across multiple UAVs is outlined. The UAV network system field demonstration comprised of 16 fixed ground nodes, 1 mobile ground node, and 2 airborne nodes connected by two routing gateways to a legacy wired network is described. Networking and communications effects, including the impact of node mobility, network partitioning and merging, as well as gateway failovers are examined through the course of the demonstration. Discussion of the networking protocols deployed, as well as analysis of performance issues are then placed in context of the overall field demonstration.

## 5.1 Unmanned Aerial Vehicle (UAV) Platform Systems

The Aerospace Laboratory for Embedded Autonomous Systems (ALEAS) Unmanned Aerial Vehicle (UAV) platform system consists of several individual Unmanned Aerial Vehicles (UAVs). Each system had a common architecture along with attendant ground stations which can be used to relay commands, as well as operational procedures, to the UAVs if needed. A modular software infrastructure was then used to enable collaborative UAV flight [361]. The main purpose of the UAS platform was to develop assurance methods to enable autonomous flight of UAVs in a shared airspace environment, such as the National Airspace System (NAS). Thus, a primary concern was making sure that the system exhibits high safety and reliability standards. The field demonstration [125], further detailed in Section 5.2.1, required that a video be streamed across a network comprised of both ground and aerial nodes. As fielded in the outlined experiment, the

* The field demonstration section in this article has been presented and published as "Experiment and Field Demonstration of a 802.11-based Ground-UAV Mobile Ad-Hoc Network," *IEEE MILCOM 2009*, Boston, MA, October 18–21, 2009 by Claudiu Danilov, Thomas R. Henderson, Thomas Goff, Jae H. Kim, Joseph Macker, Jeff Weston, Natasha Neogi, Andres Ortiz, and Daniel Uhlig.

UAS system consisted of two UAVs, continuously operating at an altitude of approximately 100–300ft. above ground level for roughly 50 minutes each.

### 5.1.1 UAV Platform System

The base UAV platform consists of several key subsystems: a standard hobby model aircraft, a gasoline engine, a PC104 embedded form factor computer, a commercial off-the-shelf (COTS) autopilot, along with other safety oriented electronics (i.e., Deadman's switch) and payload (i.e., video camera system). The model aircraft chosen was a quarter-scale Extra 300S, and the autopilot used was a Piccolo Plus manufactured by Cloud Cap Technology. Additional subsystems (e.g., the video camera, auxiliary computing, and communications packages) were integrated into the base platform as necessary to provide the desired functionality.

#### Airframe

The quarter-scale Extra 300S is a popular sport-aerobatic hobby aircraft. One of the main advantages of an aerobatic design is its ability to withstand loads that exceed the nominal operational conditions of a UAV. This is an important consideration since platforms under development are not likely to remain within nominal operation at all times, thus having a safety margin for recovery becomes crucial. In addition to being neutrally stable, the aerobatic airframe is partially decoupled in its control surfaces and degrees of freedom. This decoupled nature of the control greatly facilitates the debugging and tuning of autopilot control loops.

The aerobatic UAV platform also provides a large, easily accessible payload bay, and a generous wing-surface area in an airplane with a small (80 inch) wingspan. The wing can be disassembled into two pieces, so the UAV can be easily transported, and no significant disassembly is required in order to gain access to any of the aircraft's interior components. In addition to the aircraft meeting the objectives of the platform, it is important to choose an airframe that is well known by the R/C safety pilot. Experience with the airframe allows the safety pilot to detect problems that can be corrected early on to avoid catastrophic outcomes. Moreover, the pilot's opinion on the aircraft's behavior during autonomous flight is crucial for the process of properly tuning autopilot control loop gains.

#### Vehicle measurements and layout

Key vehicle weights and dimensions are summarized Table 5.1.

A top view photograph of the aircraft is illustrated in Figure 5.1. Here the canopy and cowling are removed, showing the general layout of the UAV test bed.

#### Power Plant

The engine that currently drives the Extra 300S is a Brillelli 46GT, which spins a Menz $20 \times 8$ propeller. The Brillelli is a gasoline engine, with a displacement of 46cm$^3$, weight of approximately 16kg, RPM range of 1800–13500, and flight time (with a full tank) of about 50 minutes. A dedicated NiMH battery powers the electronic ignition.

**Table 5.1** Weights and dimensions for vehicle platforms

| Measurement | Value |
|---|---|
| Wingspan | 2.03m |
| Wing area | $0.75m^2$ |
| Horizontal tail span | 0.914m |
| Horizontal tail area | $0.17m^2$ |
| Vertical tail span | 0.38m |
| Vertical tail area | $0.087m^2$ |
| Payload bay dimensions | 0.61m × 0.2m × 0.15m |
| Aircraft length (firewall to tail) | 1.47m |
| Distance from spinner tip to firewall | 0.34m |
| Weight of wings (w/pitot probe) | 1.4kg |
| Weight of fuselage (w/engine) | 7.3kg |
| Weight of avionics, wiring and batteries | 0.8kg |
| Total empty weight | 9.5kg |

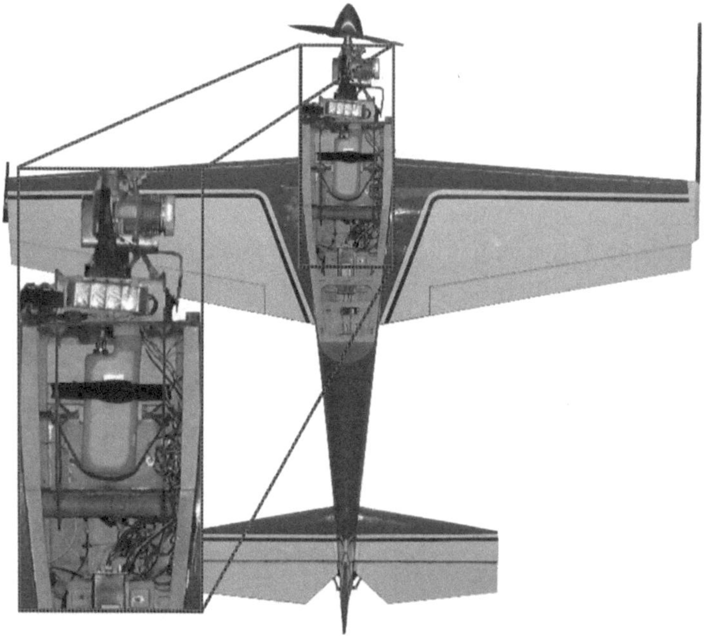

**Figure 5.1** Interior layout of the unmanned aerial vehicle

Isolation mounts were built in order to mitigate the effects of engine vibrations on the measurements from the autopilot's sensors, which otherwise are capable of compromising autonomous flight safety. The gasoline engine possesses a Hall sensor in the shaft, which plays a vital role in the monitoring of the propulsion system.

## 5.1.2 UAV Autopilot Control System

### Autopilot

The Piccolo Plus autopilot, manufactured by Cloud Cap Technology, serves as the main avionics unit. It was selected due to its proven track record in government UAV applications, widespread use in academia, and its highly configurable interface [426]. The Piccolo Plus embeds a microprocessor and a sensor suite that includes a 1Hz Global Positioning System (GPS), 20Hz Inertial Measurement Unit (IMU), and an air data sensor. A pitot-static probe is used to measure the true airspeed and altitude. The Piccolo also has a 900MHz radio to establish a communication link with the ground station [424]. The radio link provides the ability to modify the UAV platform's autopilot parameters, and relays remote pilot commands when in manual control mode (see Figure 5.2). The ground station is used to send and receive messages to and from the autopilot, display and record telemetry, and provide an interface to make changes to the autopilot in-flight [426]. One ground station is capable of monitoring up to eight UAVs (each equipped with an on-board unit).

An operator interface (OI), which can run on any computer with a Windows operating system, communicates with the ground station via an RS-232 link to enable user interaction with the avionics. The OI is intended to visualize the state of the autopilot and its sensors, as well as update configurations and issue commands. Configurations available to the user range from control loop gains and aircraft surface limits to mission decisions (e.g., what to do in the case of a loss of communication with the ground station). A set of closed loop controls is used to maintain stability, navigate waypoints, and perform other basic autopilot functions.

Externally, the Piccolo connects to a tachometer board, which itself connects to the propulsion system. Through this device the Piccolo can receive measurements of engine RPM, as well as command an engine kill when required. The Piccolo has 16 servo output channels in order to interact with the airframe. The UAV only utilizes five of these channels: one for the elevator, two for the ailerons, one for the rudder, and one for the throttle. Control surface actuation is performed by Futaba S3152 servos that

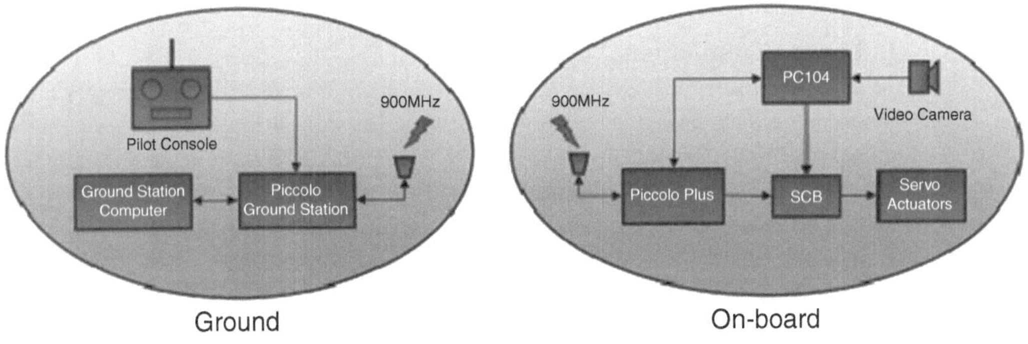


**Figure 5.2** Interconnection diagram of experimental flight setup

are powered and controlled using a three wire connection (consisting of positive 5V power, ground, and signal wires). The servos can be commanded to deflect and hold a position 45 degrees in either direction from their neutral point, by being sent a modulated square wave signal with a pulse width between 1 and 2ms. The deflection of the servo is linearly proportional to the width of the pulse starting at the neutral position (which corresponds to a 1.5ms pulse width). The pulse width commands were generated either by the Piccolo or by the Safety Pilot, depending on the mode of operation. A Futaba T9CAP R/C transmitter connects to the ground station to enable a safety pilot to take manual control of the aircraft without disabling the Piccolo Plus. Additionally, the autopilot has two serial ports that can be used to establish a communication link with payload units, such as sensors or computers. A software development kit is available in order to develop custom communication protocols with these types of units [54].

### Electrical Subsystems

A battery-based power system is used in order to meet the electrical power needs of the autopilot, on-board computing equipment, payload, servos, and engine. The use of COTS and custom components that have different power consumption and voltage requirements makes the design of the electrical subsystem challenging, especially when size and weight constitute two of the main constraints. In order to meet the desired criterion, the electrical power needs are addressed by separating the electrical systems into four main subsystems: propulsion power, servo motor power, autopilot power, and computational platform (including payload) power. Each subsystem is powered by an independent battery, chosen to minimize cost and weight, while meeting the current, voltage, and operation time requirements of the devices being powered. Electrical switches are also installed on the airplane for each subsystem in order to protect the batteries from becoming entirely drained. Finally the routing of the wiring, the type of connectors used, and the location of the batteries are also taken into consideration in order to keep the center of gravity (CG) of the aircraft in place and avoid failures due to electrical interference, vibrations, and overheating.

### On-board Computational Capability

The UAV platforms are each equipped with an x86 form factor computer to enable monitoring, redundancy, and autopilot functions. The PC104 Plus form factor computational platform – specifically the Cool RoadRunner LX-800 manufactured by Lippert Embedded Inc. – was chosen for the on-board UAV platform. This computer is based on the AMD Geode LX800 processor, which is a very low power chip that consumes about 1 watt while running at 500MHz. The trade-off between processing speed and power consumption is of particular interest in order to achieve the maximum performance without sacrificing on-board battery power. In addition to this feature, the computer has several interfacing ports that allow communication with the required onboard devices and payload, not always found in platforms of this size.

### Payload: Boeing Field Demonstration Network Hardware and Vision System

A payload bay with dimensions of 4" × 5.75" × 6.25" was available to house the Boeing Mobile Router (BMR), a.k.a., the Open Tactical Router (OTR) for a ruggedized military version. Minor modifications were made to the Extra 300S UAVs to support the field demonstration: a small USB-based video camera, which is connected to the Boeing mobile router, was mounted to the UAV, and a small rubber dipole antenna for the BMR's 802.11 radio was attached through a protrusion in the hull.

For the airborne platform, we selected a ruggedized PC-104 based, 400MHz computer from Parvus as our mobile router platform. Packed together with a MiniPCI adapter, an 802.11 card, a power board, and a GPS board, it fits into an aluminum railed card cage (also from Parvus), that fits into a cube of 4" on each dimension. We also added rubber shock absorbers (Shock Rocks from Parvus) that attach to the corners of the cube, adding about 0.5" in each direction, as seen in Figure 5.3. The entire hardware is about 1.13kg without the battery. We used the payload battery in conjunction with the power board (allowing input between 8 and 40V DC) to power the computer; it consumes about 10W.

The ground nodes have similar specifications, but are not ruggedized; we use commercial 802.11b radios and antennas. The radio model is an EnGenius EMP-8602 PLUS-S dual band 802.11 a/b/g card with up to 600mW of transmit power. During the demonstration, the nodes are set on 802.11b mode, 5.5Mbps base rate for both unicast and multicast, at either 600mW or 400mW transmit power. The antenna is a 7dBi Rubber Duck Omni RP-SMA for the 2.4GHz band. The link rate adaptation is turned off for the duration of the experiment. All mobile routers use a GlobalSat BU-353 USB GPS receiver based on the SiRF Star III High Performance GPS chipset. We use the built-in patch antenna and USB connector to the router. The video cameras used were Logitech QuickCam for Notebooks Pro, and the video rate is set to 400Kbps, for an image of 320 × 240 pixels at a rate of 15fps. We selected the Cisco 3845 Integrated Services Router for the backbone topology segment of our field demonstration configuration. Further specifications are available on the web [99].

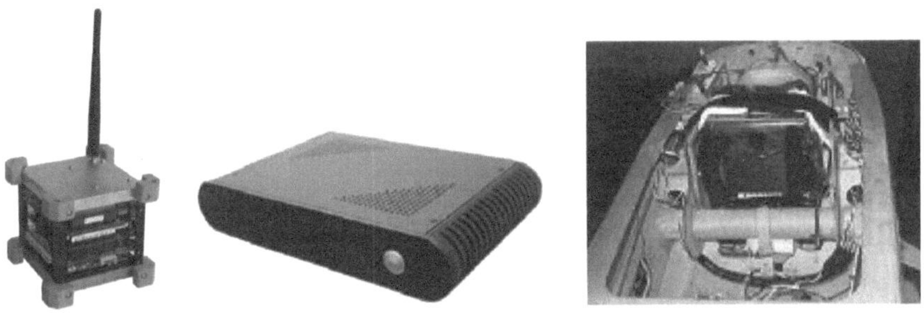

Miniaturized (Airborne) Standard (Ground) Integrated on UAV

**Figure 5.3** Boeing mobile routers: airborne, ground, UAV integrated versions

### 5.1.3 UAV Communication System

#### Piccolo Link Communication Failure: PC104 Actuated Servo Control of Vehicle

Initially, the Piccolo radio link was a possible single point of failure for the UAS. The radio link was the only path of communications between the ground station and the UAV platform. In an attempt to mitigate this hazard, a Backup Servo Control board that multiplexed servo inputs from the Piccolo and a secondary source was designed. This circuit board defaults to passing through signals from the PC-104 or alternate RC receiver to servos. When the autopilot is functioning normally, a 5V signal (high output) from the Piccolo's deadman switch output prompts the circuit into passing commands to the servos from the Piccolo servo outputs. In the event the Piccolo software realizes there is an autopilot malfunction, a watchdog timer turns the deadman switch off. Extra inputs are also present on the backup board that allow the ground station operator, or on-board fault monitoring systems that can formulate further conditions, to trigger the switch away from Piccolo control.

There are several caveats to the effective functioning of the PC-104 servo control of the UAV. First, during preliminary testing, it was noted that a standard hobby R/C receiver encountered significant interference while operating anywhere near the range of the Piccolo's 900MHz radio system. Secondly, the reliability of the software running the watchdog timer, and the electrical connections on the designed circuit board, could not be ensured to have the required degree of reliability conventionally seen for airworthiness in the NAS. There have been no documented UAV testbed failures in which the Piccolo deadman switch was triggered [425]. In addition, the sheer number of servo connectors present on the backup servo board (of which there are 24) presented a situation where elementary errors such as connecting a lead into the incorrect socket or improperly securing a connection could easily result in the loss of control of the UAV testbed.

#### Flight Termination in Absence of Network Control Commands

A basic level of flight safety is implemented in all the UAV platforms. Should the avionics lose the ability to communicate with the ground station, the autopilot will navigate the airplane to a predetermined GPS waypoint, and loiter until the communications link is restored. Should the engine also fail while in this state, the aircraft will continue to loiter, but will start losing altitude while trying to maintain the commanded airspeed. If the airspeed set by the ground station operator *a priori* is chosen properly, this will result in a semi-controlled descent at a specific location. Furthermore, the loss of both the communication and GPS links will result in the autopilot attempting to hold its specified course using its inertial navigation aids and air-data sensors. However, as time elapses, inertial sensor drift and varying winds will result in the airplane deviating from its intended path. The actual amount of deviation in the flight path could not be quantified realistically, since influencing variables, such as weather and other disturbances, were difficult to characterize. Furthermore, owing to the limited accuracy of the GPS system, and the lack of a sonic or laser altimeter on-board, the UAV testbed cannot safely autoland itself if the radio link cannot be restored.

Under the current engine, autopilot, and electrical system configuration, passive flight thrust termination is possible. The power source for the electronic ignition can be tied into the Piccolo's deadman switch through a tach/deadman interface board [424]. In the event of an autopilot malfunction, the deadman switch will no longer provide a signal to the interface board, which will terminate powered flight by turning off the engine's ignition. A modified version of the throttle closed, full up elevator, full right rudder, full right aileron flight-termination system leading to a spiral descent into terrain was implemented. The servo commands that allow for a gentle circling descent of the airplane with the engine off have been captured and, under the passive flight thrust termination mode, fed back to the servos, in the event that the communications link cannot be restored after a fixed period of time, or fuel is running out. A field-programmable gate array (FPGA) based circuit board is used to accomplish this task, in order to minimize weight and keep this system separate from the other avionics.

#### RC Link and Manual Reversion to Safety Pilot

A required safety feature was the employment of a safety pilot for each vehicle, who maintained a line-of-sight and had the ability to intervene and resume stick-to-surface control of the assigned vehicle. One of the key observations made was in the noticeable difference in the response of the control surfaces to manual input from the R/C transmitter [174]. If the manual input was relayed through the Piccolo radio system rather than the R/C receiver, there was a noticeable delay and an apparent loss of resolution in the control surface actuation. The fact that flying the UAV platform over the Piccolo radio link is quite a different experience from flying the exact same airplane over an R/C hobby 72MHz link underscores the need for an expert safety pilot.

### 5.1.4 UAV Monitoring System

The PC104 enables our software oriented approach to fault tolerance. It is a functional unit that is constantly running during flight, monitoring the performance of the physical plant (airframe and propulsion system) and the autopilot. It may also execute other mission tasks such as path planning and conflict detection and resolution if resources are available. However, in the event of a faulty mode the PC104 may be required, via software, to assume the function of the failed component(s). An example of such a situation may be a failure in the Piccolo that prevents it from generating the expected outputs for one or more of the control surfaces. In this event, the PC104 must emulate the behavior of the Piccolo and generate itself the correct outputs to the control surfaces. Further critical events, such as a complete Piccolo failure, may require the PC104 to switch from an on-board monitoring system to spare autopilot in order to bring the aircraft to the ground.

#### Reliability and Security Engine

The LX-800 PC104 has a PCI port available to interact with custom FPGA boards. This enables the addition of a customized hardware-implemented error-checking device to ensure the safety and reliability of the implemented autopilot monitoring system.

An example of such a system is the Reliability and Security Engine (RSE) developed at the University of Illinois [312]. The RSE is an architectural framework that provides security and reliability support for a processor without sacrificing its performance [268]. The framework hosts hardware modules that provide reliability and security services and implements input/output interfaces between this module and the processor's main pipeline as well as the executing software (i.e., operating system, applications). It is intended to protect the processor against security vulnerabilities as well as internal errors, thus making the system secure and reliable. An embedded implementation of the RSE was developed on a PC104 Plus form factor FPGA board and successfully integrated with the PC104 computer, as seen in Figure 5.4 [325].

### Online Monitor Implementation: Aileron Control Loop Example

Online monitoring systems were developed for the UAV platform propulsion system and the aileron control system. Hardware in the Loop simulations (HITL) were performed in order to test the efficacy of these two online monitoring systems. The advantage of this approach is that it enables the use of the actual hardware components used in real flight to test the algorithms implemented in this research. The fault detection and diagnosis algorithms run on the PC104 as they would onboard the aircraft. The fault recovery actions are then executed, and the corresponding control commands are sent from the PC104 to the required servo actuators. The experimental setup therefore is more realistic than a software simulation environment, giving a higher degree of authenticity to the results (Figure 5.5).

**Figure 5.4** LX 800 PC104 (left) and BEN One PCI-104 FPGA board (right)

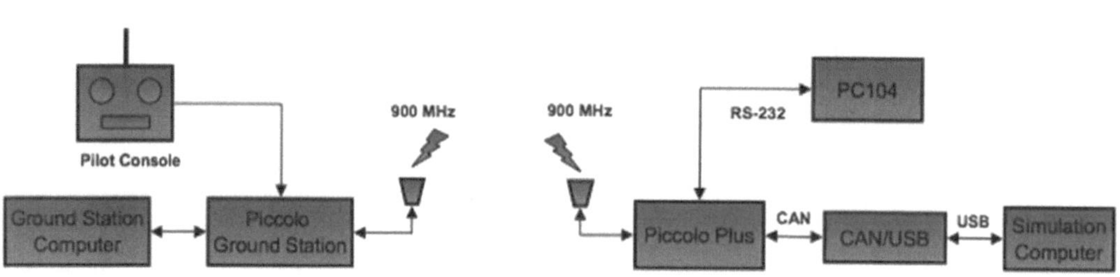


**Figure 5.5** Hardware in the loop experimental setup

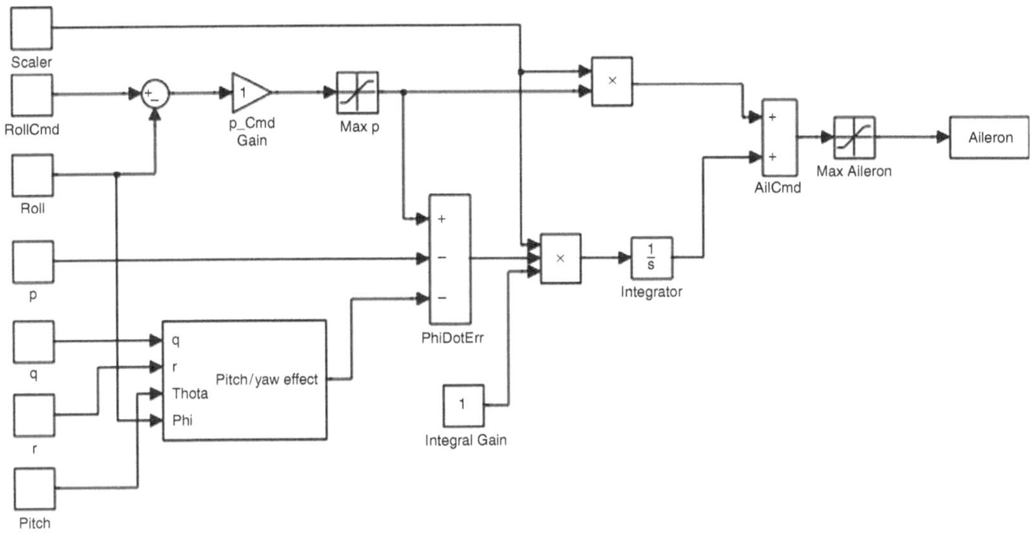


**Figure 5.6** Aileron control loop implemented in Piccolo Plus

Incorrect computation of aileron deflection is a failure that can be detected by looking at aircraft behavior or autopilot behavior. The autopilot calculates the aileron deflection based on the error between the commanded (desired) roll angle and the actual roll angle, and the angular rates (Figure 5.6). Using information from the Piccolo manufacturer [425], a control loop was designed to predict the aileron deflection. The control loop runs in the PC104 computer on-board the aircraft and is updated at 20Hz. The input to the controller comes from a telemetry broadcast from the Piccolo, which contains all the necessary sensor data to compute the aileron deflection. This broadcast is performed at 20Hz, which is the maximum update rate of the Piccolo sensors.

A similar monitor was developed and implemented for the propulsion system, and these monitors were flown during the RANGE field demonstration. In this demonstration, multiple versions of the platform, carrying implementations of the developed monitors, enabled the validation of the detection and recovery algorithms implemented through the HITL fault injection experiments [53, 327, 417].

### 5.1.5 UAV System Integration and Safety

Commercial off-the-shelf components refer to parts that are manufactured commercially and generally undergo little or no customization before they are integrated into the overall system. These systems are used frequently in UAS design, primarily to manage cost and to enable the ability for reuse [444]. However, assembling a safe and reliable UAV platform from COTS components presents many challenges. In general, it is a non-trivial exercise to assemble a safe, secure, and reliable system from potentially unsafe components [275]. The behavior of such a system cannot be easily quantified because the detailed workings of the system are usually unknown, leading to difficulty in establishing an equivalent level of safety for certification purposes [191]. The end user is only provided with the basic black-box behavior of the system, and guidelines about its intended application. In addition, each COTS component generally comes with pre-specified interfaces. Therefore limiting, quantifying, and controlling the interactions between these COTS components is difficult to ensure and assess [105].

Key systems to be integrated across the vehicle platform and ground station were the communication, computation, and control systems [245]. These three systems are described in greater detail in Sections 5.1.2–5.1.3. Similarly, the airframe was required to be integrated with the propulsion system (Section 5.1.1) and, to some extent, the electrical subsystem (Section 5.1.2). The desired level of safety was achieved through the use of a flight safety system (FSS) [54], in conjunction with an integrated vehicle health monitoring system [325], which acted to diagnose, mitigate, and control discovered faults, and terminate flight in the case of extreme failures.

The traditional purpose of an FSS is to protect people and property in the case of off-nominal vehicle flight [175]. More recently, flight safety systems also attempt to prevent catastrophic destruction of the flight vehicle. There are two major types of flight safety systems: range-centric and vehicle-centric. These systems are categorized, respectively, based on whether the decision-making activity and source of decision data originates at a ground station or in the flight vehicle's systems. This categorization is not exclusive, and range-centric FSS may have components on-board the vehicle and *vice versa*. The primary range-centric flight safety methodologies are range-containment and vehicle-destruct. The ALEAS UAS opted to employ a range-containment philosophy with a non-destructive flight termination approach [416], as expounded in Section 5.1.3.

Range-containment involves choosing the locations and trajectories of the flight vehicle such that the vehicle (or debris) can be contained in a specific area in the case of every possible malfunction. Thrust-termination involves turning off the vehicle's thrust producing system, in order to decrease the kinetic energy of the flight vehicle so that the severity of the impact is reduced. Thrust-termination is not a stand-alone flight safety methodology and is generally used in combination with flight-termination to safely end flight activities. A flight-termination system (FTS), also referred to as a vehicle recovery system (VRS), to highlight its non-destructive nature, allows the vehicle to come to a soft landing. In the ALEAS UAV, a set of control surface commands is implemented to drive the airplane into an unpowered, high-lift state on the edge of stall, to bring the airplane back to the ground quickly but not overly forcefully (Section 5.1.3) [53, 54].

In the event of a lost communication link, UAVs with a very basic autopilot may simply cease flying their course, and loiter on the spot or around a pre-determined point. A slightly more advanced UAV would fly around a few fixed points or a pre-programmed path in an attempt to reestablish its command and control link. At the most advanced level, the aircraft would be capable of flying to a pre-determined waypoint or location, and auto-landing or terminating flight [326]. Even in a fully autonomous UAV, the loss of the GPS link can further complicate matters, requiring the aircraft to fall back to a lower level of flight safety than would normally be engaged. For the ALEAS UAV, flight termination is employed in the event of a loss of control link (Section 5.1.3).

Integrated vehicle health management (IVHM) is employed in order to allow the UAS to continue operation under degraded performance, in the event of a fault. These IVHM systems have evolved into systems capable of actively managing a vehicle's health instead of simply monitoring vehicle health in a passive function [59, 287]. The two major categories of IVHM are post-flight systems and in-flight systems. Their main purpose is to promote safety by monitoring specific subsystems, and assist in the diagnosis, and possibly prediction, of failures. Truly reactive IVHM requires the timely detection, diagnosis, and mitigation of faults, accidents, and failures with real-time guarantees. The ALEAS UAV employs an inflight IVHM in order to detect and diagnose both propulsion and aileron faults (Section 5.1.4).

## 5.2 Unmanned Aerial Vehicle (UAV) Networked Systems

The UAV network system field demonstration [125] was conducted by Boeing Research & Technology (BR&T) with support from Boeing's Global Military Aircraft (GMA) division, Naval Research Laboratory (NRL), and the University of Illinois, Urbana-Champaign (UIUC) at the NASA Dryden Flight Research Center under the Office of Naval Research (ONR) sponsorship. This section describes the field demonstration of a UAV network system, focusing on the network topology, emulation, integration, experimentation, and field demonstration. The field demonstration network is composed of 16 ground nodes (not to be confused with the two Piccolo ground stations and computers used to operate the UAV system), a mobile ground vehicle and two fixed-wing UAVs, connected by two routing gateways to a legacy wired network. The network performance of an 802.11 surface (ground) UAV network along with network effects such as mobility, network partitions, network merges, and gateway failovers are demonstrated. Finally, the results for recorded data traffic and for the state of the routing protocols, with the mobile nodes participating as sources of data traffic, are presented.

### 5.2.1 UAV Internetworking Operational Concept (CONOPS)

The Operational Concept (CONOPS) in the field demonstration is based on the UAV Internetworking and range extension through multiple UAVs, as shown in Figure 5.7. It was performed for resilient mobile internetworking of two unmanned aerial vehicles (UAVs) and 16 surface (ground) nodes to extend the surveillance range and battle space connectivity, under the Robust Airborne Networking Extension (RANGE) research project, sponsored by the Office of Naval Research (ONR).

To emulate the *UAV Internetworking and Range Extension CONOPS*, we deployed a surface (ground) network of 16 nodes, and flew two small fixed-wing UAVs above this deployed site, both individually and simultaneously. Both planes were equipped with Boeing's miniaturized mobile routers and commodity video cameras. We also placed a mobile router on a ground vehicle that drove around the site and sent audio and video back to a viewing area.

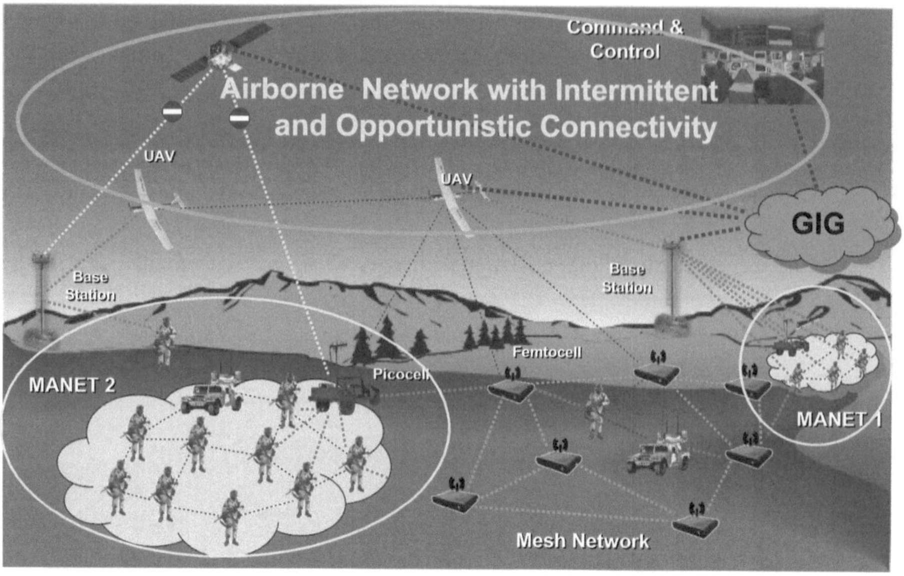


**Figure 5.7** UAV internetworking and range extension field demo CONOPS

### 5.2.2 Network Configuration

In this field demonstration, we focused on the application of Mobile Ad-hoc Network (MANET) protocols to airborne and hybrid airborne/surface scenarios; for example, we considered a use case of two UAVs supporting a surface (ground) network consisting of largely static nodes. The UAVs served as a source of data and also could be considered as disadvantaged nodes with an intermittent connectivity in the topology. The hybrid air/surface network was interconnected by two gateways to a notional backbone network running legacy protocols and devices. A key element of this field demonstration was to show how such MANETs could be interconnected to backbone networks in the non-trivial case of using multiple gateways between the backbone network and the MANET. The network topology and configuration is shown in Figure 5.8, consisting of 16 MANET routers on the ground (at NASA DFRC) and one surface (ground) and two airborne mobile routers. The MANET routing domain was connected to the backbone through two border routers that had instances of both MANET and legacy protocols running on two different interfaces.

### 5.2.3 Network Emulation

There are multiple network simulation tools, such as ns-2 [286], ns-3 [17], OPNET [22], and QualNet [336], that are typically run on a single computer, and abstract the operating system and protocols into a simulation model for producing statistical analysis of a network system. In contrast, network emulation tools, such as PlanetLab [393], NetBed [196], and MNE [348], often involve a dedicated test-bed, or connecting real systems under test to specialized hardware devices. Among these available tools that

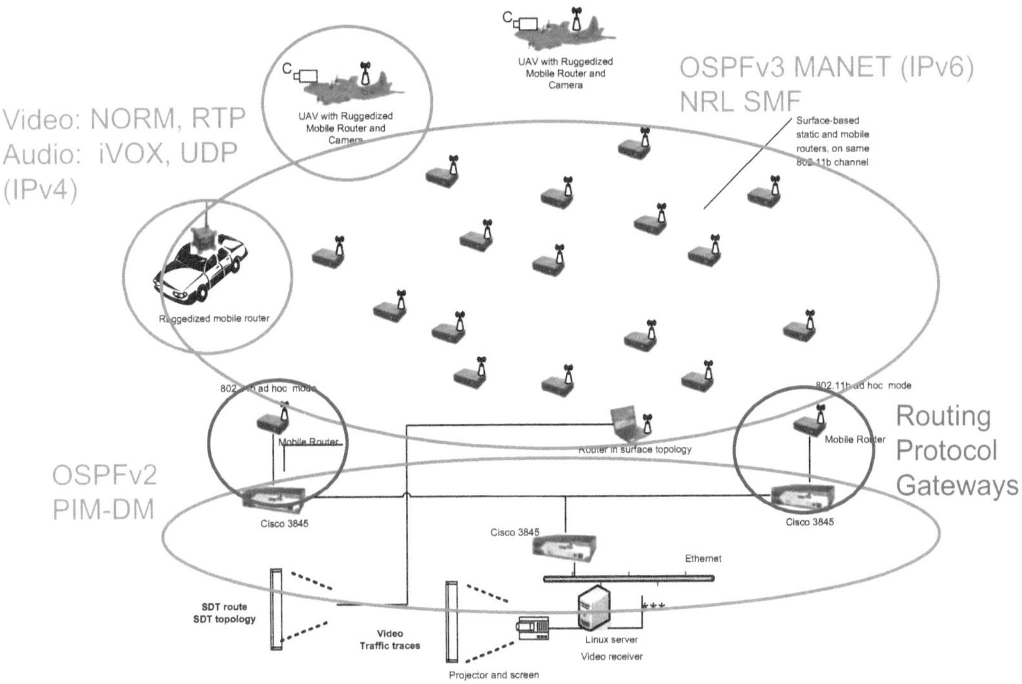


**Figure 5.8** Field demonstration network topology and configuration [125]

allow for running real applications over emulated networks, we selected the Common Open Research Emulator (CORE) in terms of scalability, ease of use, application support, and network emulation features.

CORE [22] is a framework for emulating networks on one or more PC that can emulate the network stack of routers, PCs, and other hosts as well as simulate the network links between them. Because it is a live-running emulation, these networks can be connected in real-time to the physical networks and routers. This way it can provide the realism of running live applications on an emulated network while requiring relatively inexpensive hardware. Figure 5.9 shows a snapshot of the example of air-to-ground UAV internetworking scenario network topology built by the GUI-based CORE.

EMANE, another emulator, works with layers 1 and 2 (physical and data link layers) while CORE focuses on emulating layer 3 and above (network, transport, session, and application layers). Together, the CORE/EMANE tools provide an easy-to-use graphical user interface for designing and configuring virtual networks, consisting of lightweight Virtual Machines (VMs) interconnected with pluggable MAC and PHY layer models. CORE provides a Graphical User Interface (GUI) and Python framework for building virtual networks (see example in Figure 5.9). Because CORE is using virtualization and the Linux network stack, with unmodified user protocols and user applications, the system faithfully represents networking layer 3 and above, but the link and physical layers are greatly simplified. On the other hand, EMANE provides the necessary pluggable framework for emulating link and physical layer wireless networks. More details on CORE and EMANE can be found in [22] and [221].

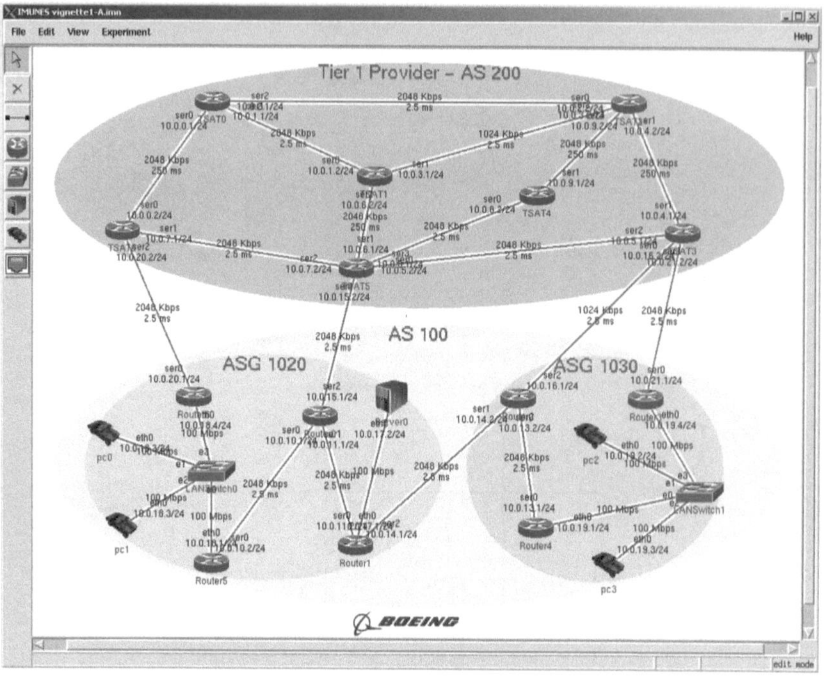


**Figure 5.9** GUI-based CORE: air-to-ground UAV internetworking scenario network topology buildup

## 5.2.4 Network Protocols

### Mobile Ad-hoc Network (MANET)

There are extensive works on MANET routing protocols, dealing primarily with packet forwarding in a mobile environment where the network is dynamically changing its topology. Most MANET routing protocols can be classified either as Proactive (e.g., OSPF MANET [322], OLSR [108]), Reactive (e.g., AODV [336]), or Flooding (e.g., SMF [285]). These types of protocols perform well in certain network conditions, which allow the selection of the right type of protocol. Several hybrid protocols [196], [348] combine both proactive and reactive approaches to organize the network such that they can either provide higher scalability, or tune network topology for better performance. A comparative study [393] suggests that there is no single protocol that works in all situations, and that different types of protocols have to be designed for different networking conditions.

**Open Shortest Path First (OSPF)** is a de facto standard routing protocol designed for wired networks. OSPFv3 MANET is an extension of OSPF for IPv6 [322] to support mobile ad-hoc networks (MANETs). The extension, called OSPF MDR, is designed as a new OSPF interface type for MANETs. OSPF MDR is based on the selection of a subset of MANET routers, consisting of MANET Designated Routers (MDRs) and Backup MDRs. The MDRs form a Connected Dominating Set (CDS), and the MDRs and Backup MDRs together form a bi-connected CDS for robustness [OSPF MANET].

Boeing has developed an implementation of OSPF MANET as an extension of the quagga routing suite. Note that while OSPF MANET is specified for IPv6, its extension exists to carry IPv4 routing information in the protocol. All of the applications in this demonstration were IPv4-based.

**Protocol Independent Multicast – Dense Mode (PIM-DM)** is a multicast routing protocol that uses the underlying unicast routing information base to flood multicast datagrams to all multicast routers. Prune messages are used to prevent future messages from propagating to routers without group membership information [15]. Boeing developed a PIM-DM software implementation as an extension to the XORP routing suite.

**Simplified Multicast Forwarding (SMF)** is a basic IP multicast forwarding protocol suitable for wireless mesh and mobile ad hoc network (MANET) use, developed by the Naval Research Laboratory (NRL). SMF specifies techniques for multicast duplicate packet detection (DPD) to assist the forwarding process. SMF also specifies DPD maintenance and checking operations for both IPv4 and IPv6. SMF takes advantage of reduced relay sets for efficient MANET multicast data distribution within a mesh topology. In the demonstration, our routing software integrated a Boeing PIM-DM implementation with NRL's SMF software, which was using the OSPF MANET CDS for a multicast relay set. More details on this integration are found in [125].

## Adaptive Mobile Ad-hoc Network (A-MANET)

To address the enumerated problems and provide an enhanced performance in the disconnected, intermittent, and low bandwidth (DIL) environments, Boeing developed Adaptive Routing (a.k.a., Multi-Mode Routing) [126], which switches between proactive, reactive, and flooding routing strategies automatically, depending on the network dynamics. The Adaptive Routing protocol uses an efficient link state routing protocol, such as OLSR [108], OSPF MANET [322], if the network topology is stable enough. When the link state routing protocol cannot keep up with the network dynamics, the route to a specific destination may enter a flooding state, where packets will be disseminated to the entire connected MANET. Whenever possible, data driven on-demand routes will be created based on the paths traversed by flooded data packets. The protocol may use an on-demand routing state as a flooding optimization when such routes become available, and until the link state protocol becomes aware of the new topology. Figure 5.10 shows the general state diagram of the adaptive routing protocol.

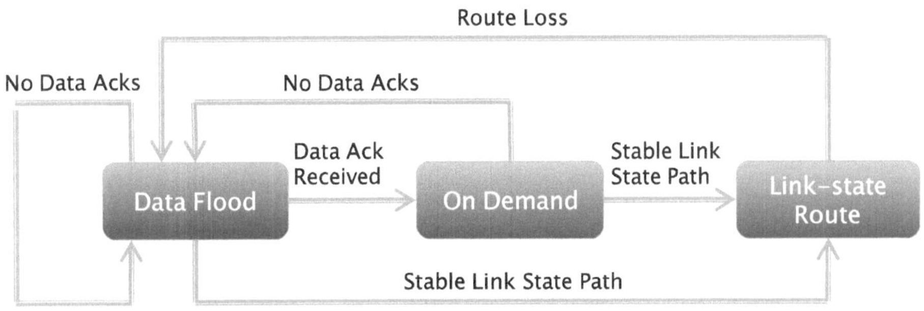


**Figure 5.10** Adaptive routing state diagram

**Table 5.2** Comparison of protocols performance

| Protocol | Delivery Ratio | Overhead Ratio |
|---|---|---|
| Flooding | 100% | 29.16 |
| OSPF MANET Default | 50.3% | 5.36 |
| RANGE OSPF MANET | 65.9% | 6.01 |
| Adaptive Routing | 92.4% | 5.12 |

Adaptive Routing [126] avoids organizing the network into nodal regions or neighborhoods, and instead operates on a per-source/destination or per-flow basis. The Adaptive Mobile Ad-hoc Network (A-MANET) based on Adaptive Routing adaptively applies different forwarding strategies based on the network conditions and the priority of the flows, on per-path or per-flow basis. A key design aspect is that synchronization required between the network nodes is reduced, because the protocol permits forwarding decisions to be taken without global or local coordination between nodes. Loop detection and avoidance is provided by standard duplicate packet detection [284]. Basically, the adaptive routing blends unicast-based flooding, on-demand (reactive) routing, and link state (proactive) routing without tight coordination (synchronization) between routers.

Our simulation and actual test results show in Table 5.2 that the packet delivery performance of broadcast-based flooding can be achieved at a level of network overhead closer to that of link state routing, in a mobile network scenario that is challenging for a pure link state protocol. This Adaptive MANET will significantly improve the airborne UAV networking connectivity and performance in the future.

**Evaluation Scenario**

(a) 30 nodes located in a 1200m by 1200m area, moving according to a random way point scenario with random speeds between 0 and 60mph.
(b) The wireless range of each node was set to 250m.
(c) Out of the 30 nodes, 15 were randomly selected to send data traffic each to one other randomly selected node, such that every node in the network was either a sender or a receiver.

### 5.2.5 Network Systems Integration

**UAV Network Software**

Software in use included the following:

(1) OSPF MANET software with extensions developed under the RANGE project for multi-gateway operation;
(2) PIM/SMF gateways for multicast integration;
(3) NRL's Scripted Display Tool (SDT) for visualization of node position and routing links;
(4) NRL's MGEN traffic generation software, including GPS integration through gpsLogger;

(5) NORM, RTP, and VLC for video transmission and reception;
(6) IVoX for voice transmission and reception;

The **Scripted Display Tool (SDT)** is open source software by NRL's PROTEAN Research Group that provides a simple visualization capability using standard image files for a background and set of overlaid nodes. A custom coordinate system can be defined for the background – in this case, the GPS coordinates of the demonstration area, as shown in Figure 5.11 – and node positions can be dynamically updated to "move" their associated icons about the background [310]. The Multi-Generator (MGEN) is open source software developed by the NRL's PROTEAN Research Group. MGEN provides the ability to perform IP network performance tests and measurements of both UDP and TCP. It also supports the inclusion of the node's current GPS position with each packet sent through the network, as well as the time the packet was sent (for latency measurements) [309].

The **NORM** protocol and software [17], developed at NRL, is designed to provide end-to-end reliable transport of bulk data objects or streams over generic IP multicast or unicast forwarding services. NORM uses a selective, negative acknowledgement (NACK) mechanism for transport reliability and offers additional protocol mechanisms to conduct reliable multicast sessions with limited *a priori* coordination among senders and receivers. A congestion control scheme is specified to allow the NORM protocol to fairly share available network bandwidth with other transport protocols such as Transmission Control Protocol (TCP). It is capable of operating with both reciprocal multicast routing among senders and receivers and with asymmetric connectivity (possibly a unicast return path) from the senders to receivers. The protocol offers a number of features that enable different types of applications or possibly other higher level transport protocols to utilize its service in different ways. NORM leverages the use of FEC-based repair and other IETF reliable multicast transport (RMT) building blocks in its design [17].

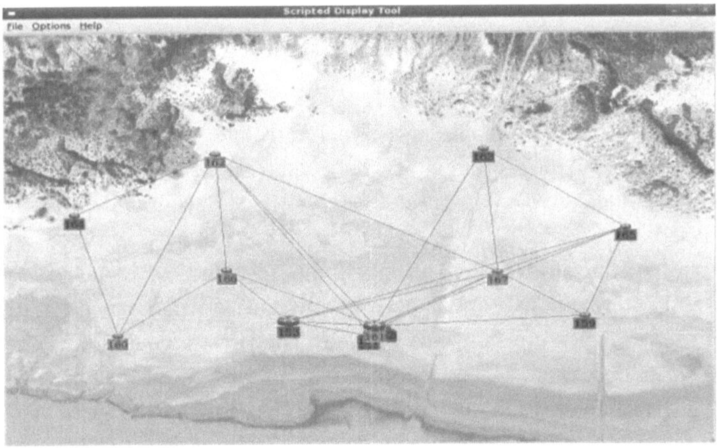


**Figure 5.11** SDT display of the deployed surface topology [125]

NRL's IVoX, the **Interactive Voice Exchange** (IVoX) application, is a voice over IP (VoIP) application that supports unicast and multicast, and also includes NORM integration for reliable communications. IVoX supports a number of voice encoding algorithms with data rates extending from as low as 600bps. We have instrumented our mobile routers to store a variety of logs. The experiment logging and data collection framework is based on Python and shell scripting. It includes sending MGEN beacons (including location information) to the visualization node; logging GPS information (latitude, longitude, altitude, and time); logging signal strength information from up to eight other wireless peers (the iwspy statistic limit); monitoring kernel route changes using rtmon; saving a full TCP dump from each specified network interface; using athstats to record wireless statistics; and saving the output of quagga, XORP, and SMF log files. Logging can be configured to start automatically at boot time, or at the time of acquiring a GPS fix. Scripts have been developed to process the multicast experimental results to generate end-to-end outage statistics and traffic graphs at each gateway.

## Visualization

We integrated our code with GPS logging and NRL's scripted display tool (SDT) for visualization (Figure 5.11), used a Boeing custom traffic trace plotter to show OSPF overhead, and used the video LAN client (VLC) for video display. We used SDT in two ways during the demonstration. The first use was to show the dynamic OSPF topology. In Figure 5.11, the geographic layout of the surface nodes, as well as links between them, are rendered against an aerial photograph of the lakebed. We modified the quagga OSPFv3 code to log network links to a file in a format compatible with NRL's CMAP tool. The log file and update interval can be configured using a quagga sty command either interactively or from a configuration file. Specifically, nodes were color coded as follows: (1) Purple nodes were active OSPF MANET MDR routers that were selected as MDR forwarders (also SMF forwarders in the multicast topology). (2) Green nodes were active OSPF MANET routers that were not MDRs. (3) Red nodes illustrated nodes for which FPS reporting is absent, such as the node 152 (airplane node) in the screenshot after it was returned to ground and powered off (Figure 5.11). The red lines between nodes displayed the links advertised by OSPF MDR routers in Router-LSAs. Note that in OSPF MANET MDR, this set of links does not represent the full topology but instead represents a pruned routing topology designed to given early shortest paths without the need to report all neighbor adjacencies. Therefore, the usable RF topology was actually greater than that depicted in Figure 5.11. In addition, we configured another display of SDT to show the active unicast route between the surface mobile router node and the gateways.

During the course of the demonstration, the SDT displays dynamically updated the network topology display as node position and connectivity changed. When the planes were airborne, they were shown as fast movers against the rest of the network on the map. We captured and displayed real-time plots of the OSPF traffic both in the MANET and in the backbone, on similar vertical scales. The displays illustrated that the MANET routing protocol overhead was largely contained within the MANET routing domain, and the routes redistributed by the gateway nodes did not contribute much to the backbone overhead [158].

### 5.2.6 Field Demonstration and Analysis

The field demonstration [125] was conducted with an 802.11 ground-UAV network composed of 16 ground stations, a mobile vehicle, and two fixed-wing UAVs, connected by two routing gateways to a legacy wired network. The network effects demonstrated include mobility, network partitions, network merges, and gateway failovers. This section presents experimental results and analysis for recorded data traffic and for the state of the routing protocols, with the mobile nodes participating as sources of data traffic.

We extracted the packet delivery ratio (excluding duplicates) of multicast traffic sent from a flying UAV, received at a host computer within the legacy network sitting behind the two gateways running PIM-DM/SMF. As the UAV was streaming video at a rate of 400Kbps, we sent two streams of multicast traffic in parallel, each at a rate of 10 packets per second, each packet carrying a 100byte payload. One of the streams was received natively at the host computer, while the other multicast stream was forwarded through the NRL NORM implementation. NORM was configured with a buffer of 75kB at the sender (about 1.5 sec of video), and with a 25% FEC redundancy. Figures 5.12 and 5.13 show the trajectory of the streaming UAV and the ground mobile node, respectively, during the experiment. Detailed analysis can be referred to reference [125].

The flight test was divided into five different phases, over which each network performance was analyzed. Phase 1 had both planes airborne (Nodes 150,152), with one plane sending multicast video and data, and the other plane forwarding opportunistically depending on its MDR status. Phase 2 involved turning off the second plane's router (Node 150), so that the plane streaming video and data (Node 152) had to rely on the ground network for forwarding. In Phase 3 the gateway used for data forwarding between SMF and PIM was turned off, so that the multicast routing protocol had to

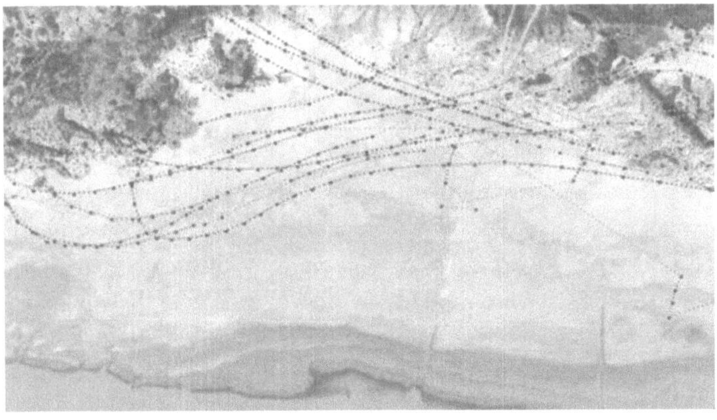

**Figure 5.12** Trajectory of the UAV that streamed video [125]

**Figure 5.13** Trajectory of the ground mobile node [125]

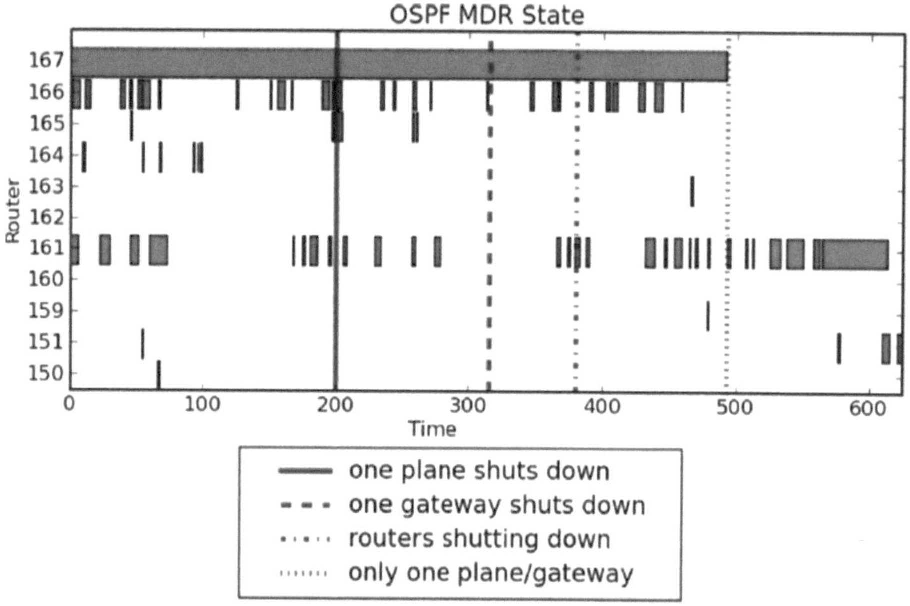


**Figure 5.14** OSPF MDR status of the network nodes [125]

fail over to the second gateway. Phase 4 involved successively failing ground nodes, until only the second gateway remained. Phase 5 involved monitoring the connectivity between the plane and second gateway, which were the only nodes remaining.

As can be seen from Figure 5.14, for the majority of the demonstration, Node 167 (ground node) functioned as an MDR, as due to its geographical placement in the network it possessed both excellent connectivity as well as a high router ID number. Nodes 164 and 165 were placed at the eastern and western extremities of the geographical ground network area. Node 161 functioned as the second gateway, and was the last ground node left alive in Phase 5. The multicast loss rates in Phases 1–2 are nearly identical (about 24–28% without NORM, and 14% with NORM), as the second plane only opportunistically functioned as an MDR. In phase 3, we experienced a short disconnection event (over 90% NORM multicast loss rate) as we failed from the primary to secondary gateway. As successive nodes are shut down in Phase 4, we begin to experience more disconnection events, where sudden loss rates evince localized spikes to 40–70% with NORM. In Phase 5, connectivity, and thus loss rates, are correlated with the position of the plane with respect to the secondary gateway. For a further in depth analysis of the results, see [125].

During the entire demonstration we noticed the wireless link quality was highly variable during the day, corresponding with changing environmental conditions (e.g., wind speed, temperature, and dust particles). We noticed a general degradation of the wireless channel quality, at times rendering the network topology completely unstable. Although we did not measure wind speed and dust particles on the desert area (lakebed), we observed some correlation between the increased wind speed and reduced stability of the network topology. We suspect this may be due to wind kicking up dust particles that can cause problems for the wireless signal, particularly given the low antenna heights of the fixed ground nodes.

## 5.3 Related Works


1. *Optic Flow: A Computer Vision Approach to Object Avoidance on UAVs*, by A. Ortiz and N. Neogi, IEEE Digital Avionics Systems Conference, Portland, OR, October 2006.
2. *A Safety-Oriented Approach to Designing UAVs Using COTS Technology*, by D. Uhlig, K. Bhamidipati, and N. Neogi, IEEE Digital Avionics Systems Conference, Portland, OR, October 16-19, 2006.
3. *Engineering Safety and Reliability into UAV Systems: Mitigating the Ground Impact Hazard*, by K. Bhamidipati, D. Uhlig, and N. Neogi, AIAA Guidance, Navigation and Control Conference and Exhibition, Hilton Head, SC, August 2007.
4. *A Dynamic Threshold Approach to Fault Detection in Uninhabited Aerial Vehicles*, by A. Ortiz and N. Neogi, AIAA Guidance, Navigation, and Control Conference and Exhibition, Honolulu, HI, August 2008.
5. Fault Tolerant Triangulation in Distributed Aircraft Networks with Automatic Dependent Surveillance Broadcast, by D. Uhlig, N. Kiyavash, and N. Neogi, *International Journal of Systems, Control and Communications*, eds. C. Hadjicostis, C. Langbort, N. Martins, and S. Yuksel, June 2009.
6. *CORE: A Real-Time Network Emulator*, by Jeff Ahrenholz, Claudiu Danilov, Thomas R. Henderson, and Jae H. Kim, IEEE MILCOM 2008, San Diego, CA, November 2008.
7. *Experiment and Field Demonstration of a 802.11-based Ground-UAV Mobile Ad-hoc Network*, by Claudiu Danilov, Thomas R. Henderson, Thomas Goff, Jae H. Kim, Joseph Macker, Jeff Weston, Natasha Neogi, Andres Ortiz and Daniel Uhlig, IEEE MILCOM 2009, Boston. MA, October 18–21, 2009.

8. *Comparison of CORE Network Emulation Platforms*, by Jeff Ahrenholz, IEEE MILCOM 2010, San Jose, CA, November 2010.
9. *Integration of the CORE and EMANE Network Emulators*, by Jeff Ahrenholz, Tom Goff, and Brian Adamson, IEEE MILCOM 2011, Baltimore, MD, November 2011.
10. *Adaptive Routing for Tactical Communications*, by Claudiu Danilov, Thomas R. Henderson, Thomas Goff, Orlie Brewer, Jae H. Kim, Joseph Macker, and Brian Adamson, IEEE MILCOM 2012, Orlando, FL, November 2012.
11. *Analysis of Mobility Models for Airborne Networks*, by Junfei Xie, Yan Wan, Jae H. Kim, Shengli Fu, and Kamesh Namuduri, IEEE MILCOM 2013, San Diego, CA, November 2013.

## 5.4 Summary

This chapter provides two main contributions. The first is a process to enable the integration of multiple critical UAV systems in order to assure the emergent property of global system safety. The second is a description and analysis of the networked ground-to-UAV field demonstration, under the range extension example use case of the UAV Internetworking CONOPS. Several key lessons were learned in the course of this work.

1. Correct requirements on controls, communications, and computational systems (especially software requirements) are crucial to ensuring UAS global system safety.
2. Resource allocation over the UAV platforms (both inter-UAV and intra-UAV) between propulsion, power, and monitoring systems are key points in achieving desirable emergent behavior, and directly impact the endurance of the UAVs.
3. Multiple failure condition scenarios revealed that connectivity can be maintained under several simultaneous faults subject to service degradation.
4. Under the failure scenarios, the network exhibited effects under node mobility such as network partitioning, network merging, and gateway failovers.

It should be noted that the ground-UAV airborne internetworking configuration and field demonstration is specifically described under the range extension scenario. The network emulation was based on the Common Open Research Emulator (CORE) tool. The detailed specification of the network protocols deployed and network system integration enabled the analysis of the network performance under the outlined five failure scenarios.

This demonstration, which took place at the NASA Dryden Flight Research Center in April 2008, was the first airborne field demonstration that combined OSPF MANET and NRL SMF protocols with OSPFv2 and PIM-DM legacy protocols, in a multiple gateway scenario using UAVs in the MANET. While the ability to connect MANET unicast and multicast to larger legacy networks was proven by this demonstration, several future directions presented remain to be investigated, such as improving the MDR selection process and using link quality information in network configuration.

## Acknowledgements

The author (Jae H. Kim) would like to thank BR&T and NRL researchers, Tom Henderson, Tom Goff, Joseph Macker, and Jeff Weston for their technical support. This work was supported by Office of Naval Research Contract N00014-05-C-0012 and also by Boeing Independent R&D and capital funds. The authors also would like to thank Dr. Santanu Das (ONR Program Manager) for his support.

The author (Natasha A. Neogi) would like to thank the UIUC ALEAS researchers, Keerti Bhamidipati, Daniel Uhlig, Ross Allen, and Christina McQuirk, as well as Miguel Frontera from Hobbico Inc. for all of their research efforts. This work was partially conducted by funding provided by the NSF through grant ITR 0325716, and the author wishes to thank Dr. Helen Gill and Dr. Timothy M. Pinkston for their support. This author also wishes to thank the Boeing Company for providing funding through the grant Boeing ITI RPS No.15. She is also grateful to NASA Langley Research Center for their support while writing this review article.

# 6 Integrating UAS into the NAS – Regulatory, Technical, and Research Challenges

Natasha Neogi and Arunabha Sen

This chapter provides the background and context for integrating unmanned aircraft systems (UAS) into a civilian airspace system. This chapter will cover regulatory concerns, social issues, and technical challenges with respect to integration efforts for UAS. As the airspace system continuously evolves, there is a migration towards a state of higher risk. When there is a set of standard designs for aircraft, procedures, and infrastructure, it becomes possible to use well-established frameworks for analyzing and mitigating risk. However, as change is introduced, the conventional frameworks may no longer yield the results in practice that are predicted in theory. These frameworks may thus prove misleading, and it becomes necessary to examine them in the context in which they were developed, in order to adapt them to accommodate changing factors. In the following section, the current regulatory systems for the certification of conventionally piloted aircraft (CPA) in civil airspace is examined. The applicability of this framework to the sea change posed by the introduction of UAS is discussed. The issue of privacy is briefly touched upon, as the common perception of this notion will be challenged by the integration of UAS into civil airspace. Gaps in current techniques used for assuring aviation systems, as well as necessary enabling technologies that will be required for safe UAS integration, are outlined. Finally, several major technical challenges produced by the introduction of UAS into civil airspace are outlined. A discussion of how their proposed solutions will interact with the current system is also tackled.

## 6.1 Regulatory Framework For Civil Aviation – Past and Present

Before any newly designed aircraft can enter into operation in civilian airspace over almost any country, it must receive an airworthiness certificate from the responsible aviation authority in the country in which the aircraft is registered, thereby demonstrating that it is airworthy. Receipt of an airworthiness certificate is only one step in the operation of an aircraft in the airspace; certificates relating to pilot and crew qualifications, operational certificates, and matters relating to other regulatory issues, such as continuing airworthiness, must also be addressed (see Figure 6.1). A strict adherence to these regulations result in a safe airspace system.

However, the functional purpose executed by unmanned aircraft systems (UAS) in civil airspace does not necessarily fit the conventional operational model of today's aircraft, as conveying people from an origin to a destination – or carrying people at

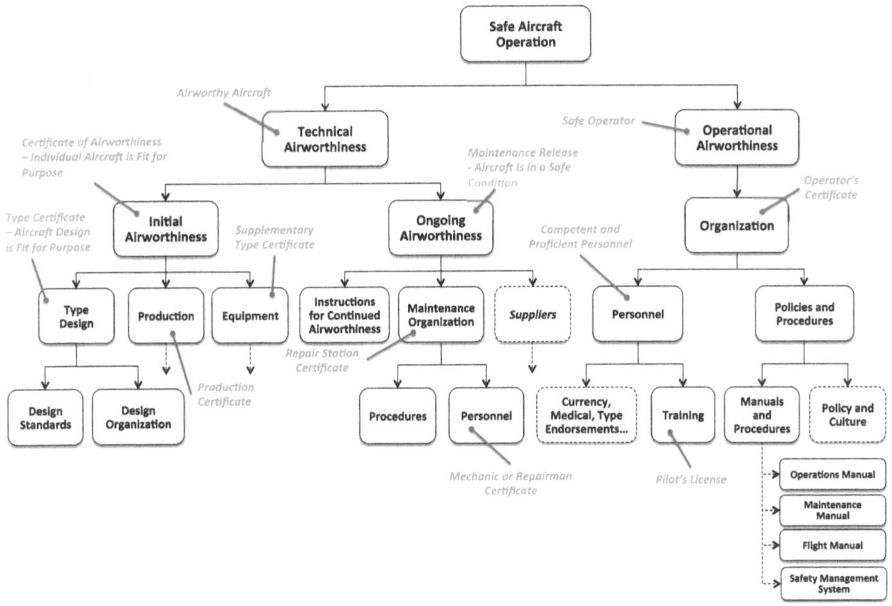


**Figure 6.1** Abstraction of framework for regulation of aircraft

all – is not a primary goal. It is unclear if enforcing the current airworthiness regulations for conventionally piloted aircraft (CPA) stringently, for all UAS, will be the best means to achieve safety in the airspace. In this section, we examine the general approach used by several prominent aviation regulatory bodies to regulate CPA, thereby enabling the safe operation of CPA in civil airspace. Regardless of whether the conventional regulations are applied to all UAS, basic knowledge of the current practices regarding certification is beneficial. The class of vehicles to which these regulations already apply will be, in essence, the fellow users of the airspace that UAS will have to contend with, if full, unfettered access is to be granted.

### 6.1.1 Airworthiness Certification

Airworthiness describes an aircraft's fitness for flight operations, in all possible environments and foreseeable circumstances for which the aircraft or device has been designed [194]. From a regulatory standpoint, airworthiness refers to an aircraft's ability to meet the established standards for safe flight. Demonstration of compliance to these standards is often achieved through adherence to a given set of regulations (e.g., US Title 14 Code of Federal Regulations (14CFR) [418], Regulation (EC) No. 216/2008 of the European Parliament and Council [118] etc.), created by a national civil aviation authority (CAA). In the following, we shall discuss concepts related to the regulation of CPA in the general framework of a CAA, to avoid focusing upon one particular country's regulatory framework.

The CAA is responsible for assessing whether or not an applicant demonstrates that a proposed aircraft design complies with the relevant regulation. Guidance materials (often called "*soft law*"), including acceptable means of compliance to regulations, are issued by the CAA via Advisory Circulars (AC) [165], Acceptable Means of Compliance (AMC) [122], or Temporary Guidance Leaflets (TGL) [248]. These, along with Minimum Operating Performance Standards (MOPS) and industry standards, developed by consortia such as the Radio Technical Commission for Aeronautics (RTCA) and the European Organization for Civil Aviation Equipment (EUROCAE), aid in the development of an aircraft designed to meet the requisite regulatory requirements.

The spectrum of aircraft designs covers a wide variety of characteristics and attributes, and it is difficult to come up with a single set of rules that, when applied to all aircraft designs, renders them airworthy without imposing undue regulatory (and cost) burdens upon certain aircraft types. In order to provide a fitting set of regulations to an aircraft design in order to promote airworthiness without imposing an undue burden, it is desirable to subdivide aircraft designs into groups, and thus *classify* them. Each aircraft design in that group possesses a set of common attributes, thereby enabling the application of fitting regulation in order to achieve the overall airworthiness of all designs in that group. Some examples of attributes that have been used for classification of aircraft are aerodynamic aspects (e.g., fixed wing, rotary wing, etc.) or mode of operation (e.g., passenger aircraft, cargo transport aircraft, etc.). Similarly, there are separate regulations that apply for different propulsion mechanisms (e.g., reciprocating engine, turbine engine, etc.). Regulations applicable to each group, relevant to its characteristics, can then be applied so that they are commensurate to the risk implied by the design. The first step in establishing airworthiness is often accomplished by developing the set of regulations which will apply to an aircraft manufacturing design, which has been classified according to the CAA's regulatory framework (the type certification basis). Many CPA fall within predetermined groups, which correspond to *parts* of the regulatory code that have a prearranged set of regulations that forms the initial basis for the aircraft *type certificate*. For example, Normal, Utility, Acrobatic, and Commuter Aircraft have airworthiness regulations specified in 14CFR Part 23 [420], EU-CS Part 23 [156], CASR Part 23 [104], etc.

### Type Certificate

The type certificate (TC) is a design approval issued by the CAA when the applicant demonstrates that an aircraft or aircraft product complies with the set of applicable regulations deemed necessary by the CAA. A type certificate can be issued for an aircraft, an aircraft engine or propeller, or other aircraft appliance [419, 120]. The type certificate confirms that the product is properly designed, performs suitably, and meets the required standards and regulations. The TC includes the type design (i.e., manufacturing design), the operating limitations, the type certificate data sheet (TCDS), the applicable regulations, and other conditions or limitations prescribed by the CAA. The set of applicable regulations forms the *type certification basis* for the aircraft. If operational restrictions or limitations are used in order to limit the risk posed by the aircraft, the vehicle can receive a type certificate for restricted operations. The TC is the foundation upon which all other approvals are granted, including the airworthiness certification. Traditionally, the CAA issues an airworthiness certificate for an aircraft when it finds that the aircraft conforms to its type certificate and, after inspection, is in a condition for safe operation.

Once a type certificate is issued, the design of the aircraft is fixed, and any changes require further approval, often through a supplementary type certificate (STC). Note that an airworthiness certificate can be achieved by means other than a type certificate: however, possessing a type certificate greatly simplifies and enhances the ability to demonstrate the airworthiness of an aircraft.

### Aircraft Production Certificate

Any aircraft part must be approved in order to be installed in a type-certificated aircraft [419, 120]. This approval can come via a production certificate (PC)/production organization approval (POA), a technical standard order (TSO)/European technical standard order (E/TSO) Authorization, or a parts manufacture approval (PMA). An aircraft production certificate or production organization approval enables the production of duplicate parts that conform to the given type certificate for which the PC or POA has been granted. This enables the production of multiple aircraft which conform to a type certificate. Each aircraft still requires a certificate of airworthiness; however, the type certificate and production certificate associated with their manufacture is an essential part of the airworthiness case.

An E/TSO is a minimum performance standard used to evaluate an article (e.g., material, part, component, process, or appliance). An E/TSO authorization is obtained for a specific article, when a manufacturer demonstrates that the manufactured article meets the E/TSO standards. An E/TSO authorization serves as both a design approval and a production approval. Receiving an E/TSO authorization means that the article meets a minimum performance requirement independent of the article's intended installation on an aircraft. In all cases, the installer must apply for an additional installation approval to install the article on-board the aircraft.

Parts manufacture approval is a US mechanism to enable the installation of a replacement or modified part onto a type certificated aircraft. It also covers both design approval and production approval. Approval of an application for PMA requires approval of the design by the Aircraft Certification Office (ACO) and quality system approval by the Manufacturing Inspection District Office (MIDO). The design approval phase certifies that the replacement or modified part complies with the airworthiness standards of eligible products (e.g., aircraft, engine, or propeller). PMA production approval allows the production and sale of approved aircraft parts that are eligible for installation on type certificated products. PMA is issued to the principal manufacturing facility that controls the design and quality of the article(s) for which the approval was granted. PMA is not transferable, and is valid until surrendered, withdrawn, or terminated.

### Summary

The use of the type certificates and production certificates/approvals to assure aircraft quality is a valid model for conventionally piloted aircraft. The translation of this concept to UAS, especially small remotely piloted aircraft systems (RPAS), is not quite as clear, as not only is there no longer a pilot on-board at risk, but the conventional points of failure between the pilot controls and aircraft control surfaces is now encapsulated in the wireless command and control (C2) link. There are no provisions under this current system to handle this issue. Furthermore, the ability to control the design and manufacture of UAS to the same degree of quality as conventionally piloted aircraft may act to remove the short time-to-market many UAS applications have. Current UAS tend to use unpedigreed parts for engines, sensors, actuators, and radios, due to size, weight, and power (SWAP) constraints, as well as cost, and may have their design modified on the fly. This does not easily lend itself to the current structures in place to assure airworthiness.

### 6.1.2 Regulations for Continuing Airworthiness

Once an airworthiness certificate has been granted for an aircraft, and the aircraft registered with the CAA, it becomes necessary to ensure the continuing airworthiness of the aircraft as it operates. Continuing airworthiness activities encompass all of the processes ensuring that, at any time in its life, an aircraft complies with the technical conditions fixed to the issue of the certificate of airworthiness and is in a condition for safe operation [224]. Continuing airworthiness can apply to an aircraft, engine, propeller, or part, and involves illustrating that it complies with the applicable airworthiness requirements and remains in a condition for safe operation throughout its operating life.

The process of continuing airworthiness flows from the initial type certification, maintenance, and operational regulatory authority approvals. Instructions for continuing airworthiness (ICA) are provided as compliance documents during certification activities, and are prepared in accordance with the applicable type certification. The ICA form the basis for the operators' approved maintenance data and enables inspection, adjustment, lubrication, removal, or replacement of parts and appliances. The ICA are seen as a means of ensuring the airworthiness of the aircraft through its operational life cycle. They touch on aspects of safety, maintaining conformance to the original type design, as well as delegating responsibility for ICA tasks. These tasks include operational maintenance, record keeping, incident and defect reporting and analysis, documenting design criteria which provide the necessary accessibility for inspection, and compliance with parts life limits (structural inspection programs), among other things. Continuing airworthiness also involves compliance with any airworthiness directives issued by the CAA, which carry the force of law [421]. Airworthiness directives are issued by a CAA in order to address safety concerns with certified aircraft currently being operated that must be corrected. Consequently, no person may operate an aircraft to which an AD applies, except in accordance with the requirements of that AD unless otherwise specified by the CAA.

### 6.1.3 Certification for Crew and Operators

Aircrew or airmen certificates are issued by the CAA in order to regulate the interactions between humans with very specific roles and the aircraft. The CAA sets the technical requirements and administrative procedures related to certification of civil aviation aircrew as well as acceptable means of compliance along guidance material on the licensing of the aircrew [422, 119]. Aircrew for CPA are subdivided further into pilots, airborne crew (excluding pilots), and ground crew.

### Pilot

Much like aircraft, pilots must be classified in order to apply the relevant regulation. Pilots can obtain different *privilege levels* of pilots' certificate (e.g., student, sport, recreational, private etc.). At a given privilege level, pilots are then *rated* to fly aircraft of a specific category (e.g., airplane, rotorcraft, glider etc. ). If a category is further divided into classes (e.g., rotorcraft are divided into helicopters and gyroplanes), the pilot must obtain a class rating. Pilot training is comprised of theoretical knowledge tests as well as practical tests, and may require the pilot to possess a certain number of flight hours with the specified category or class of aircraft referenced in the desired certificate. Furthermore, a medical certificate commensurate to the privilege being exercised may be required, and must be maintained in good standing. Flight instructors must also undergo certification.

### Aircrew

An aircraft's aircrew is comprised of those individuals who operate the aircraft while it is in flight. Similarly to pilots, supplementary flight crew must also be certified in order to operate aboard a type of certified aircraft. Regulations pertaining to the aircrew encompass the length of time the crew may spend aboard the aircraft before rest is required to knowledge of practical tests concerning the aircraft that must be administered. Aircrew can run the gamut from flight engineers to flight navigators, who were common on aircraft up to the 1970s and present in the cockpit. Aircraft type-specific training and conversion training between aircraft types are the responsibility of the air operator. Crew training so that the necessary qualifications are achieved are therefore requirements directed by the CAA to the operator who will have to train the cabin crew members accordingly, in topics such as safety procedures, record keeping, and incident reporting.

### Ground crew

Ground crew encompasses all those individuals who are ground-based and support the aircraft's operation. There are certification requirements for key members of the ground crew, such as maintenance technicians and flight dispatchers. Maintenance technicians or mechanics can be rated for the different aircraft systems they are qualified to work upon, such as the vehicle airframe or powerplant. These roles have significant reporting requirements, and can figure greatly in the development of aircraft and flight operation manuals, as well as influence continuing airworthiness certification. Ground school instructor and control tower operator certification is also considered under the purview of ground crew activity.

### Summary

The skills required for a remote pilot or operator may differ greatly from those of a pilot physically present in the cockpit. Furthermore, the ground and maintenance crew will play a more integral part in maintaining the overall airworthiness of the UAS system, as there may be little conventional situational awareness communicated via feedback to the remote operator, especially regarding vehicle health. The roles of the UAS operator and crew will require a sea change in the current practice of regulation concerning certification for crew and operators.

### Air Operator's Certificate

Air Operation Certification establishes the technical requirements and administrative procedures required for the commercial operation of an airworthy aircraft. The commercial operation of an aircraft is defined as the operation of an aircraft, in return for remuneration or other valuable considerations, which is available to the public or, when not made available to the public, which is performed under a contract between an operator and a customer, where the latter has no control over the operator [423, 121]. The commercial operation of an aircraft can involve the transport of passengers, cargo, or mail (referred to as commercial air transport), as well as applications such as aerial surveying, advertising, photography, and agricultural application, along with aero-medical and flight-training activities.

An air operator's certificate involves the operator of the aircraft (or fleet of aircraft) demonstrating to the CAA that the operation possesses sufficient certified crew members to support the operation, as well as having safety processes in place in order to assure quality of the operation. This may comprise activities such as providing training programs for crew certification, development of operations and maintenance manuals, as well as implementation of system safety processes to ensure regulations are followed. These activities directly influence continuing airworthiness for the aircraft. Financial considerations are also a part of this certification process, as the operator must demonstrate that they possess sufficient infrastructure, assets, and insurance in order to commence and sustain the operation.

### Summary

It seems likely that some form of an air operator's certificate will be required to operate a UAS greater than a pre-determined size. It is likely that the air operator would be responsible for maintaining the fixed infrastructure required to field the UAS, which possibly includes multiple groundstations, as well as data links. Configuration management may come to play a much greater role in the operational manuals and safety processes required to gain an air operator's certificate.

## 6.2 Regulatory Bodies and UAS Legislation – Present and Future

Any new regulation developed in order to facilitate safe integration of UAS into a continuously functioning civil airspace must balance a host of needs. A regulatory regime needs to set levels of safety and environmental protection acceptable to the society as well as provide protection of other public interests such as privacy and security. However, regulation must not place an undue burden on this fledgling industry, thereby allowing it to evolve, innovate, and mature. The regulatory framework should not simply transpose the system put in place for manned aviation; the rules must express objectives that clearly act to mitigate risk, and be complemented by industry standards. This regulatory framework must act as an enabler and not as an impediment. Therefore, a balance must be sought between innovation and the societal concerns about safety, environmental protection, privacy, and security. The following subsection provides a sample of several nascent efforts in this direction.

### 6.2.1 European Union (EU)

The European Aviation Safety Agency (EASA) is responsible for the type certification of aircraft, engines, and parts across EU member nations [118]. EASA also drafts aviation safety regulations, conducts inspections to ensure their uniform implementation, and performs equivalent certifications (i.e., evaluates applicant airworthiness type certifications received from foreign regulatory bodies for acceptability). EASA is responsible for the issuance of type certificates to aircraft registered within the European Union community, along with Joint Aviation Authorities (JAA) member states for products under JAA procedures.

EASA is mandated to regulate UAS, and more particularly RPAS, which are at or above 150kg weight and are being used for civil applications throughout the EU [118]. RPAS are comprised of a set of configurable elements consisting of a remotely piloted aircraft, its associated remote pilot station(s), the required command and control links, and any other system elements as may be required at any point during flight operations. Experimental or amateur-built RPAS, military and non-military governmental RPAS flights, civil RPAS below 150kg, as well as model aircraft are regulated separately by the CAAs in each member nation.

EASA has come up with a concept of operations for drones [157], whereby three categories of operations are proposed along with an associated regulatory regime: open, specific, and certified. The open operation category would not require an authorization by an aviation authority for flight, but would require that the aircraft stay within defined boundaries for the duration of the operation, as well as maintain other safety properties (e.g., maintain distance from aerodromes and people). The specific operation category would require a risk assessment of the operation, in order to gain an operations authorization with specific limitations adapted to the operation. The certified operations category would be required for operations with a higher associated risk, or might be requested on a voluntary basis by organizations providing services, for example remote piloting or equipment such as *detect and avoid* functionality. This would generally involve the attainment of a TC for the UAS or RPAS.

EASA has published the policy document entitled "Airworthiness Certification of Unmanned Aircraft Systems (UAS)," which establishes general principles for type certification (including environmental protection) of an unmanned aircraft system

(UAS) [155]. This policy specifically addresses procedures for the issuance of a type certificate, or a restricted type certificate for a UAS; operational regulations pertaining to UAS are not addressed within the policy. This policy outlines the approach to be taken to establish an airworthiness certificate (via means of a type certificate or restricted type certificate) for UAS over 150kg. As outlined by this policy, these UAS will, in general, have to comply to standards for manned aviation, namely EC Part 21 [120]. EASA has already published guidance material to support approved design organizations to select the appropriate certification specifications (among the ones applicable to manned aviation) from which to build the certification basis for RPAS design. Furthermore, the development of Standard European Rules of the Air (amendment 43 of ICAO Annex 2 [227]), which will apply equally to UAS as conventionally piloted aircraft, should help to integrate UAS fully into civilian airspace.

## United Kingdom

The UK's Civil Aviation Authority regulates all UAS which have mass less than 150kg. The guidance document CAP 722 [364], compiled by the Flight Operations Policy Department, has been recently revised to introduce a concept of operations approach, which aligns with EASA directions [157]. The document outlines safety requirements that have to be met, in terms of airworthiness and operational standards, in order to operate in the UK.

Under the current policy, UAS are divided into two categories based on weight: (1) 20kg and under, and (2) over 20kg but under 150kg. Article 166 of the Air Navigation Order (ANO, 2009) [363] includes specific regulations for small unmanned aircraft (20kg) and Article 167 includes additional regulations for small unmanned aircraft that are "equipped to undertake any form of surveillance or data acquisition."

Provided the UAS has a mass of 20kg or less, airworthiness approval and registration is not required under the following conditions:

1. The UAS must be kept within the visual line of sight (normally taken to be within 500m horizontally and 120m vertically) of its remote pilot (i.e., the person in charge of it).
2. The UAS cannot be operated commercially (for remuneration or compensation for hire).
3. The UAS many not fly in Class A, C, D, or E airspace, or within an aerodrome traffic zone, or at a height of more than 400ft above the ground.
4. The UAS operator may not cause anything to be dropped from the aircraft, and must ensure that the flight can be safely made before take-off.

If the UAS is being flown away from persons, properties, and congested areas, and is not being used for surveillance purposes, operating permission for the UAS is not required.

Furthermore, if the UAS is being used for surveillance purposes without permission (irrespective of the UAS' weight being under 20kg), additional restrictions to those above also apply. The UAS must not operate over or within 150m of any congested area, over or within 150m of an organized open-air assembly of more than 1000 persons, within 50m of any vessel, vehicle, or structure which is not under the control of the person in charge of the aircraft, or within 50m of any person. Otherwise, permission must be obtained in order to operate a UAS for surveillance purposes within these minima [100].

All UAS operators are required to demonstrate piloting qualifications to the CAA. Pilots must demonstrate a basic understanding of applicable regulations (e.g., Air Navigation Order, Rules of the Air Regulations). There are a variety of means of demonstrating pilot competence, the most common being the completion of a course where the applicant demonstrates the necessary skills and knowledge by passing a ground exam and flight test. The aviation authority does not run these courses directly, but instead approves commercial National Qualified Entities (NQEs) to conduct the training and assessment on the CAA's behalf.

UAS with an operating mass of more than 20kg are subject to regulation as though they are CPA. However, it may be possible to obtain an exemption from certain regulations with which it is impossible for unmanned aircraft to comply. Article 138 of ANO 2009 states that "a person shall not recklessly or negligently cause or permit an aircraft to endanger any person or property" [363]. This is equally applicable to all manned and unmanned aircraft. UAS exceeding 20kg or being operated commercially require a *Permit to Carry Out Aerial Work*. Attaining one of these permits requires considerable effort on the behalf of the operator, and addresses issues such as pilot qualification to design and construction certificates. As of 2015, there are 1005 groups or companies that have permission to fly UAS in UK airspace.

### France

In France, the Direction Generale de l'Aviation Civile (DGAC) regulates all UAS which have mass less than 150kg. On April 11, 2012, the Ministry of Transport's Civil Aviation Authority (DGAC) enacted two decrees: the Aircraft Decree [142] relating to the design and classification of drones, and the Airspace Decree [143] relating to the use of drones under various operational scenarios in France's airspace.

The DGAC Aircraft Decree classifies UAS into seven categories (A to G) by mass and function. Categories A and B encompass model aircraft, category C is relevant to tethered UAS, and categories D–G deal with UAS. Category D refers to UAS that are neither model aircraft nor tethered, with a maximum mass of less than 2kg, at take-off. Category E is comprised of UAS whose maximum take-off mass is less than 25kg but more than 2kg. Category F handles UAS whose maximum take-off weight is less than 150kg and does not meet the requirements of category C, D, or E. Category G refers to UAS whose maximum take-off weight is greater than or equal to 150kg, and are regulated by EASA.

The Airspace Decree explicitly outlines four operational scenarios. Scenario 1 refers to UAS operations taking place in the direct sight of the remote pilot, held outside populated areas, with a maximum horizontal distance of 100m from the remote pilot. Scenario 2 applies to operations taking place outside a populated area, in a volume of maximum horizontal dimension of 1km radius and height less than 50m above ground or above artificial obstacles. Furthermore, people may be present in the operational volume, with whom the remote pilot must deconflict the operation. Scenario 3 concerns operations taking place in urban areas or near gatherings of people in direct view of the remote pilot and at a maximum horizontal distance of 100m from the remote pilot. Scenario 4 is relevant to operations involving aerial surveillance, aerial surveys occurring outside populated areas and which do not meet the criteria of scenario 2, where the flying height is less than 150m above ground or above artificial obstacles. These operations can be for renumeration or compensation.

The category and scenario combination for a given UAS operation is then used to determine the level of oversight and regulation to be applied. Depending on the category of UAS and the types of operation to be carried out, the DGAC may require an airworthiness certificate for operation. For example, a UAS used for commercial purposes would fit under category E, even though it might weigh under 4kg and operate within line-of-sight and at most 100m from the operator, or beyond-line-of-sight for up to 1km at maximum altitude of 50m in an unpopulated zone. The UAS vehicle must then certify to certain requirements, such as maximum impact energy. All category D and E UAS are required to have a barometric sensor and barometric altitude limit device to ensure the aircraft never exceeds its maximum flight ceiling. Category D and E UAS must also have a flight termination device, enabling removal of the aircraft from the airspace when it exceeds its predetermined operational boundaries, as well as a device that limits the impact of the aircraft upon descent into terrain, thereby protecting people on the ground. All of these devices must function under lost link conditions.

In addition, there are specific requirements on pilots of certain categories of UAS used in certain scenarios; they must have training, minimum flight time, and possibly an aircraft or helicopter pilot license. Operational permission may have to be obtained depending on the scenario and UAS category. Furthermore, the UAS operator must also have adequate insurance. Currently, for any UAS flying beyond the remote pilot's line-of-sight, specific avionics equipment and security devices are required, along with extensive certification and operational permissions. Of note, there are significant administrative procedures that must be obeyed, before operations may commence. For instance, all UAS must have, on file with the DGAC, an operator's manual, a maintenance manual, and a certificate of conformance. The remote pilot must have completed the theoretical part of a private pilot's license, and the operator must have, on file, a manual of the particular activity, which details the type of operational scenario to be performed, as well as an air activity statement and a request for flight.

There over 1000 registered UAS in France in 2015, and over 40 manufacturers.

## Germany

In Germany, the Luftfahrt-Bundesamt (LBA) regulates all UAS which have mass less than 150kg. Following an amendment to the German Aviation Act (Luftverkehrsgesetz, LuftVG), civilian UAS are now recognized as aircraft; operational use of UAS are regulated in the German Aviation Order (Luftverkehrs-Ordnung, LuftVO) [283]. A license is required to operate every UAS weighing greater that 5kg. Every commercial UAS flight requires a license, and, unless a special permit has been obtained, requires that the UAS be under 25kg. Licenses are issued by the state in which the UAS is operating. The operator must file a comprehensive flight plan, along with documentation regarding the aircraft's design, along with proof of insurance in order to obtain a license. Permission can be sought for multiple flights simultaneously, as long as the flights are in the same location and timeframe. There are specific blanket prohibitions for the operation of UAS throughout Germany [449]. The operation of any UAS over 25kg, the operation of UAS beyond line-of-sight, and the operation of the UAS over 100m above ground level (AGL) are all categorically prohibited. All UAS must be operated by a remote pilot, and are required to have a manual control mode (where inner loop control is flown by a remote pilot). Flight over people not associated with the UAS operation is also prohibited.

As of June 2012, UAS weighing less than 5kg can receive a limited permit for operation up to 100m above ground, within line-of-sight of the operator for either a single flight or repeated use. However, the Air Traffic Act stipulates that these small UAS cannot be flown over crowds of people, accident sites, disaster zones, and other areas where police and security personnel are being deployed. Prisons, industrial facilities, power plants, and military complexes are also no-fly zones. Note that these regulations have no stipulation on the nature of the activity being carried out, as commercial operation is not prohibited. German states have issued more than 500 special permits for UAS, as of 2015.

### 6.2.2 United States of America

Currently, the Federal Aviation Authority (FAA) is responsible for the regulation of civil aviation activities in the US. At present, the intent is to enable certification of UAS through the use of 14CFR 21.17(b) or 21.25 [419], which allows for special classes of aircraft to be designated or restricted type certificates to be issued. These non-conventional aircraft can then have their airworthiness determined on a case-by-case basis, in the context of their operational concepts.

In July 2013, the FAA issued Part 21.25 restricted-category type certifications to AeroVironment's Puma AE and the Insitu Scan Eagle, both small UAS, in order to enable their operation in Alaska. They are permitted to be used for commercial purposes by operators who have obtained a Certificate of Waiver or Authorization (COA). The certificates were limited to aerial surveillance only over Arctic waters. Insitu and energy company ConocoPhillips conducted the first commercial UAS flight over water on September 12, 2013. The FAA modified Puma's data sheet in June of 2014 for its restricted category type certificate to allow commercial operations over land. The FAA issued a Certificate of Waiver or Authorization for BP to survey pipelines and infrastructure over land in Alaska, and flights over land were occurring as of 2015.

The FAA also introduced an exemption process. By law, any aircraft operation in the national airspace requires the aircraft to possess a certificate of airworthiness and be registered, as well as requiring a licensed pilot, and operational approval. Section 333 of the FAA Modernization and Reform Act of 2012 (FMRA) [11] grants the Secretary of Transportation the authority to determine whether an airworthiness certificate is required for a UAS to operate safely in the NAS. This authority was leveraged to grant case-by-case authorization for certain unmanned aircraft to perform commercial operations prior to the finalization of the Small UAS Rule, Part 107, which is the primary method for authorizing small UAS operations.

The FAA has also established an interim policy to speed up airspace authorizations for certain commercial unmanned aircraft (UAS) operators who obtain Section 333 exemptions. Under this policy, the FAA will grant a Certificate of Waiver or Authorization (COA) for flights at or below 200 feet to any UAS operator with a Section 333 exemption for aircraft that weigh less than 25kg, operate during daytime visual flight rules (VFR) conditions, operate within visual line-of-sight (VLOS) of the pilots, and stay within given fixed distances between certain airports. The blanket 200-foot COA allows flights anywhere in the country except restricted airspace or other restricted areas, such as major cities, where the FAA prohibits UAS operations.

Part 107, the small UAS rule, covers the class of UAS under 25kg, with maximum airspeed and altitude of 100mph and 500ft, and flown in direct line-of-sight of the pilot-in-command [167]. These aircraft must be flown under daylight only visual meteorological conditions, and yield right-of-way to other aircraft. Visual observers may be utilized in order to assist in aircraft avoidance and deconfliction, with the assignment of one observer to one aircraft; but first-person camera views cannot be used. There is a minimum visibility requirement of three miles from the ground control system. Operations are allowed, with permission from ATC, in class B, C, D, E airspace. Permission is not required to operate in class G airspace, and operations are not permitted in class A airspace. Operators are required to pass an initial aeronautical knowledge test at an FAA-approved knowledge testing center. Under this small UAS rule, an airworthiness certification is not required. However, the operator would have to maintain the small UAS in a condition for safe operation, and prior to flight would have to inspect the UAS to ensure that it is in a condition for safe operation. Aircraft registration is still required under Part 107 rules.

As of 2015, there are over 1500 Section 333 exemptions granted to enable UAS operation, though all operations are currently required to occur within visual line-of-sight of the UAS remote pilot. This is in addition to the two restricted category type certificates, granted for Arctic operations to AeroVironment and Insitu.

### 6.2.3 Canada

Transport Canada's Advisory Circular AC 600-004 [413], issued November 27, 2014, entitled "Guidance Material for Operating Unmanned Air Vehicle Systems" under an Exemption establishes two classes of UAS which can be operated under exemption in the airspace system: (1) vehicles under 2kg and (2) vehicles between 2 and 35kg. For vehicles in the 2–35kg range, the maximum calibrated airspeed must be 87 knots or less. Both of these exemptions require that the vehicle remain within visual line-of-sight of the operator, who must be at least 18 years old. The remote pilot may operate a single UAV (from a single control station) and must maintain continuous unaided visual contact with the UAV sufficient to be able to maintain operational control of the UAS. The remote pilot must know the location of the UAV at all times, and be able to scan the airspace in which it is operating to see and decisively avoid other air traffic or objects. No relay stations may be present to extend the range of the UAS (nor is the use of visual observers or video feed permitted to extend line of sight).

The operation of a UAS requires the possession of liability insurance of at least \$100 000. The UAS may only be operated during daylight hours, at or below 300ft altitude in Class G airspace, and at least 5 nautical miles from any aerodrome or urban area. A minimum clearance of 500ft must be maintained from all structures, buildings, vehicles, and persons. The remote pilot must be able to intervene and take control of the vehicle at any time. Furthermore, the UAS operator must perform an assessment of the risks posed by loss of any communications links. This assessment must address when it would be safe for the vehicle to commence a pre-programmed "return home" maneuver (including routes and altitudes) or, alternatively, when the pilot would initiate flight termination. The remote pilot operating the UAS system under this exemption also has to have successfully completed a pilot ground school program.

All other UAS operations are subject to the requirement of obtaining a Special Flight Operations Certificate (SFOC) as required by section 602.41, and the requirement to comply with the conditions of an SFOC as required by section 603.66 of the Canadian Aviation Regulations (CARs) [414]. The above-mentioned two exemptions only apply to the recreational operation of a UAS; in order to operate a UAS for commercial purposes, one must apply for a Special Flight Operations Certificate.

### 6.2.4 Australia

The Civil Aviation Safety Authority (CASA) is the national statutory authority in Australia responsible for the regulation of aviation safety and the promotion of safety standards in the aviation community. The Civil Aviation Safety Regulation (CASR) Part 101 [102], released in 2002, was the first operational regulation for unmanned aircraft. This regulation dealt primarily with remotely piloted aircraft, where a human remains in the control loop as the decision maker for the air vehicle. Currently, CASA is in the process of performing updates to CSAR Part 101. Updates include a suite of advisory circulars providing guidance on the topics of training and licensing, operations, manufacturing and initial airworthiness, maintenance and continuing airworthiness, safety management and human performance, operator certification, and operations in a controlled airspace [101] (AC101 1-10). CASA has undertaken the development of a notice of proposed rulemaking, entitled Remotely Piloted Aircraft Systems Terminology and Weight Categorisation of Remotely Piloted Aircraft (NPRM-1309OS) [103]. This will help to update CASR Part 101 to CSAR Part 102, which will entail a substantive rewrite of the current regulation.

Currently, CASR 101 divides fixed-wing UAS into two classes: small (1–150kg) and large (>150kg). Rotary wing aircraft have the same partition, but with a weight break point of 100kg. CASR Part 102 intends to introduce the micro class of UAS. At present, all large UAS require CASA approval in order to operate. CASR 101 dictates that UAS must not be flown over a Restricted or Prohibited area without written approval from the relevant authority. In order to operate a UAS above 400ft, or less than 3 nautical miles (nmi) from an airfield, approval for the flown volume must be received. When operating in a controlled airspace (i.e., other than class G), area approval is required, as is compliance with all air traffic control instructions. Operating a UAS near a runway, airfield movement area, and aircraft approach or departure path is prohibited unless exceptional grounds for approval are presented. Thus, a small UAS may be operated up to 400ft above ground level (AGL), in uncontrolled airspace, without an area approval, given that they meet all other CASA requirements, and are at least 3 nautical miles from any airfield.

Additionally, a UAS must not be operated within 30m of a person not directly associated with the operation of that UAS. A person or company may only operate a UAS for hire or reward if they hold a UAV Operators Certificate. If an operator owns more than one RPAS or UAS, a chief controller and maintenance manager must be appointed for the operation, and have their details on file with CASA. As well, in order to operate a UAS for hire or reward, the remote pilot (or remote operator) must also hold an RPA Controller's Certificate. A remote pilot or operator may only operate in controlled airspace if they hold an Aircraft Radio Operators Certificate of Proficiency (AROCoP). Furthermore, a UAS must not drop or discharge an object that poses a risk to another aircraft, persons, or property while in operation. The UAS must only be operated in Visual Meteorological Conditions, in line of sight of the remote pilot, unless prior approval and training are sought. Moreover, unless the remote operator of a UAS is provided with sufficient visual cues to enable the acquisition and avoidance of other air traffic, UAS flights in controlled airspace are treated as Instrument Flight Rules (IFR) flights, subject to ATC control. CASA may require a large UAS to be equipped with an SSR transponder, a collision avoidance system, or forward-looking vision system as appropriate for the type of operation. Finally, for operations beyond the visual line-of-sight, the filed UAS flight plan must include information and procedures regarding pre-planned emergency flight profiles in the event positive data link control of the UAS is lost. Dependent on system capabilities, these profiles will include: (a) UAV autonomous transit to a pre-designated recovery area followed by an autonomous recovery; (b) UAV autonomous transit to a pre-designated recovery area followed by activation of a flight termination system (FTS).

Communication requirements for UAS performing visual line-of-sight operations are as required by CASA, based on the class of airspace in which the flight will occur. If the remote pilot or UAS operator is not co-located with the RPAS/UAS launch controller, the launch and recovery control station as well as the primary ground control station must establish communications with the ATC authorities responsible for the area of flight prior to commencement of flight. For operations beyond visual line-of-sight, specific abort and flight termination procedures must be developed by the UAS operator, and are provided to the ATC as required. The ground control station must utilize a communications architecture that interfaces with existing ATC communications equipment and procedures, so that the fact that the remote pilot or UAS operator is on the ground is transparent to ATC personnel. Upon check-in with ATC personnel, the remote pilot or UAS operator must request a direct telephone number for the ATC for contingency use in case the radio communications fail. At a minimum, the UAS operator for beyond-line-of-sight operations must have completed the ground training applicable to the issue of an instrument rating in order to operate UAS in controlled airspace under an IFR clearance. Generally, the requirement for certification will limit flights over populous areas to large UAS; however, the designer of a small UAS still may apply for a type certificate, subject to the requirements of CAR 1998 Part 21 and accompanying advisory material.

### 6.2.5 Brazil

The Agencia Nacional de Aviacao Civil (ANAC) is the civil aviation authority in Brazil, and the Regulamentos Brasileiros de Aviacao Civil (RBAC) [18] comprises the regulations that govern aviation. There are no direct laws that infringe upon the free civilian use of UAS in Brazilian airspace under these regulations, as the current regulation is not properly written in order to deal with UAS airworthiness. This has led to a thriving UAS industry in Brazil, with commercial use for photography, filmmaking, and surveys being popular applications. ANAC has attempted to curtail this unregulated commercial use of UAS by introducing a roadmap for the regulatory integration of UAS into Brazilian airspace.

ANAC has introduced two paths for the use of UAS in the airspace. The first involves receiving an experimental certificate, issued under RBAC 21.191, which encompasses operations such as research and development, or model testing. These experimental certificates are issued under a restricted operation, as defined in the Brazilian Regulation of Aeronautics Ratification Part 91 (RBHA 91). These certificates are granted on a case-by-case basis. Alternatively, a Special Flight Permit can be obtained under RBAC 7.565/86, Article 20; however, the operations are still subject to restriction under Part 91. The path forward to a new rule for the commercial operation of UAS is being developed under the ANAC-DECEA Committee, which is a Brazilian UAS Industry forum. Currently, all UAS activity must occur under non-commercial operations, removed from populated areas, and within visual line-of-sight of the remote operator. No-fly zones include airports and military installations.

### 6.2.6 South Africa

South Africa has a rich history of using UAS for commercial purposes, but recent changes in the law have curtailed future use of UAS for remuneration or compensation. In May 2014, a moratorium was issued on the use of UAS platforms for photography, film, and surveillance operations by the South African Civil Aviation Authority (SACAA), and they ceased issuing operational permits [388]. The controversial move was in response to the lack of regulations governing the use of UAS; the SACAA intends to issue more appropriate regulations governing the commercial use of UAS, but a timeframe has not been given for accomplishing this task.

Thus, the commercial operations of unmanned aircraft systems are no longer approved in South Africa by the Director of Civil Aviation Authority, nor has any exemption been issued for any UAS operations in South Africa since 2014. The current regulation governing the use of UAS in South Africa derives from the Civil Aviation Act, 2009 (Act No. 13, of 2009), which imposes conditions on UAS similar to those imposed on conventionally piloted aircraft [387]. The SACAA is in the process of developing guidance material for interim operations and the subsequent integration of UAS in national airspace.

### 6.2.7 Japan

Air transportation and aviation safety activities are regulated by the Japan Civil Aviation Bureau, which is administered under the Ministry of Land, Infrastructure, Transportation, and Tourism. Japan currently has no separate regulatory material relating purely to UAS operation. The only aviation regulations that cover UAS operation in Japan require that they fly below 150 meters and at least nine kilometers away from airports.

The industrial use of unmanned helicopters in an agricultural context has been prevalent in Japan since the 1980s [371]. The helicopters are remotely piloted, and they generally operate at low speeds and altitudes (20km/h, 10m). For agricultural use, the Japan Agricultural Aviation Association (under the auspices of the Ministry of Agriculture, Forestry and Fisheries of Japan) sets standards governing the use of unmanned helicopters. These standards cover a diverse range of airworthiness, operational, maintenance, and reporting issues (including pilot training and aircraft registration), but are limited to agricultural applications. Note that these UAS operations occur under the oversight of the Japanese department of agriculture, and not under the auspices of the Civil Aviation Bureau (CAB). Nonetheless, these agricultural helicopters still meet safety standards regarding structures and flight performance. The operators must demonstrate compliance with regulations concerning inspections, have a valid Maintenance Operator License, and be registered under the Operator Registration System. Over 2000 unmanned helicopters are currently used for chemical spraying operations.

Even though there are no formal regulations (outside the agriculture industry) governing the use of UAS in Japan, there is extensive work being done to establish industry standards. These industry standards often have great impact on the eventual development of, or interpretation of, compliance with regulation. The Japan UAV Association (JUAV) represents the Japanese industries which develop, manufacture, and operate UAS in Japan under diverse applications, including agriculture, monitoring, surveillance, and disaster support. This body has developed industry standards [239] for both fixed-wing and rotary-wing vehicles operating over uninhabited areas. JUAV is now examining the establishment of industry standards for small electrically powered fixed-wing aircraft [240].

### 6.2.8 Summary

Clearly, there is a diverse set of approaches being taken across a multitude of nations in order to develop regulatory standards to safely integrate UAS into the airspace system. A common theme seems to be leaning towards tailoring standards to be commensurate with the risk posed by both the vehicle and its operational concept. Integration of the concept of operations into establishing the correct regulatory basis is not a necessarily new idea; however, the vast range of operational models for UAS and RPAS will lead to a larger design and risk space to consider when developing the requisite regulations.

We next examine the influence of standards organizations and the industry standards they produce on the integration of UAS into civilian airspace.

## 6.3 Standards Organizations

Modern aviation is a complex interaction between aircraft, humans, software, hardware, operational procedures, manufacturing and maintenance processes, along with environmental conditions and a host of other factors. The uniformity in procedures and systems used in civil aviation enables its safe and efficient operation. This is made possible by the universal adoption and interpretation of standards and recommended practices by all stakeholders. Below, we examine several bodies that are in the process of developing standards that would aid in the path to integrating UAS into civil airspace.

### 6.3.1 International Civil Aviation Organization (ICAO)

The International Civil Aviation Organization (ICAO) is a specialized agency of the United Nations, and is comprised of 191 member states. ICAO develops and maintains universally accepted standards that govern civil aviation, known as standards and recommended practices, or SARPs. SARPs cover all technical and operational aspects of international civil aviation, such as safety, personnel licensing, operation of aircraft, aerodromes, air traffic services, accident investigation and the environment. ICAO is currently developing SARPs to aid in the integration of UAS into civil airspace. ICAO uses the term unmanned aircraft systems to refer to aerial vehicles that do not have a pilot on-board, and the term remotely piloted aircraft system (used in ICAO circular 328-AN/190 [226]) to refer to vehicles who have an operator, and are thus not autonomous.

ICAO has postulated two possible paths to certification for RPAS. The first notion is similar to conventional notions of certification [225], where the RPAS is documented with a type certificate issued to the remotely piloted aircraft (RPA). The configuration of the RPAS as a whole would be included in the TC of the RPA, under the responsibility of one unique TC holder. The remote pilot station associated with the aircraft would be a separate entity, likely to be treated in a manner similar to engines and propellers, in that it could have a TC issued by the CAA in which the remote pilot station is designed. The configuration of RPA and the remote pilot station(s) it uses would be certified in conjunction with the RPA by the CAA in which the RPA is designed, and documented in the TC data sheet. The remote pilot station then becomes an integrated part of the RPAS. This would give the CAA in which the RPA is designed responsibility for certifying the overall system design. The RPA's CAA would also have responsibility for providing any mandatory continuing airworthiness information. The State of Registry would have responsibility for determining continuing airworthiness of the RPAS in relation to the appropriate airworthiness requirements. More than one remote pilot station could be associated with the RPA as long as the configuration is described in the TC. A certificate of airworthiness would be issued for the RPA, and it would remain the responsibility of the operator to control the configuration of the RPAS (e.g., RPA, remote pilot station, and data links). SARPs for the design standard of the remote pilot station would have to be developed in the context of the airworthiness of the aircraft [225].

The second option envisaged would require not only new SARPs to be developed for the airworthiness of RPA/UAS, but also new certificates comparable to the existing type certificate and certificate of airworthiness for the remote pilot station(s). This option diverges significantly from the traditional approach in that the design configuration of the RPAS would be defined separately for the RPA and remote pilot station. This means that the airworthiness of the RPA and the comparable certification for the remote pilot station would be dealt with individually. An RPAS designer would have the responsibility for verifying the RPA, and the remote pilot station(s) could be configured into an *airworthy* system. It is not clear yet what the exact RPAS design process approval (similar to what is currently called type certificate) and RPAS production process approval would be, but they would require a fundamental change to the approach to certification [225].

Regardless of which approach is selected, ICAO believes that the RPAS/UAS should have some form of a certificate of airworthiness. In the first option, the remote pilot station associated with the aircraft will be linked to the RPA Certificate of Airworthiness, either through the certificate of airworthiness directly or through configuration control mechanisms per flight (e.g., RPA logbook). In this option, only the RPA will be registered. In the second option, the remote pilot station will have a separate certificate, similar to the RPA certificate of airworthiness, and there must be an operator-controlled system document with which the RPAS (i.e., RPA and remote pilot station) configuration is controlled. In this option, requirements for registration of the RPAS elements will have to be clearly defined. In both options, a method will need to be developed to certify the adequacy of the connection between the remote pilot station(s) and the RPA. Traditionally, the equipment only is certified; the data link(s) are not. In this new scenario, the data link will likely replace the traditional wiring and cables connecting pilot operated flight controls to control surfaces. As such, the appropriate CAA authority will need to consider the data link performance as part of the RPA/RPAS certification process.

ICAO has postulated the introduction of a UAS operator certificate (UOC), similar to the structure of the current air operator certificate. This would enable the operator to conduct operations, provided they have demonstrated an adequate organization, method of control, supervision of flight operations, and training program, as well as ground handling and maintenance. The air operator would then be responsible for meeting the quality of service requirements on the voice and data links for the remote pilot stations.

### 6.3.2 Radio Technical Commission for Aeronautics: SC-228

SC-228 is a working group established under the auspices of the Radio Technical Commission for Aeronautics (RTCA), in order to determine what the minimum operational performance standards for unmanned aircraft systems should be. Some of the topics being addressed by the working group are: (1) developing the MOPS for detect and avoid (DAA) equipment, and (2) establishing command and control (C2) data link MOPS for

L-band and C-band solutions. The initial phase of the standards development is focusing on civil UAS equipped to operate in Class A airspace under IFR flight rules [372]. The assumed operational environment for the MOPS is the transitioning of a UAS to and from Class A or special use airspace, while traversing Class D and E, and perhaps Class G airspace. A second phase of MOPS development is envisaged to specify DAA equipment to support extended UAS operations in Class D, E, and perhaps G, airspace. SC-228 has made two white papers available on their website: the first addresses the SC-228 initial position on DDA issues [374], and the second outlines relevant C2 issues [373] in the context of the SC-228 operational environment. A Phase I preliminary MOPS for DAA and C2 are expected to be released to the public in 2015 with a finalized version of the MOPS planned for 2016.

### 6.3.3 European Organization for Civil Aviation Equipment: WG 73/WG 93

The European Organization for Civil Aviation Equipment (EUROCAE) addresses technical problems dealing with on-board avionics and ground equipment relevant to air transportation applications. EUROCAE deals exclusively with aviation standardization (airborne and ground systems and equipment) and related documents as required for use in the regulation of aviation equipment and systems. EUROCAE Working Group (WG) 73 was created to analyze and develop standards to facilitate the integration of UAS in civil airspace [250]. The working group strategy involves enabling incremental access to airspace, based on defined safety considerations. This approach was adopted in order to focus on specifically selected fundamental capabilities that would allow UAS operations with certain restrictions in a tractable fashion. Current work has been focused on two main scenarios: (1) Entry into, flight in and exit from airspace classes A, B, and C under IFR, and (2) UAS operations of all kinds and in all locations within the limits of visual range conditions. In the course of exploring the first scenario, a safety and performance requirements (SPR) document is under development, which will serve to derive communication, command and control (C3) and see and avoid (S&A) requirements. Furthermore, WG 93, a new working group within the EUROCAE structure, is being established to focus on small/light UAS. WG 93 will work on seeing if the standards created by WG 73 can be adapted to small UAS or whether new standards need to be derived [447].

### 6.3.4 Joint Authorities for Rulemaking on Unmanned Systems

The Joint Authorities for Rulemaking on Unmanned Systems (JARUS) is composed of a group of experts from a consortium of civil aviation authorities and regional aviation safety organizations. The primary purpose of JARUS is to recommend a single set of technical, safety, and operational requirements for the certification and safe integration of unmanned aircraft systems (UAS) into civil airspace. The objective of JARUS is to provide guidance material aimed at facilitating each CAA to write their own requirements, which will hopefully harmonize with one another, while avoiding duplication of efforts. JARUS considers regulation to cover all aspects of UAS operations, such as personnel, organizational, and operational issues, along with technical (airworthiness) requirements. JARUS is currently cooperating with ongoing efforts from ICAO, RTCA, and EUROCAE. JARUS has published several documents, including the JARUS CS-LURS, which is a civil certification specification for a remotely piloted rotorcraft [246]. They have made available documents for review relating to organizational regulations for RPAS, acceptable means of compliance, C2 link concept of operations (CONOPS), and C2 required communications performance (RCP) [247].

### 6.3.5 Summary

It is clear that there is a plethora of approaches being employed by diverse organizations in order to develop uniform standards that will govern the design, deployment, and operation of UAS. However, most of this work is preliminary in nature, and is stymied by the scope of the problem. Clearly, issues posed by large UAS which transit between multiple airspaces at high speeds and altitudes, and have lengthy mission durations, will be different from those of small, electrically powered UAS whose flight time is measured in minutes, but who operate in urban areas in close proximity to large groups of people. Standards must be tailored to reflect operational models as well as vehicle characteristics, and should be able to accommodate the rapidly evolving enabling technologies that will facilitate near-term and far-future applications.

The expanding application space of UAS, and their ability to operate in proximity to urban areas leads us into our next section, which addresses a crucial barrier to gaining public acceptance of UAS, and their integration into civil airspace: the issues of privacy and security.

## 6.4 Social Implications – Privacy and Security

Two key issues driving the ongoing debate regarding the integration of UAS into civil airspace, and general public acceptance of the risks posed by their operation, are those of privacy and security. The topic of security is dealt with in depth in subsequent chapters; in this chapter we only consider security so far as it impacts privacy and safety. Current regulation governing the civil airspace system focuses clearly on safety-related aspects, and little thought is given to any privacy issues incurred in the safe operation of the air transportation system. UAS will enable an entirely new model of operation, where concerns regarding emerging issues such as privacy will be brought to the forefront.

### 6.4.1 Privacy

Defining privacy in the context of the possible proliferation of UAS in civil airspace is a non-trivial matter. Traditional definitions, relating to the concepts of control over personal information and secrecy, are inadequate in capturing the full scope of issues that will be encompassed by the potential for pervasive visual surveillance [332].

UAS operations will have implications upon personal autonomy and anonymity, which must be considered in order to gain a broader picture of the socio-technical concerns emerging around the issue of privacy-aware UAS policies and regulations. Additionally, a separate set of privacy interests are raised through the subsequent aggregation, use, and retention of UAS-obtained information.

## Regulation

Currently, there does not exist a definitive body of regulation surrounding privacy in the direct context of UAS operations. Several countries possess data protection and privacy acts, which serve to limit the terms surrounding general surveillance of the public and the use of the information thereby collected. Furthermore, regulations may come from myriad sources, some statutory, some regulatory, and some practical. Aviation authorities, intra-governmental commerce authorities, domestic security authorities as well as law enforcement and public safety authorities, along with the legislative branch of government, will all likely have some role to play in assessing what regulation is appropriate to achieve the desired level of societal privacy that will be legally conferred upon the public.

At present, there are three types of regulation pertaining to UAS. Firstly, there are longstanding statutory and common-law protections against non-governmental intrusions [50]. For example, trespass, whereby an individual performs an incursion onto another person's property without permission, is both a crime and a tort. The prohibitions against invading privacy, intruding upon seclusion, publishing private facts, and stalking may all likely be just as applicable to actions carried out by UAS. Secondly, there are laws relating to specific forms of surveillance, which preclude the recording of images or conversations without both parties' consent. Similarly, there are anti-voyeurism laws, which bar unauthorized viewing into homes under certain circumstances. It is relatively clear how UAS may be employed in a manner that may violate these laws, by gaining unlawful access to an individual via aerial means. Finally, there is a growing body of civil and criminal laws designed specifically to block unwanted aerial surveillance from public or privately owned UAS [50]. This is an emerging field, and it is as of yet uncertain how efficacious these laws will prove to be, as usage models for UAS continuously evolve. There are two main points of note: (1) there is currently a substantial set of privacy regulation which will have to be examined and interpreted in the context of UAS; and (2) the regulations being formulated to specifically deal with UAS concerns are diverse in nature, often enacting different standards of privacy. That is, not all regulations define trespassing or UAS surveillance in the same way, or apply identical privacy protections to identical places.

## Aspects of Privacy Impacted by UAS Use

The primary aspect of privacy that is impacted by the use of UAS in civil airspace is the initial collection of information about people. The nature of aerial surveillance has changed radically, as the introduction of low cost UAS, coupled with the miniaturization of hi-fidelity camera technology and significant storage capacity, act to enable pervasive surveillance, not only by governmental agencies, but by private citizens. UAS are capable of operating at varying altitudes, as well as in close proximity to people. Many applications of UAS, both commercial and recreational, involve the attachment of an imaging device, such as a camera, thereby enabling the capture of imagery or information regarding all of the overflown region. The use of these UAS in public forums, or even over private property, presents many challenges to society's current expectations of privacy, not to mention the impact on other aspects of privacy, regarding the storage and use of personal images and data.

Personal control is one of the central tenets of privacy. Privacy has previously been defined as "the claim of individuals, groups, or institutions to determine for themselves when, how, and to what extent information about them is communicated to others" [446]. However, this approach to privacy may prove to be restrictive to the future applications enabled by UAS, if the permission of every individual overflown on a camera-equipped vehicle's path must be obtained. Furthermore, the expectation of privacy in public places is not necessarily easily defined.

Similarly, the notion of secrecy is often used to define privacy restrictions. Under this idea, a person's privacy has been violated if information previously concealed has become known, or part of the public domain. Unfortunately, many countries currently possess case law or precedents that state that any information obtained by overflight in the civil airspace is already in the public domain. That is, information obtained by overflight of manned aircraft can be distributed without violation of an individual's privacy. Thus, a person is not entitled to privacy if they have revealed their activities to the general public by rendering them viewable via overflight. This is a clearly worrying definition as UAS possess the ability to perform surveillance activities for great durations, without making their presence known to the general public; two things that conventionally piloted aircraft are incapable of doing.

Personal autonomy implies that people are able to make decisions free from interference or control from both government or private actors. It has been stated that: "Surveillance of human behavior is in place to control human behavior, whether by limiting access to programs or institutions, monitoring and affecting behavior within those arenas, or otherwise enforcing rules and norms by observing and recording acts of compliance and deviance" [181]. It is apparent that a pervasive degree of UAS surveillance may pose potential risks to personal autonomy. The key idea here is the pervasive nature of the surveillance; currently when in public, the average individual is under surveillance from a multitude of sources at any given time. However, traffic cameras, cellphone cameras, surveillance at stores and banks, as well as overflight from news cameras is often location specific and of limited duration. Even though it is possible to track an individual's progress throughout a city using a variety of camera sources from public agencies, commercial institutions and private individuals, the ability to assemble this footage is non-trivial, and requires a significant outlay of resources and specific access. A single UAS unobtrusively following an individual would render this ability accessible to any member of the general public.

Anonymity is a "state of privacy" that "occurs when the individual is in public places or performing public acts but still seeks, and finds, freedom from identification and surveillance" [446]. Pervasive surveillance over urban landscapes or large public events does not pose a great threat to anonymity, as there must be sufficient resolution over individuals in order for identification to be performed, as well as a necessary intent to seek out specific individuals. This can be likened to the notion of cellphone footage of large gatherings of people being taken and posted on the Internet, whereby the identification of people in the group is then crowdsourced to all those who view the image, and pick out individuals they recognize. However, UAS surveillance targeted at a specific individual, by government agencies, commercial entities, or private individuals, for example through the use of facial recognition technology or cellphone-targeted GPS location, will greatly degrade or remove one's anonymity while in public.

### Data Use and Retention

An additional class of privacy risks concerns the use of the data that are collected by UAS in civilian airspace. The collection, processing, and analysis of large quantities of data over large periods of time, and thus the manipulation and storage of these data, can have direct privacy implications that may not be easily seen from the collection of any single piece of data.

Under the privacy theory of aggregation, it is assumed that, while the collection of bits of data about a person may not violate his or her privacy interests, extensive collection of information about him or her can rise to the level of a legal privacy intrusion [223]. This extensive collection of information may be as a result of continuous surveillance via a single source, or may be due to multiple means of surveillance or data collection being aggregated in order to reveal patterns of behavior that would otherwise not be readily apparent. Thus, while the collection of each individual piece of data may, singly, not be a violation of the individual's privacy (or implicate the privacy interests of secrecy, anonymity, or autonomy), the aggregation of the data yields a violation of privacy. Furthermore, the potential ubiquity of UAS will dramatically alter the quantity of information that can be gained regarding an individual over time, especially in conjunction with alternative sources of personal information, such as utility, banking, telephone, or insurance records. Thus, the issue of data aggregation, in the context of surveillance (sustained or otherwise) by single or multiple UAS, poses a direct challenge to current expectations of privacy.

As individuals share more and more information with both governmental agencies and commercial corporations, it becomes paramount that standards be fixed regarding the appropriate use of collected data. Data which may be authorized and collected for a specific context may subsequently be used in an unauthorized way that breaches an individual's privacy. Many countries possess Data Privacy Acts, which require federal agencies that maintain a database of personal records to inform each individual about whom it collects information of the principal purpose, or purposes, for which the information is intended to be used [223]. The application of current frameworks for data collection and privacy need to be examined in the context of UAS data collection through potentially prolonged surveillance. Additionally the disclosure of this information must also be carefully considered, as the ability to make large swathes of data searchable, which were formerly under the capability of government agencies or large corporations, is now within the realm of private individuals. A potential solution is to place regulation not on how data are collected, but to restrict how data are used: thus, the deployment of UAS would not necessarily be curtailed, but stringent data management and protection schemes would have to be enacted a posteriori on any aggregated information.

Due to the decreased cost of data storage, there is a near limitless ability to store and retain vast databases of information about individuals indefinitely. This issue of data retention has been one of the drivers behind Europe's "Right to be Forgotten Law", which includes a provision requiring that data be retained "for no longer than is necessary for the purposes for which it was collected" [117]. The privacy impact of retention of UAS-derived data will hinge upon whether specific individuals can be identified through those data. Indefinite retention of data that contain personally identifiable information, and are subsequently locatable based on such personal information, would clearly impact an individual's privacy via the notions of anonymity and autonomy.

## Privacy Implication of Public Use of UAS

Presently, UAS are used by governments in many roles. Primary uses include customs and border patrol, immigration and customs enforcement, law enforcement (including drug enforcement), search and rescue, and fugitive apprehension. This is in addition to military use, as well as by intelligence agencies. There is some use of UAS by land bureaus, as well as geological surveying and ecological and herd monitoring. UAS operated by governmental agencies should, in all likelihood, be subject to all of the legal restrictions placed on operations regarding the protection of privacy of the state's citizens. All information that is amassed via public (government owned and operated) UAS must be collected, used, retained, and disseminated consistent with the constitutional, criminal, civil, and common law of the land, along with all other applicable regulations and policies. Furthermore, the public use of UAS for the collection of information must be subject to accountability and transparency processes.

As today's UAS enhanced information collection systems outstrip the nature and amount of data that could previously be collected, new regulations must be developed to ensure that the expectations of privacy previously enjoyed by the general public are not circumvented. Furthermore, actions taken by governmental agencies, such as law enforcement or tax collection agencies, predicated on data collected via UAS platforms must be audited carefully. Public scrutiny processes should be in place to ensure that misuse of UAS, or the data collected therefrom, is not without consequence.

## Privacy Implications of Commercial or Private Use of UAS

Similarly, there is a need for a framework to be developed that will encompass issues of privacy, accountability, and transparency for commercial and private UAS use. Careful note should be taken that the collection of images of identifiable individuals, even inadvertently, when using surveillance cameras mounted on a small unmanned surveillance aircraft, will be subject to any Data Protection Act in the state of operation of the UAS. Data protection acts contain requirements concerning the collection, storage, and use of such images as apply to private citizens. In the absence of such an act, there may be native requirements on the fair use and sharing of imagery captured over private property, or without permission of the owner.

The proliferation of commercial or private UAS could also have legal implications that make government surveillance more pervasive. For instance, it is unclear to what standard of due process of law accessing a cloud server, which serves as storage device for information captured by commercial (or private) UAS, would require. That is, it is generally accepted that the government should not be able to compel a private entity to turn over large quantities of non-specific third-party data. Additionally, information generated by location tracking of individuals could be valuable to advertisers and data brokers; commercial data collectors may soon have the capacity to continuously track the movements of individuals in public spaces just as they now track online activity. Regulation of private UAS use must be crafted with a recognition that UAS not only make possible new types of crime, but make it much easier to commit others. Offenses such as stalking, harassment, blackmail, and invasions of privacy could be committed more easily, covertly, and anonymously with the use of UAS.

## 6.5 Gaps between Regulatory Needs and Technical State-of-the-Art

It is clear to see that there are significant barriers to the safe integration of UAS into civilian airspace. There is a clear question regarding what standards, if any, are necessary in order to assure the airworthiness of a UAS, before it can be allowed to operate in civil airspace without degrading the overall safety of the airspace system. Furthermore, the mode of operating UAS will differ significantly from the way that aircraft currently operate in airspace systems. Operational regulations will be required to maintain safety standards for organizational structures, such as safety management systems, accident and incident reporting, training and facilities or infrastructure maintenance. Harmonization of regulation across international boundaries will be a challenge, as much of the regulation currently being designed has been done in an ad hoc or piecemeal fashion. In addition to the issue of safety, notions such as security and privacy will also dictate policy and regulation governing the integration of UAS into civil airspace systems.

These barriers are socio-technical in nature; that is, they carry a strong technical component in relation to their socio-regulatory nature. Airworthiness standards for UAS in civilian airspace will be prominently impacted by the ability of these aircraft to detect and avoid other CPA. The quality of the C2 links between the vehicle platform and UAS ground station, in terms of link security (and confidentiality or privacy) will have a direct effect on the safety of the vehicle, especially in terms of "lost link" protocols and behavior. The ability to correctly and accurately locate the UAS vehicle platform at all times when it is operational in the civilian airspace system, whether through conventional means such as radar or ADS-B, or through operational means, will enable access to high traffic airspace, such as Class A or B. Coordinated or cooperative networking solutions and infrastructure interoperability will be required to transition UAS across different airspace sectors, as well as international boundaries.

In the next section, we focus on the specific technical issues of networked UAS localization, whereby multiple UAS are simultaneously localized, in order to maintain complete network or C2 coverage over a given area.

## 6.6 Technical Challenges

This section first provides a glimpse into the vastly heterogeneous world of the UAVs. Such heterogeneity of purpose, function, and technical capabilities of the UAVs make integration of the UAVs in the national airspace all the more challenging. As presented in Table 6.1, the two UAVs at the two ends of the spectrum, the RQ4 Global Hawk and the Nano Hummingbird could not be more different. Their wingspans range from 131 feet to 6.3 inches, weight spans from 32,250 pounds to 6.3 ounces, and flight time from 28 hours to 11 minutes. Accordingly, the technical and research challenges posed by them are quite different and a *one size fits all* type of approach is unlikely to be effective in such a diverse environment. In the following section, we first list a number of research challenges in this domain and then discuss two of them in greater detail.

**Table 6.1** The wide world of UAVs

| | 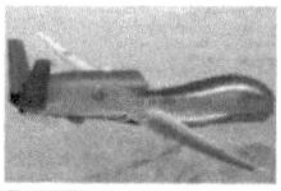 |  |  |  |
|---|---|---|---|---|
| UAV Name | RQ-4 Global Hawk | MQ-9 Reaper | MQ-1 Predator | MQ-8 Fire Scout |
| Manufacturer | Northrop Grumman | General Atomics | General Atomics | Northrop Grumman |
| Wingspan | 131 feet | 66 feet | 55 feet | 28 feet |
| Gross Weight | 32250lbs. | 10500lbs. | 2250lbs. | 3150lbs. |
| Flight Time | 28 hours | 24 hours | 24+ hours | 6+ hours |
| Range | 12300 nautical miles | 1000 nautical miles | 675 nautical miles | 110 nautical miles |
| Communication Range | – | – | – | – |
| Primary Users | Armed Forces | Armed Forces | Armed Forces | Armed Forces |
| Primary Use | Surveillance | Surveillance and combat | Surveillance and combat | Situational awareness |

| |  |  |  |  |
|---|---|---|---|---|
| UAV Name | RQ-7 Shadow | Phantom 2 Vision+ | U818A | Nano Hummingbird |
| Manufacturer | AAI | DJI | UDI | AeroVironment |
| Wingspan | 14 feet | 18 inches | 16 inches | 6.3 inches |
| Gross Weight | 375lbs. | 10.8lbs. | 3lbs. | 0.67oz. |
| Flight Time | 6 hours | 25 mins. | 10 mins. | 11 mins. |
| Range | 31 miles | 700 meters | 30 meters | – |
| Communication Range | – | – | – | – |
| Primary Users | Armed Forces | Amateur and professional photographers | Flight enthusiasts, amateur photographers | Research and Development personnel |
| Primary Use | Reconnaissance | Aerial photography | Recreational flight, aerial photography | Prototype for precision flight |

### 6.6.1 Research Questions

1. What is the safe density of a set of homogeneous UAVs to be operating in a fixed volume (a spherical or cubic volume)? A first step can be made by analyzing on a two-dimensional plane.
2. What is the safe density of a set of heterogeneous UAVS to be operating in a fixed volume (a spherical or cubic volume)? A first step can be made by analyzing on a two-dimensional plane.
3. What is the minimum transmission range needed by the UAVs, flying on a known trajectory (predictable flight path) so that the network formed by the UAVs will always remain connected?
4. What is the minimum transmission range needed by the UAVs, flying on a known trajectory (predictable flight path), so that the network formed by the UAVs will always remain connected, even when all the UAVs in a specific subarea (or subvolume or spatially correlated region) become ineffective due to malicious jamming or failure due to some natural cause or targeted attack?
5. What is the minimum transmission range needed by the UAVs, flying on a known trajectory (predictable flight path), so that the network formed by the UAVs will always remain connected, even when all be UAVs in a specific subarea become ineffective due to malicious jamming or some other event?
6. What is the minimum transmission range needed by the UAVs, flying on an unknown trajectory (unpredictable flight path), but in a confined space (say a spherical or cubic area), so that the network formed by the UAVs will always remain connected?
7. If the UAV network has to operate in a disconnected mode at times (if the minimum transmission rate needed to keep the network always connected is more than the transmission range of the on-board transmitter of the UAVs), what type of protocol will ensure that the data can be transferred from the source to the destination in the earliest possible time?
8. If a set of UAVs are used to monitor a set of suspected mobile targets that seem to be traveling in a predefined trajectory, what is fewest number of UAVs that will be needed to monitor every suspect mobile target for the entire duration of flight of the mobile targets? It may be noted that a UAV can monitor a suspected mobile target only if the mobile target is within the "sensing range" of a UAV.

In the following two subsections we will discuss the research challenges 3 and 8 in further detail.

### 6.6.2 Minimum Transmission Range Needed by the UAVs to Keep the Airborne Backbone Network Connected at all Times

An Airborne Network (AN) is a mobile ad hoc network that utilizes a heterogeneous set of physical links to interconnect a set of highly mobile airborne platforms. UAVs can play a major role in airborne networks as the mobile airborne platforms. Efforts are currently underway, in both the civilian as well the as military domain, for the creation of an airborne network. In the civilian domain, both Google and Facebook are in the process of creating an airborne network to bring Internet services to remote parts of the world. In the military domain, air forces of several nations are in the process of creating an airborne network. The design, development, deployment, and management of a network where the nodes are mobile are considerably more complex and challenging than a network of static nodes. This is evident by the elusive promise of the Mobile Ad Hoc Network (MANET) technology, where, despite intense research activity over the previous years, mature solutions are yet to emerge [74, 114]. One major challenge in the MANET environment is the unpredictable movement pattern of the mobile nodes and its impact on the network structure and dynamics. In case of an AN, there exists considerable control over the movement pattern of the mobile airborne networking platforms. A designer can specify the controlling parameters, such as the *location, flight path, and speed* of the Airborne Networking Platforms (ANPs), to realize an AN with desired functionalities. Such control provides the designer with an opportunity to develop a topologically stable network, even when the nodes of the network are highly mobile.

It is increasingly being recognized in the networking research community that the level of *reliability* needed for continuous (interruption-free) operation of an AN may be difficult to achieve through *a completely mobile, infrastructure-less network* [298]. In order to enhance *reliability* and *scalability* of an AN, Milner *et al.* in [298] suggested the formation of a *backbone network* with ANPs. In order to meet the reliability and scalability requirements in an AN, we have proposed an architecture for an AN where a set of ANPs form the *backbone* of the AN. This set of ANPs may be viewed as *mobile base stations* with *predictable and well-structured flight paths* and whose clients may be stationary or mobile. The clients may include individuals, aircrafts, ships, cars, or any other entities that might need networking services. From the reliability standpoint, it is imperative that the backbone network formed by the mobile ANPs remains connected all the times even though the topology of the backbone network changes with the movement of the ANPs.

Network connectivity can be easily achieved by making the transmission range of the ANPs very large. However, large transmission range also implies large power consumption. In order to minimize power consumption and hence extend network lifetime, one would like to operate with the smallest transmission range that ensures that the backbone network remains connected at all times. We define the *critical transmission range* (*CTR*) to be the minimum transmission range of the ANPs to ensure that the dynamic network formed by the movement of the ANPs remains connected at all times. We present an algorithm to compute *CTR* when the flight paths are known. As a part of the design of this algorithm, we develop techniques to compute the *dynamic topology* of the AN at any instance of time.

In our proposed architecture for an airborne network, the nodes of the backbone networks (ANPs) are viewed as *mobile base stations* with *predictable and well-structured flight paths*, and the clients, both stationary and mobile, can be anything such as an individual, an aircraft, or a ship. A schematic diagram of this architecture is shown in Figure 6.2. In the diagram, the UAVs are the ANPs forming the infrastructure of

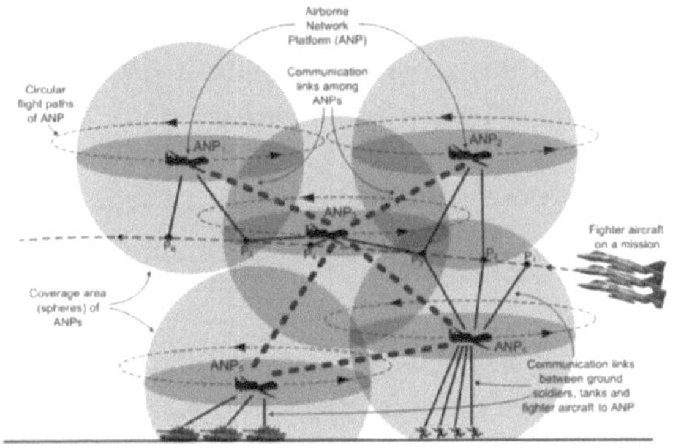


**Figure 6.2** A schematic view of an airborne network. (Reproduced from [432] with permission from the IEEE)

the AN (although in Figure 6.2, only UAVs are shown as ANPs; other entities such as aircraft and satellites can also be considered as ANPs). We assume that the ANPs follow a circular flight path. The circular flight paths of the ANPs and their coverage area (shaded spheres with ANPs at the center), are also shown in Figure 6.2. Thick dashed lines indicate the communication links between the ANPs. The figure also shows three fighter aircraft on a mission passing through space known as the *air corridor*, where network coverage is provided by ANPs 1 through 5. As the fighter aircraft move along their flight trajectories, they pass through the coverage area of multiple ANPs and there is a smooth hand-off from one ANP to another when the fighter aircraft move from the coverage area of one ANP to that of another. At points P1, P2, P3, P4, P5, and P6 on their flight path in Figure 6.2, the fighter aircraft are connected to the ANPs (4), (2, 4), (2, 3, 4), (3), (1, 3), and (1), respectively.

We make a simplifying assumption that two ANPs can communicate with each other whenever the distance between them does not exceed the specified threshold (transmission range of the on-board transmitter). It is true that successful communication between two airborne platforms depends not only on the distance between them, but also on various other factors such as (i) the line of sight between the platforms [411]; (ii) changes in the atmospheric channel conditions due to turbulence, clouds, and scattering; (iii) the banking angle, the wing obstruction, and the dead zone produced by the wake vortex of the aircraft [150]; and (iv) the Doppler effect. Moreover, the transmission range of a link is not a constant and is impacted by various factors, such as transmission power, receiver sensitivity, scattering loss over altitude and range, path loss over propagation range, loss due to turbulence, and the transmission aperture size [150]. However, as the distance between the ANPs is a very important parameter in determining whether communication between the ANPs can take place, we feel such an assumption is justified. Once the basic issues of the problem are well understood, factors (i)–(iv) can be incorporated into the model to obtain a more accurate solution.

For simplicity of analysis, we make two more assumptions. We assume that (i) all ANPs are flying at the same altitude and (ii) they follow a circular flight path. The first assumption allows us to reduce the problem from a three-dimensional space to two.

However, neither of these two assumptions is critical and our analysis technique can easily be extended to scenarios where the ANPs are not flying at the same altitude and they are not following a circular flight path. As a consequence of assumption (i), we can view the $n$ backbone nodes (ANPs) as moving points on a two-dimensional plane. Let $(x_i(t), y_i(t))$ be the coordinates of the node $i$ at time $t$. The network of flying ANPs gives rise to a *dynamic graph* $G(t) = (V, E(t))$, where $V = \{1, 2, \ldots, n\}$ is the set of nodes indexed by the ANPs and $E(t)$ is the set of edges at time $t$. There is an edge between two nodes if their Euclidean distance, $s_{ij}$ is less than the transmission range $Tr$ at time $t$, that is $E(t) = \{(i,j) | s_{ij}(t) < Tr\}$. It may be noted that the dynamic graph $G(t) = (V, E(t))$ is completely defined by the following five controlling parameters:

1. A set of points $\{c_1, c_2, \ldots, c_n\}$ on a two-dimensional plane (representing the centers of circular flight paths).
2. A set of radii $\{r_1, r_2, \ldots, r_n\}$ representing the radii of circular flight paths.
3. A set of points $\{p_1, p_2, \ldots, p_n\}$ representing the initial locations of the platforms.
4. A set of velocities $\{v_1, v_2, \ldots, v_n\}$ representing the speeds of the platforms.
5. Transmission range $Tr$ of the transceivers on the airborne platforms.

In the following, we explain the computation of the dynamic topology of graph $G(t) = (V, E(t))$ when all five controlling parameters are provided. We also provide a technique to compute the minimum transmission range needed by the ANPs to keep the resulting dynamic graph connected at all times.

Suppose that two ANPs, represented by two points $i$ and $j$ (either in two or in three-dimensional space, the two-dimensional case corresponds to the scenario where the ANPs are flying at same altitude) are moving along two circular orbits with centers at $c_i$ and $c_j$ with orbit radius $r_i$ and $r_j$ as shown in Figure 6.3a with velocities $v_i$ and $v_j$ (with corresponding angular velocities $\omega_i$ and $\omega_j$), respectively.

A moving node $i$ is specified by the radius vector $\vec{R}_i(t)$ directed from some origin point $O$, and similarly $\vec{R}_j(t)$ for point $j$. Therefore, the distance $s_{ij}(t)$ between the nodes $i - j$ at time $t$ is given by:

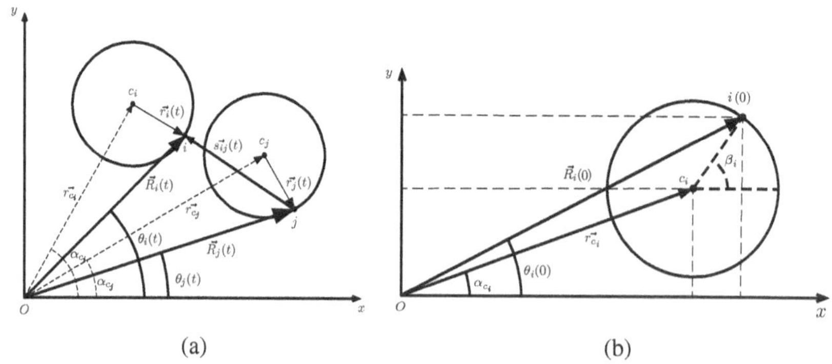


**Figure 6.3** (a) Initial phase angle $\beta_i$ of point $i$; at time 0 point is shown as $i(0)$, (b) Vector representations ($\vec{R}_i(t)$ and $\vec{R}_j(t)$) of two points $i$ and $j$ at time $t$ moving along two circular orbits: $r_{c_i} = 15$, $r_{c_j} = 27$, $\angle c_iOx = \alpha_{c_i} = \frac{\pi}{3}$, $\angle c_jOx = \alpha_{c_j} = \frac{\pi}{6}$ (right). (Reproduced from [432] with permission from the IEEE)

$$s_{ij}^2(t) = (\vec{R}_i(t) - \vec{R}_j(t))^2 = R_i^2(t) + R_j^2(t) - 2\vec{R}_i(t) \cdot \vec{R}_j(t). \quad (6.1)$$

As mentioned earlier, we have assumed that the communication between the ANPs is possible if and only if the Euclidean distance between them does not exceed the communication threshold distance $Tr$. This implies that the link between the nodes $i$ and $j$ is alive (or active) when

$$s_{ij}(t) \leq Tr. \quad (6.2)$$

In the analysis that follows, we have assumed that ANPs are flying at the same altitude, that is, we focus our attention on the two-dimensional scenario. In this case, we can view the ANPs as points on a two-dimensional plane moving along two circular orbits, as shown in Figure 6.3a. In Figure 6.3a, the vectors from the origin $O$ to the centers of the orbits $c_i$ and $c_j$ are given as $\vec{r_{c_i}}$ and $\vec{r_{cj}}$. The cartesian co-ordinates of the centers can be readily obtained as $\vec{r_{c_i}} = (r_{c_i} \cos\ \alpha_{c_i}, r_{c_i} \sin\ \alpha_{c_i})$ and $\vec{r_{c_j}} = (r_{c_j} \cos\ \alpha_{cj}, r_{c_j} \sin\ \alpha_{c_j})$. Accordingly, $\vec{R}_i(t)$ can be expressed in polar coordinates: $(R_i(t), \theta_i(t))$ with respect to origin point $O$, as shown in Figure 6.3a, and similarly for $\vec{R}_j(t)$. The initial location of the points $\vec{R}_i(0)$ and $\vec{R}_j(0)$ are given. From Figure 6.3b, the phase angle $\beta_i$ for node $i$ with respect to the center of orbit $c_i$, can be calculated as (by taking projection on the axes):

$$\tan\ \beta_i = \frac{R_i(0) \cos\ \theta_i(0) - r_{c_i} \cos\ \alpha_{c_i}}{R_i(0) \sin\ \theta_i(0) - r_{c_i} \sin\ \alpha_{c_i}}. \quad (6.3)$$

From Figure 6.3a:

$$\vec{R}_i(t) = \vec{r}_{c_i} + \vec{r}_i(t) \quad (6.4)$$

where $\vec{r}_i(t) = (r_i \cos\ (\beta_i + \omega_i t), r_i \sin\ (\beta_i + \omega_i t))$ (since angle made by $i$ at time $t$ w.r.t. $c_i$ is given by $(\beta_i + \omega_i t)$). Therefore, the angle between $\vec{r}_i(t)$ and $\vec{r}_{c_i}$ is $(\beta_i - \alpha_{c_i} + \omega_i t)$. Hence:

$$R_i^2(t) = r_{c_i}^2 + r_i^2 + 2r_{c_i} r_i \cos\ (\beta_i - \alpha_{c_i} + \omega_i t). \quad (6.5)$$

Now, taking the projection of $\vec{R}_i(t) = \vec{r}_{c_i} + \vec{r}_i(t)$ on the $x$ and $y$ axes, we get:

$$R_i(t) \cos\theta_i(t) = r_{c_i} \cos\ \alpha_{c_i} + r_i \cos\ (\beta_i + \omega_i t), \quad (6.6)$$
$$R_i(t) \sin\ \theta_i(t) = r_{c_i} \sin\ \alpha_{c_i} + r_i \sin(\beta_i + \omega_i t). \quad (6.7)$$

Recalling $\cos(A - B) = \cos A \cos B + \sin A \sin B$, and simplifying we get:

$$\begin{aligned} R_i(t)R_j(t) \cos(\theta_i(t) - \theta_j(t)) &= r_{c_i} r_{c_j} \cos\ \alpha_{c_i c_j} + r_i r_j \cos(\beta_{ij} + (\omega_i - \omega_j)t) \\ &\quad + r_{c_i} r_j \cos(\alpha_{c_i} - \beta_j - \omega_j t) \\ &\quad + r_{c_j} r_i \cos(\alpha_{c_j} - \beta_i - \omega_i t) \end{aligned} \quad (6.8)$$

where $\alpha_{c_{ij}} = \alpha_{c_i} - \alpha_{c_j}$ and $\beta_{ij} = \beta_i - \beta_j$. Combining equation (6.1) with equations (6.5) and (6.8), we have:

$$\begin{aligned} s_{ij}^2(t) &= r_{c_i}^2 + r_i^2 + 2r_{c_i}r_i\cos(\beta_i - \alpha_{c_i} + \omega_i t) \\ &\quad + r_{c_j}^2 + r_j^2 + 2r_{c_j}r_j\cos(\beta_j - \alpha_{c_j} + \omega_j t) \\ &\quad - 2r_{c_i}r_{c_j}\cos\alpha_{c_ic_j} + r_ir_j\cos(\beta_{ij} + (\omega_i - \omega_j)t) \\ &\quad + r_{c_i}r_j\cos(\alpha_{c_i} - \beta_j - \omega_j t) + r_{c_j}r_i\cos(\alpha_{c_j} - \beta_i - \omega_i t). \end{aligned} \tag{6.9}$$

An example of a plot of equation (6.9) (generated using MATLAB) is shown in Figure 6.4a with communication threshold distance $Tr = 18$. This implies that the link between the nodes $i$ and $j$ exists, when the distance between them is at most 18 and the link does not exist otherwise. This is shown in Figure 6.4b. The dark gray part indicates the time interval when the link is *inactive* (or dead) and the light gray part indicates when it is *active* (or live).

Thus using equation (6.9) and comparing the distance between any two nodes with the communication threshold $Tr$, we can determine active/inactive times of all links. This can be represented as intervals on a time line as shown in Figure 6.5. By drawing projections from the end-points of the active/inactive times of each link on the time line, we can find all the links that are active during an interval on the time line. As shown in Figure 6.5, links 1, 2, and 3 are active in interval 1; links 1 and 3 are active in interval 2; links 1, 2, and 3 are active in interval 3, and so on. Once we know all the links that are active during a time interval, we can determine if the graph is connected during that interval using any algorithm for computing graph connectivity [141]. By checking if the graph is connected at all intervals, we can determine if the graph is connected at all times, when the ANPs are moving at specified velocities.

We note that based on equation (6.9), $s_{ij}$ is periodic if every pair of velocities $\omega_i$ and $\omega_j$ are *commensurate*, that is $\omega_i/\omega_j$ is a rational number [323]. Therefore, the network topologies will be repeated periodically and it is enough to check network connectivity in one period.

If the problem parameters (1) through (5) are specified, we can check if the dynamic graph is connected at all times following these two steps. In the first step, we determine the lifetime (active/inactive intervals) of every link between every pair of nodes $i$ and $j$

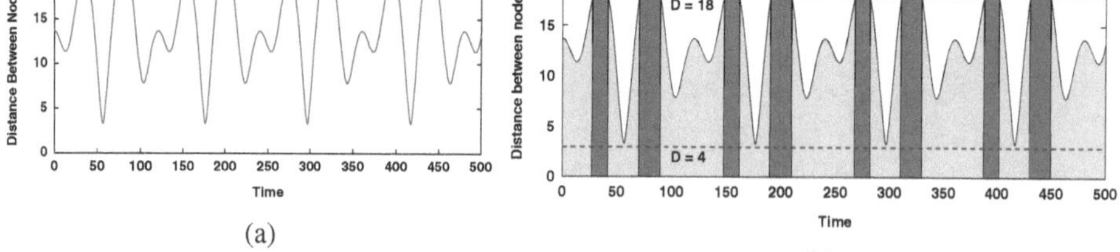


(a)

(b)

**Figure 6.4** Effect of the distance between nodes on the existence of the communication link between them; (a) Distance between two points $i$ and $j$ as a function of time. (b) Active (light gray)/Inactive (dark gray) times of the link between $i$ and $j$ with transmission range $Tr = 18$

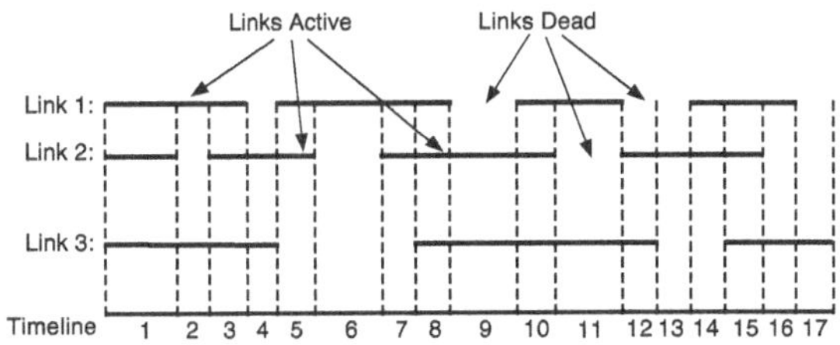


**Figure 6.5** Active/inactive time interval of each link and interval intersection projections on the time line. (Reproduced from [432] with permission from the IEEE)

by comparing $s_{ij}(t)$ with $Tr$ and finding the time points that the state of a link changes. Let $L(Tr) = \{e_1, e_2, \ldots, e_l\}$ denote the set of events $e_i$s that the state of a link changes when transmission range is $Tr$; $L(Tr)$ is sorted in increasing order of the time of the events. Hence, between two consecutive events $e_i$ and $e_{i+1}$ that happen at times $t_i$ and $t_{i+1}$ the set of active links is unchanged. Algorithm 1 shows the details of computing $L(Tr)$. In the second step, we check the graph connectivity in each interval $(t_i, t_{i+1})$ for all $0 \leq i \leq l - 1$ using the connectivity checking algorithm of [115], where $t_0$ shows the current time (starting point). Step 2 is described in detail in Algorithm 2.

**Algorithm 1** Link lifetime computation

*Input*: (i) a set of points $\{c_1, c_2, \ldots, c_n\}$ representing the centers of circular flight paths, (ii) a set of radii $\{r_1, r_2, \ldots, r_n\}$ representing the radii of circular flight paths, (iii) a set of points $\{p_1, p_2, \ldots, p_n\}$ representing the initial locations of the platforms, (iv) a set of velocities $\{v_1, v_2, \ldots, v_n\}$ representing the speeds of the platforms.
*Output*: $L(Tr)$: an ordered set of events that the state of a link changes from active to inactive or inactive to active.

```
L(Tr) ← Ø
for all pairs i,j do
    Compute l to be the set of time points t such that s_ij(t) = Tr (equation 6.9) over a period of
    time, to find the instances of times t where the state of the link (i,j) changes. If s_ij(t) = Tr
    and s_ij(t) is increasing at t, it implies that the link dies at t, and if s_ij(t) is decreasing at t, it
    implies that the link becomes active at t.
    for all l_k ∈ l do
        Find the position of l_k in L(Tr) using binary search and add the event into L(Tr). (L(Tr)
        is sorted in increasing order.)
    end for
end for
```

**Algorithm 2** Checking connectivity of airborne network

*Input*: $L(Tr)$
*Output*: *true* if graph is connected all the time; otherwise *false*.

```
for all l_i ∈ L(Tr) do
    Check if the AN graph is connected with the set of live links during interval (l_i, l_{i+1}). This
    can be done with the connectivity testing algorithm in [115].
    if AN graph is not connected, return false.
end for
return true
```

Let $n$ be the number of ANPs. The first loop of Algorithm 1 is executed for $O(n^2)$ times. The number of iterations of the inner loop depends on the number of the solutions of $s_{ij}(t) = Tr$. For the case where the ANPs move at the same velocity, that is $\omega_i = \omega_j = \omega$ it is obvious that equation (6.9) is periodic and the length of one periodic interval is $2\pi/\omega$. So, it is enough to execute Algorithm 1 for one period $(t_0, t_0+2\pi/\omega)$. In this case, equation (6.9) can be written as $A\cos(\omega t) + B\sin(\omega t) = \sqrt{A^2+B^2}\sin(\phi + \omega t)$ where $A, B$, and $\phi$ are constants and can easily be obtained from equation (6.9). In this case, the equation $s_{ij}(t) = Tr$ can have at most two solutions and the solutions can be found in constant time. Therefore, for every link, the timeline is divided into at most three segments in one period and the size of the set of intervals, $|L(Tr)|$ is $O(n^2)$; also, the time complexity of the binary search is $O(\log n^2)$. So, the total time complexity of Algorithm 1 is $O(n^2 \log n)$. Even when the velocities of the ANPs are different, $s_{ij}$ remains periodic if every pair of velocities $\omega_i$ and $\omega_j$ is *commensurate*, that is $\omega_i/\omega_j$ is a rational number [323]. In this case, we also need to solve $s_{ij}(t) = Tr$ for one period only. Otherwise, equation (6.9) is not periodic and we need to consider a period of time between $t_0$ and a finish time $t_f$ and find the solutions in that period. For the sake of simplicity, we assume that the ANPs move at the same speed. The running time of connectivity testing algorithm in [115] is $O(n^2)$. Also, as $|L(Tr)| = O(n^2)$, time complexity of Algorithm 2 is $O(n^4)$.

As we would like to determine *CTR*, the problem can be specified in the following way. Given controlling parameters 1, 2, 3, and 4, what is the minimum transmission range of the ANPs so that the resulting graph is *connected* at all times?

Earlier we explained how we check network connectivity when all five parameters are given. The maximum transmission range of an ANP $Tr_{max}$ is known in advance. In order to compute *CTR*, we can conduct a binary search within the range $0 - Tr_{max}$ and we can determine the smallest transmission range that will ensure a connected AN during the entire operational time when all other problem parameters have already been specified. The binary search adds a factor of $\log Tr_{max}$ to the complexity of Algorithms 1 and 2.

### 6.6.3 Minimum Number of UAVs Needed to Monitor all Suspect Mobile Targets at all Times

Motivated by the importance of target tracking in military and civilian environments and widespread use of UAVs in target tracking, considerable research has been conducted on target tracking problems using UAVs and mobile sensors [468], [465], [448], [318], [457], [469]. There also exists a large body of research on target tracking problems using sensor networks. However, in most of these studies the sensor nodes (trackers) are static and as such the issue of path planning of trackers does not exist. The authors in [311] provide a survey of these studies. Typically, the target tracking problem using mobile trackers has two components: (*i*) estimation of target positions using sensor data, and (*ii*) mobility management of trackers (sensors). Most of the studies on target tracking using mobile sensors focus on the quality of detection of the mobile targets with a given set of UAVs (mobile sensors). In these studies, one or more mobile trackers are used to track a single or multiple targets. The authors in [14] study multi-target tracking using multiple UAVs and develop a decentralized approach for target location estimation and UAV mobility management. In spite of extensive studies on the target tracking problem, there exists only a handful of studies on the problem of continuous time tracking of mobile targets with an optimal number of mobile trackers.

In this section, we discuss the scenario where there are multiple mobile targets whose trajectories are known in advance. Accordingly, estimation and prediction of target location is not an issue here and the focus of this target tracking problem is on *finding the minimum number of UAVs with bounded velocity and their trajectories so that every target is tracked (covered) by at least one UAV during the entire period of observation.* Although it may appear that the assumption regarding complete knowledge of the trajectories of the targets makes the problem very simple (if not trivial), it can be shown that even with this assumption, the problem of computation of minimum number of trackers and their trajectories for continuous coverage of the mobile targets remains computationally hard, that is NP-complete.

**Target tracking problem (TTP)**: The objective of TTP is to minimize the total number of trackers needed to track all the targets during the entire period of observation. Figure 6.6 depicts an example with three targets. The locations of three targets (squares) on their trajectories (curves) are shown at time instances 0 to 5. Dashed curves show the trajectories of trackers and the disks centered at the locations of trackers depict the sensing area of trackers at different time instances.

The target tracking problem can be solved by using network flow techniques and integer linear programming. We first discretize time into equal time intervals of length $\delta$. We consider a set of $n$ targets $\mathcal{A} = \{a_0, \ldots, a_{n-1}\}$ moving on two-dimensional space over time instances $0, \ldots, T$.[1] Let $p(a_i, t) = (x(a_i, t), y(a_i, t))$ be the location of target $a_i$ at time instance $t$ where $x(a_i, t)$ and $y(a_i, t)$ denote the x-coordinate and y-coordinate of $a_i$ at time $t$. We assume that a target is covered by a tracker $b_j$ if the distance between

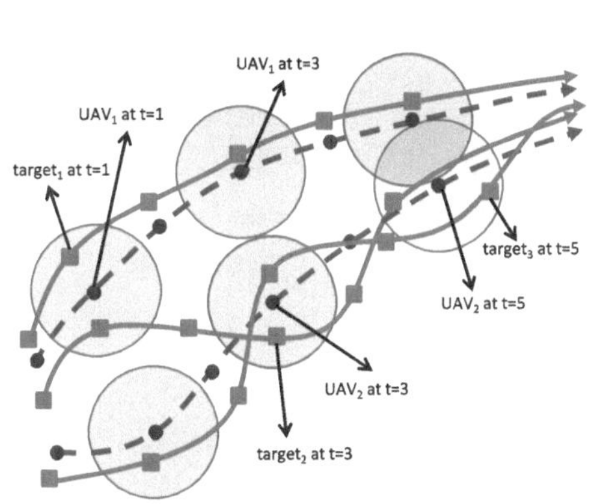


**Figure 6.6** Trackers and their trajectories for tracking all the three targets. (Reproduced from [433] with permission from Springer)

[1] We present the formulation in two dimensions for brevity. Extensions to higher dimensions is straightforward and is discussed in Section 6.6.3

them is less than the sensing radius $r$. Let $|p_1 \sim p_2|$ denote the distance between two points $p_1$ and $p_2$ on the two-dimensional space.

In TTP our goal is to find a smallest set of trackers $\mathcal{B}$, such that:

1. Coverage – For any target $a_i$ and any time instance $t$, there is a tracker $b_j \in \mathcal{B}$ whose location in time $t$, denoted by $p(b_j, t)$, is in distance $r$ from $p(a_i, t)$; that is, $|p(a_i, t) \sim p(b_j, t)| \leq r$.
2. Mobility – For any tracker $b_j$ with specified velocity $d$ and any timeslot $t \geq 1$, $|p(b_j, t-1) \sim p(b_j, t)| \leq d$.

The target tracking problem can be shown to be computationally hard, that is NP-complete. Consider a special case of TTP where targets should be covered at just one time instance $t = T = 0$. This special case of the TTP problem is equivalent to the Geometric Disk Cover Problem [172]. The objective of the Geometric Disk Cover Problem is to find the fewest number of disks (of prescribed radius) that can be used to cover a set of points on a two-dimensional plane. That is, given points in the plane, identify a minimally sized set of disks (of prescribed radius) that covers all points.

**Solution technique** In order to solve the TTP, first we model it by a flow network, by *also discretizing space*. Next, we find the solution of the problems by computing (a modified version of) *minimum flow* on a directed graph $G = \langle V, E \rangle$. The flow is then split into paths, where each path represents movement of a tracker. Before explaining our graph construction, we give the definition of the classical *minimum flow* problem [158].

**Minimum flow problem:** Given a capacitated network $G = \langle V, E \rangle$ with a non-negative capacity $c(i,j)$ and with a non-negative lower bound $l(i,j)$ associated with each edge $(i,j)$ and two special nodes, a source node $S$ and a sink node $D$, a flow is defined to be a function $f : E \rightarrow \mathbb{R}^+$ satisfying the following conditions:

$$\sum_{j \in V} f(i,j) - \sum_{j \in V} f(j,i) = \begin{cases} F, & i = S \\ 0, & i \neq S, D \\ -F, & i = D \end{cases}$$

$$l(i,j) \leq f(i,j) \leq c(i,j)$$

for some $F \geq 0$ where $F$ is the value of the flow $f$. The minimum flow problem is to determine a flow $f$ for which $F$ is minimized.

**TTP Graph Construction**

We construct a directed graph $G = \langle V, E \rangle$ representing the positions of the targets and the possible trackers' movements. This construction involves space discretization (in addition to the time discretization described earlier). Specifically, we consider a *grid* over the two- (or three-) dimensional space and restrict our trackers to move between points on that grid. The granularity of our grid is denoted by $\varepsilon$, which represents a trade-off between the accuracy of our solution and its running time. Let $\mathcal{N}$ be the set of all points on the grid; namely, for the two-dimensional space, $\mathcal{N} = \{i \cdot \varepsilon, j \cdot \varepsilon \mid i, j \in \mathbb{Z}\}$.

We note that based on the coverage constraint, for a target $a_i$ to be covered by a tracker at a time instance $t$, there should be at least one tracker in the disk of radius $r$ centered at

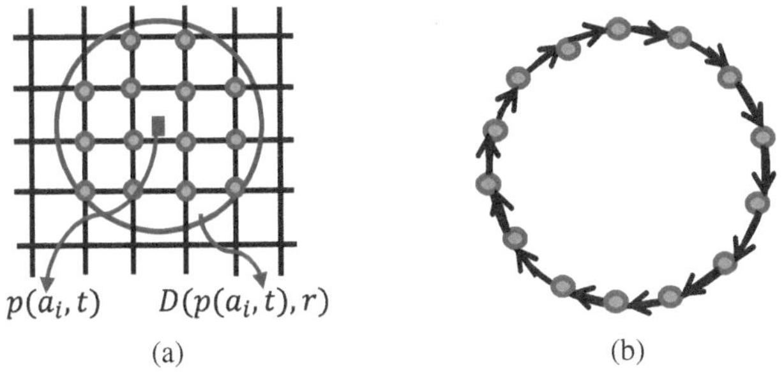


**Figure 6.7** (a) A disk centered at the location of target $a_i$ at time $t$. (b) The ring corresponding to the discrete points in the disk shown in (a). (Reproduced from [433] with permission from Springer)

$p(a_i, t)$. Let $D(p(a_i, t), r)$ denote that disk. Thus, a tracker should be located at one of the points in $\mathcal{N} \cap D(p(a_i, t), r)$. For every such potential location, at any given time, we add a vertex to the graph, which is represented by the triplet $\langle$location, target id, timeslot$\rangle$, where *location* corresponds to the coordinates of the points on the grid (namely, the same location can be added multiple times for many targets and/or timeslots). Figure 6.7a depicts a disk $D(p(a_i, t), r)$. Circles show $\mathcal{N} \cap D(p(a_i, t), r)$ and the square represents the center of the disk.

In addition, we add one supersource vertex $S$ and one supersink vertex $D$. In other words, the set of vertices will be

$$V = \{S, D\} \cup \bigcup_{t=0}^{T} \bigcup_{i=0}^{n-1} \langle \mathcal{N} \cap D(p(a_i, t), r), a_i, t \rangle .$$

We will have four types of edges:

1. Intra-target edges: For every target $a_i$ and time slot $t$, we construct a directed ring connecting all the vertices with target id $a_i$ and time instance $t$ (Figure 6.7b). We note that the order of the nodes in the ring is arbitrary.
2. Mobility edges: $\{(\langle p, a_i, t\rangle, \langle p', a_{i'}, t'\rangle) | i \neq i', |p' \sim p| \leq d|t' - t|\}$. Note that if $t' = t$, then $p' = p$. These edges have capacity 1 and demand 0. The direction of these edges goes from the node with lower timeslot to the node with higher timeslot, where ties are broken by nodes' target id. A mobility edge represents that a tracker can move from a location $p$ to $p'$ during time interval $|t' - t|$.
3. Supersource edges: All vertices are connected to the supersource $S$ with edges of capacity 1 and demand 0. These edges originate in the supersource.
4. Supersink edges: All vertices are connected to the supersink $D$ with edges of capacity 1 and demand 0. These edges terminate in the supersink.

We note that space discretization comes with a price, as it rules out solutions in which the trackers are not at a grid point in each time slot. As $\varepsilon \ll d$ (namely, the granularity of the grid is much finer than the maximum velocity of the tracker), these differences are negligible.

### 6.6.4 Modified Minimum Flow Problem

In Figure 6.8 we illustrate an example on a line (one dimensional) with three targets and we show their locations on the $X$ coordinate at two timeslots. In this example, $r = 2$, $\epsilon = 1$, and $d = 1$. Figure 6.9 depicts graph $G = \langle V, E \rangle$. For the sake of clarity, not all edges in $E$ are shown. The number on each node shows the location on the $X$ dimension. In this example, we note that the minimum flow value is zero while the value of $f$ of the intra-target edges is 1 and all the other edges have flow zero.

More specifically, all the lower bound constraints on the intra-target edges are satisfied while there is no flow starting from $S$. Hence, we have modified the minimum flow problem such that the lower bound constraints on the intra-target edges in graph $G = \langle V, E \rangle$ are satisfied if the flow starts from source node $S$. In this regard, we add additional constraints to make sure there is a path from $S$ to every node in $V \setminus \{S, D\}$ such that the flow $f$ on the edges of the path is 1. Let $g_u$ be a flow function defined with respect to node $u \in V \setminus \{S, D\}$. A closer look at our construction shows that if there is a path from $S$ to one node in a ring, there is a path to all nodes in that ring. Hence, for every ring in $G$, corresponding to a target $a_i$ at a time instance $t$, we select an arbitrary node $u$ (which is of form $\langle p, a_i, t \rangle$) and add the following constraints to the minimum flow problem

$$\sum_{j \in V} g_u(i,j) - \sum_{j \in V} g_u(j,i) = \begin{cases} 1, & i = S \\ 0, & i \neq S, D \\ -1, & i = D \end{cases}$$

$$g_u(i,j) \leq f(i,j).$$

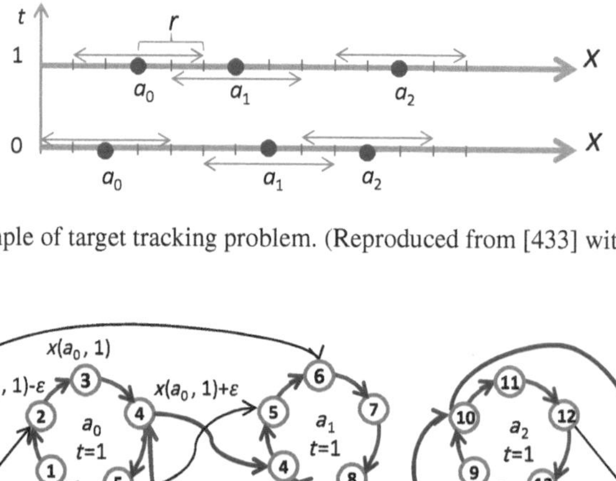


**Figure 6.8** An example of target tracking problem. (Reproduced from [433] with permission from Springer)

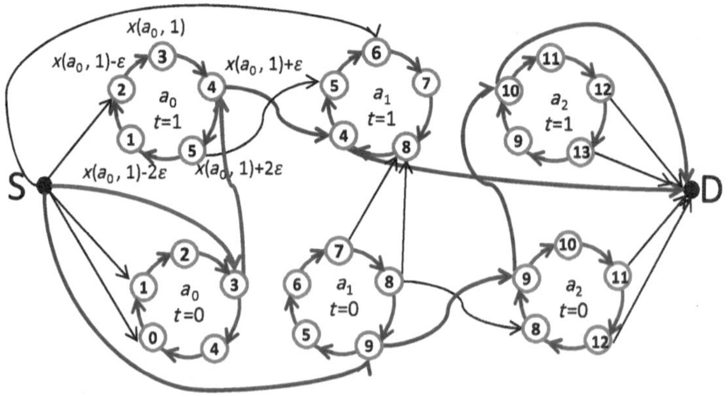


**Figure 6.9** The modified minimum flow solution of the example in Figure 6.8. (Reproduced from [433] with permission from Springer)

**Solution to the Target Tracking Problem**

Considering that a tracker covers a target only if it is placed at a discrete point in the disk centered at the location of the target at each time instance, we can show that the minimum number of trackers required to cover the targets at all time instances is equal to the value of the modified minimum flow in graph $G = \langle V, E \rangle$, and the trajectories of the trackers can be computed from the solution of the modified minimum flow problem.

In Figure 6.9, the edges show the edges with flow 1 in the modified minimum flow solution. We can see that the value of the modified minimum flow is 2, meaning that 2 trackers are required to cover the three targets in two timeslots. The first tracker at time 0 is at location $X = 3$ ($x(a_0, 0) + \epsilon = 3$), covering target $a_0$ and moves to $X = 4$ at time 1, covering both $a_0$ and $a_1$ ($x(a_0, 1) + \epsilon = 4$ and $x(a_1, 1) - 2\epsilon = 4$). The second tracker at time 0 is at location $X = 9$ covering targets $a_1$ and $a_2$, where $x(a_1, 0) + 2\epsilon = 9$ and $x(a_2, 0) - \epsilon = 9$, and moves to $X = 10$, ($x(a_2, t_1) - \epsilon = 10$) covering $a_2$.

**Extension to higher dimensions**: In two-dimensional space, we use $O((r/\varepsilon)^2)$ nodes connected in a ring to represent the disk around a target. In order to extend this solution to three dimensions, it is enough to discretize space in all three dimensions and use $O((r/\varepsilon)^3)$ nodes to represent the ball around a target and connect the nodes as a ring. Similarly, we can extend the model to higher dimensions.

## 6.7 Summary

This chapter provides a review of the current regulatory framework used for airworthiness certification of conventionally piloted aircraft, as well as the implications of these processes on potential avenues for UAS regulation. A survey of global UAS regulation is performed, highlighting the trend towards operational and risk-based approaches; and the impact of international and industry oriented standards organizations on emerging regulation is highlighted. An in-depth examination of privacy in the context of UAS use and regulation is examined. Finally, two technical problems, related to maintaining UAS networks and interconnectivity are explored in detail. Further work needs to be done as the landscape of UAS regulation is continuously evolving, and technical solutions along with acceptable means of compliance to these solutions, for communication and controls link and protocols for UAS, need to be formalized for the true potential of UAS to be realized.

## 6.8 Acknowledgements

The author (Natasha A. Neogi) would like Kelly Hayhurst (NASA Langley), Jeff Maddalon (NASA Langley), and Reece Clothier (RMIT) for all of their help and advice in information this research effort. She is also grateful to NASA Langley Research Center for their support while writing this article.

# 7 Safety, Security, and Privacy Aspects in UAV Networks

Kamesh Namuduri and Damien Sauveron

## 7.1 Introduction

As the airspace is being opened up for UAS, the UAS market is expected to see a tremendous growth over the next two decades. One projection is that by 2035, there may be as many as 250,000 UAVs including commercial, public, state, and federally owned UAVs operating in the airspace providing various kinds of services in the United States alone [321]. Inclusion of such a large number of UAVs in the airspace creates many challenges in terms of safety, security, and privacy, which are the three dimensions for integrating UAVs in the airspace and for designing real-world applications. The three factors are not completely independent of each other as one factor influences the other two.

Safety of human lives, undoubtedly, is the most critical aspect for any manned or unmanned aircraft. Safety is achieved through advances in aviation technology and statutory regulations that will help to prevent potential collisions in midair or on the ground. Factors that influence decisions related to safety include price, performance, payload, and power needs among others. Trade-offs among these factors need to be carefully considered while making statutory regulations. For example, the type of collision avoidance technology that should be made mandatory through the federal regulation for aircraft should take into account the impact of the choice on the performance of the aircraft.

Information security is critical in the sense that revealing essential flight information to hackers may jeopardize the safety of the aircraft and people on the ground. Thus, information security is not independent of safety. For example, hackers can fabricate messages impersonating or spoofing a plane, creating ghost planes in the air. Information security measures are needed to distinguish false messages from true messages intended for safety and situational awareness. Similarly, measures for aviation security may have an impact on the privacy of citizens as our experiences at airports demonstrate. While citizens may accept to forego a certain amount of comfort for security, any measures that cause privacy concerns should be minimized.

In the light of UAV integration into airspace, information security and privacy issues that surround the use of UAVs are being re-investigated. Legal definitions of individuals' privacy and information security are being re-examined in order to understand the trade-offs between the two. What are the implications of a UAV deployed by a law enforcement agency taking video footage of an individual's private property? What is the legitimacy and admissibility of evidence collected by a UAV in a court of law?

Such questions are being considered in courts leading to discussions on expectations of privacy in public and private places when UAVs fly in civilian airspace. Further, careful auditing of the use of data collected by government entities such as the law enforcement agency using UAVs is critical to protecting civil liberties. Dialogues, discussions, and answers to these questions will determine the landscape of the regulations and legislations that are to come in the future.

Although security and privacy appear to have adverse effects on one another, it is possible to develop measures that will enhance both security and privacy at the same time. This requires careful dialogue among the stakeholders, including aviation experts, civil liberty unions, and citizens. Increased awareness in society may alleviate fears surrounding UAVs in airspace. For example, informing citizens ahead of UAV deployment in their neighborhoods, regulating the type of sensors that a UAV is equipped with, and providing transparency and auditing of information collected by UAVs will help balance security and privacy.

Each country owns its own national airspace. For example, in the United States, the National Airspace System or NAS consists of the procedures, policies, regulations, infrastructure, aircraft, and personnel that compose the national air transportation system of the United States [443]. While specific regulations may differ from one country to another, there exists a common ground among the countries in terms of the fundamental concepts of safety, security, and privacy issues related to UAVs. Thus, the rules of engagement of safety, security and privacy for integrating UAS in the airspace around the world bear lots of resemblance. While all three elements, that is safety, security, and privacy, are necessary, their relative importance may vary based on the application context.

The most important tenet of UAS integration in the airspace is the safety of human lives. Safe integration of UAS in the NAS is a critical Federal Aviation Authority (FAA) milestone mandated by US Congress. At its core, safe integration means preventing collisions either in mid-air or on the ground. Thus, it is important to examine the technologies that are in use today and those proposed for future use for aircraft collision detection and avoidance as UAVs must possess similar capabilities. The metrics for assessing collision avoidance technologies from a safety perspective include reliability, longevity, and efficiency.

## 7.2 Safety in the Sky

In this section, the technologies, standards, and regulations that are aimed at improving human safety and situational awareness of aircraft will be discussed. Some of these technologies such as the Automatic Dependent Surveillance – Broadcast (ADS-B) [403] exist today, some others such as the L-Band Digital Aeronautical Communications System (LDACS)[376] are currently under consideration, and some others such as the self-organizing networks are being proposed for future applications [220, 222]. The objective of this discussion is to provide a glimpse of the state-of-the art in safety and situational awareness technologies and what is on the horizon.

### 7.2.1 Automatic Dependent Surveillance – Broadcast (ADS-B)

ADS-B increases the safety of flight operations by allowing the Air Traffic Control (ATC), other aircraft, and third party personnel to track aircraft that are flying nearby. ADS-B is the more precise, satellite-based successor to the radar systems used for Air Traffic Management (ATM). The use of ADS-B enhances the situational awareness of aircraft systems and hence their collision avoidance capability. ADS-B avionics enable aircraft to fly in weather conditions under ATC. All aircraft systems which fly in the regulated airspace where a Mode C Transponder is now required are expected to be equipped with the ADS-B by 2020. ADS-B Out uses GPS technology to determine an aircraft's location, airspeed, and other data, and broadcasts that information to a network of ground stations, which relay the data to air traffic control displays and to nearby aircraft equipped to receive the data via ADS-B In. The FAA mandated only the ADS-B Out capability for aircraft systems. ADS-B In capability, while optional, provides aircraft operators with the weather and traffic position information delivered in real-time. As of July 2015, more than 13,000 general aviation aircrafts and 425 commercial aircrafts have been equipped with ADS-B [160].

As of February 2015, the FAA has installed 634 ADS-B ground stations (Figure 7.1) throughout the US, completing baseline deployment of the network [160] (ADS-B has been integrated into automation platforms at 22 of 24 ATC facilities which control high-altitude traffic. ADS-B traffic and weather broadcasts are now available nationwide throughout the US.

A transponder used for civilian applications may operate in one of the interrogation modes A, C, or S. The interrogation itself may come from secondary surveillance radar or other ADS-B. FAA's ADS-B Out requirement can be satisfied by using either a dedicated 978MHz Universal Access Transceiver (UAT) or a 1090MHz Mode S Extended Squitter (ES) transponder with an approved GPS navigation source. 1090 ES format is internationally accepted whereas the use of UAT format is restrictive as it is acceptable only within the US.

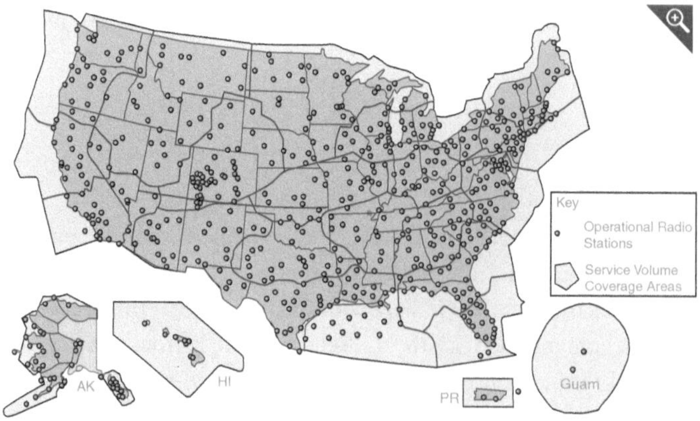


**Figure 7.1** FAA installed 634 ADS-B ground stations throughout the US as of February 2015 completing the baseline deployment of the network (Courtesy FAA, NextGen)

ADS-B also has an important role in airport ground operations via airport surface detection system-Model X (ASDE-X). ASDE-X is a ground surveillance system that uses radar and satellite technology to allow ATC to track surface movement of aircraft and airport ground vehicles. It alerts the ATC of any potential runway and taxiway conflicts by using data from ADS-B and other sensors. ADS-B may be too expensive to install for small UAVs. For example, ADS-B transceivers that are available today from companies such as SageTech, L-3 Aviation, and Freeflight cost in the range of $2000 to $8000. Some other companies such as Google are working towards designing lightweight and less-expensive versions of ADS-B transceivers that are suitable for UAVs.

### 7.2.2 FLARM

FLARM is a less-expensive solution used to alert pilots of potential collisions between aircraft. FLARM receives position and movement information from an internal 16 channel GPS receiver with an external antenna. FLARM estimates the flight path of an aircraft in which it is installed and transmits this information to other aircrafts on radio at one-second intervals. Aircraft within the communication range of the FLARM receive this information provided they are also equipped with the FLARM. The incoming signal is compared with its own flight path estimated by the second aircraft. FLARM also compares the predicted flight path with known obstacle data, including electric power lines, radio masts, and cable cars. Due to its affordability, FLARM is considered as a viable option for small UAVs as well as gliders.

### 7.2.3 ADS-B Versus FLARM for Gliders

In 2015, the FAA issued an "Advanced Notice of Proposed Rule Making" (ANPRM) soliciting public comments on whether the glider exemption for transponder use should be revoked and whether gliders should be required to comply with the rules for the use of ADS-B devices. This notice is intended to address midair collisions such as the one that happened in 2006 between a glider and a small commercial jet. FAA issues ANPRMs such as this before formulating a specific change in federal regulations or airspace rules. Organizations such as the Soaring Society of America (SSA), EAA, and Aircraft Owners and Pilots Association (AOPA) bring the awareness within the community with issues such as this. The SSA submitted a formal response to this notice and encouraged others who had personal experiences with transponders on-board gliders [258].

The key concerns in equipping a glider with ADS-B is the power drain caused by the use of a transponder and the limited battery life that supports the electronic equipment on a glider. The current technology does not permit the glider to use any existing transponder for the entire duration of the flight, which is a requirement from FAA if the glider is equipped with any transponder. King [258] argued that a transponder may not be the "magic bullet" for addressing the midair collisions. This discussion brings up the need for lightweight, less-expensive, and low-power anti-collision technology that can be used by small-scale UAVs, gliders, and other manned and unmanned flights. While it is not possible to come up with a "one size fits all" solution, there is a need for investigating the minimum operational requirements and interoperability of collision avoidance systems. In addition, there is a need for reliable and secure communication and networking strategies and protocols for sharing essential flight information among the flights that are within the vicinity.

### 7.2.4 L-Band Digital Aeronautical Communications System (LDACS)

The LDACS is foreseen as a potential standard for aeronautical communications within the coming years. Being a high-performance, high-rate data link, LDACS enables modern Air Traffic Management (ATM) applications that are currently being developed within the two international ATM projects SESAR (Europe) and NextGen (US). The LDACS has the potential to be extended towards an integrated communications, navigation, and surveillance solution for next generation aviation. It includes both Air-to-Ground communications and Air-to-Air communications. Integration of LDACS with satellite-based and terrestrial-based communications and networking technologies will lead to a redundant aeronautical network with multiple links. Such a heterogeneous and redundant aeronautical network will enhance the connectivity of aircraft with other aircrafts and ground control stations, which, in turn, will enhance the overall safety of an aircraft. As the LDACS allows direct air-to-air communications, it serves as an enabler for future ADS-B technology. The LDACS also facilitates the creation of an ad hoc network in the air in which each aircraft serves as a node, sharing information with neighboring nodes.

### 7.2.5 Aeronautical Mobile Aircraft Communication System (AeroMACS)

AeroMACS is based on the IEEE 802.16 WiMAX mobile communications standard. It is the technology that the air traffic industry has selected to support on-the-ground communication exchanges at airports. It is also being considered to support other aviation applications when the aircraft is on the ground. As the aircraft lands, it will establish a connection to the AeroMACS ground network to exchange data with the ground infrastructure and to use ATM applications. AeroMACS operates in the aeronautical C (5GHz) band allocated for Aeronautical Mobile Route Service (AMRS) by the International Telecommunication Union (ITU). AeroMACS supports both fixed and mobile connectivity. It is the wireless solution for the next generation standard in ATM selected by EUROCONTROL and the FAA and ratified by the International Civil Aviation Organization (ICAO) for the provision of dedicated aeronautical communication services for safety and regularity of flight on the airport surface globally.

### 7.2.6 Self-organized Airborne Network (SOAN)

SOAN offers a possible solution to enhanced situational awareness to manned and unmanned aircraft systems. An airborne network offers the ability for an aircraft to

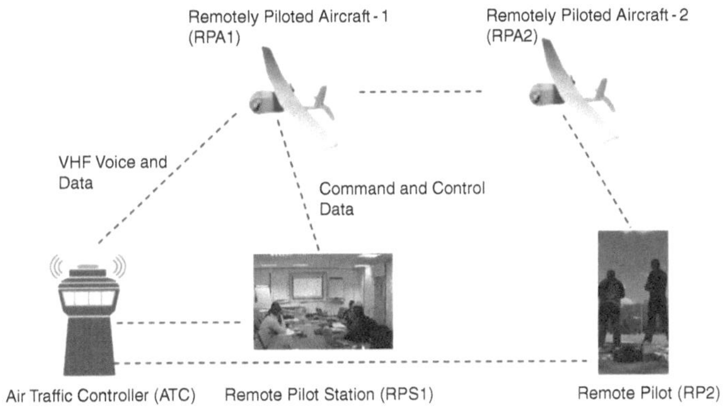


**Figure 7.2** Providing a direct communication link between two aircraft allows for the creation of self-organized airborne network

be constantly in touch with other aircraft providing a way for the ATC to keep track of aircraft continuously and globally. In order to understand the concept of SOAN, let us review the terminology defined by the ICAO [220, 222].

- Remotely Piloted Aircraft (RPA) – an unmanned aircraft such as a UAV which is piloted from a remote pilot station.
- Remotely Piloted Aircraft System (RPAS) – a remotely piloted aircraft, its associated remote pilot station(s), the required command and control links, and any other components as specified in the type design.
- Remote Pilot Station (RPS) – the component of the RPAS containing the equipment used to pilot the RPA.
- Remote Pilot (RP) – a person charged by the operator with duties essential to the operations of an RPA and who manipulates the flight controls, as appropriate during flight time.

In a typical scenario, each RPA is controlled by one RPS. In most scenarios, an RPA is expected to be always within the Radio Line of Sight (RLoS) of the RPS. As shown in Figure 7.2, the following data links are needed for managing communications when RPAs are integrated in civilian airspace.

- Command and Control (C2) data link between RPS and RPA to transmit commands from RPS to RPA and to receive RPA status information at the RPS
- VHF two-way voice communications links between ATC and RPA, and between ATC and RPS

One key aspect of integrating RPAs (UAVs) in the civilian airspace is the requirement of a reliable communication link between a UAV and its ground control station (remote pilot station or RPS). As long as the RPA–RPS communication is reliable, RPS knows the whereabouts of the UAV and can control the movement of the UAV. In a critical scenario, the RPA can be brought down gracefully or forcibly if needed. An RPS may interact with both the RPA and ATC in Radio Line of Sight (RLoS).

### 7.2.7 Beyond the Radio Line of Sight (BRLoS)

When an RPA is beyond the radio line of sight, satellite or airborne links are necessary for RPS and ATC to maintain communication. There are certain regions such as oceanic airspace and Siberian areas in which aircrafts often fly for long durations in the absence of surveillance and contact with ATC. Absence of surveillance takes place when the aircraft is more than 400km away from an ATC. While satellite links offer a solution to BRLoS communications, they introduce operational challenges that include signal delays caused by the unpredictable characteristics of communication links between RPS and the satellite, including the number of earth-to-satellite signal hops. Regulation and oversight of satellite communication service also pose a challenge to relying on satellite communications. Airborne networks offer a better alternative and back-up for satellite communications as they satisfy the requirements of C2 data links. The number of air-to-air hops and signal delays can be predicted better in airborne networks than satellite communications.

### 7.2.8 Benefits of Self-organized Airborne Networks

Self-organized airborne networks will facilitate storage of information on-board every aircraft, and exchange of information among the aircraft as long as some aircrafts are flying within the vicinity. Each RPA serves as a node in the network not only carrying its own information to others, but also serving as a relay to carry information from other RPAs. An RPS may send a command to its RPA even if the RPA is beyond its radio line of sight. When an RPA receives information from its neighboring RPAs, it can use this information for its own situational awareness and for implementing its own "detect and avoid" functionality. Each RPA receives its control signals from its own RPS, via nearby RPAs or from other aircrafts. A SOAN allows for the implementation of a robust functioning of RPAs in civilian airspace under BRLoS conditions. The information exchanged among the RPAs is received without much delay. If an RPA is within the vicinity of another RPA, a network is formed, and information is exchanged and used for collision avoidance. If there are no nodes within the vicinity of an RPA, there is no danger of any collision.

## 7.3 Privacy on the Ground

The availability of UAVs and the success stories of using UAV missions by federal entities such as the Customs and Border Patrol have prompted many state Government agencies such as the police department to acquire UAVs for law enforcement purposes. Almost every state has purchased UAVs and uses them for applications such as watching a crime scene, tracking drug dealers, and finding missing children among others [81]. Many individuals and private organizations are also acquiring UAVs for various personal and commercial applications, including monitoring farm lands, filming movies, and taking aerial photographs.

Given such opportunities, it is expected that there will be a widespread use of UAVs in real-world applications in the near future. UAVs will be flying in our own neighborhoods carrying out life-saving tasks such as searching for people, protecting critical infrastructure, and delivering medical supplies. UAVs may be flying in our neighborhoods for non-critical applications such as pizza delivery, mail delivery, and monitoring. As the Certificate of Authorization (COAs) approved by the FAA increase and the UAVs crowd the civilian airspace, serious consideration needs to be given to citizens' privacy.

### 7.3.1 Fourth Amendment in the Context of UAVs

The Fourth Amendment of the US Constitution guarantees "the right of the people to be secure in their persons, houses, papers, and effects, against unreasonable searches and seizures, shall not be violated, and no warrants shall issue, but, upon probable cause, supported by oath or affirmation, and particularly describing the place to be searched, and the persons or things to be seized."

An individual's right of privacy may conflict with Government's measures of safety and security. Whether a particular type of search is considered reasonable in the eyes of the law is determined by balancing an individual's Fourth Amendment rights and legitimate Government interests, such as public safety. There is no clear-cut answer to balancing the two sides of the coin. "While individuals can expect substantial protections against warrantless intrusions into their homes, the Fourth Amendment offers less robust restrictions upon Government surveillance occurring in public places including immediately outside the home, such as driveways and backyards" according to a study conducted by the Congressional Research Service in 2013 [262].

Although cases relating to the Fourth Amendment and privacy concerns are quite common in front of the Supreme Court, there were hardly any cases involving drone use that the US Supreme Court has addressed. However, there were several relevant cases involving surveillance by law-enforcement in general that the Supreme Court addressed. The Court has held the view that the application of the Fourth Amendment depends on whether there is a legitimate and reasonable expectation of privacy. Carr [81] refers to Justice Harlan's two-prong test to determine when Fourth Amendment protection is appropriate. First, a person must have exhibited an actual expectation of privacy. Second, the expectation must be the one that the society is prepared to recognize as reasonable. In the following discussion, few cases illustrate the Court's view of legitimate use of surveillance in law-enforcement which serve as precedents for proper use of drone surveillance.

**Is a warrant always needed for collecting evidence from a private property?** The answer is not a simple "yes" or "no" as the following cases demonstrate. In July 2012, in the first legal case involving surveillance by drones in the US, a state district judge in North Dakota allowed the evidence collected by a drone in investigating a crime scene without a warrant. The attorney questioned the legitimacy of the warrantless drone surveillance by citing *Kyllo* v. *The United States* as evidence [81]. In the Kyllo case from 2001, the Supreme Court decided that the use of infrared detection-based surveillance to identify marijuana cultivation inside a suspect's house is a violation of the Fourth

Amendment. On the contrary, in *California* v. *Ciraolo's* case from 1986, the Supreme Court ruled that as long as an aircraft is in navigable airspace, an individual's private property is not always protected by the Fourth Amendment. The defendant in Ciraolo's case built a private fence around his property. A pilot flew over his house and observed marijuana plants. The Court held the view that "The Fourth Amendment does not require police traveling in the public airways to obtain warrant in order to observe what is visible to the naked eye."

**Is it legal to track vehicles without a warrant?** In *Jones* v. *United Sates* case discussed in 2012, the Supreme Court ruled that a tracking device placed on a car to monitor the vehicle's movement patterns without a warrant is a violation of an individual's privacy. In this case, the judge cited society's expectation of law enforcement, stating that the agents are not expected to secretly monitor and catalog every movement of an individual vehicle for a long period.

## 7.4 Information Security

In many applications, UAVs may be operating in an adverse environment and thus are susceptible for attacks. For instance, in a military context, adversaries may want to know what information the owner of UAVs has collected or they may want to change this information. Adversaries may want to compromise UAVs so that they can use them for their own advantage. In a civilian context, an attacker may want to know why some UAVs are flying around and may try to access the data collected for fun or for profit.

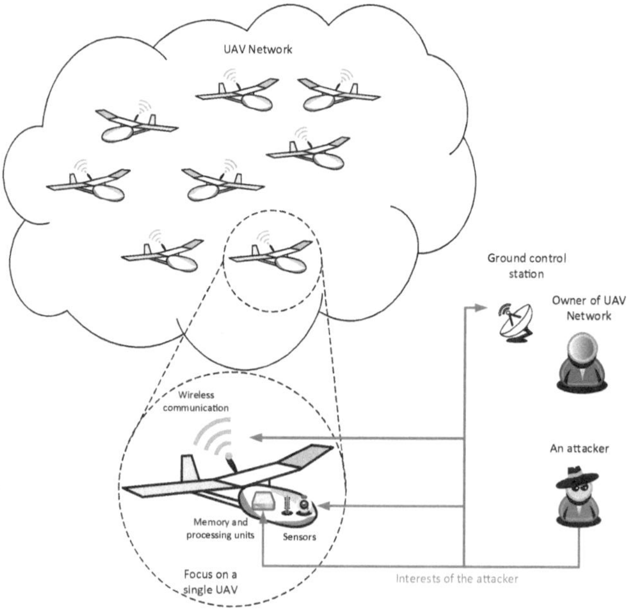


**Figure 7.3** UAV network and interests of an attacker

According to the level of classification of the information accessed, an attack may lead to serious privacy concerns. In order to avoid such leakages, appropriate information security methods need to be embedded within the UAV applications.

As illustrated in Figure 7.3, although the capabilities of the adversaries are different in civilian and military contexts, a UAV network as well as an individual UAV within a network are susceptible to attacks. The likelihood of an attack is even higher if a UAV is captured by an adversary.

Section 7.5 outlines the security requirements for a UAV. Section 7.6 provides a comprehensive overview of the security requirements at the network level for standalone UAV swarms and for ground-controlled UAV fleets. Section 7.7 focuses on related research and products.

## 7.5 Security Requirements at UAV Level

Our threat model assumes that the adversary has access to a UAV to perform advanced attacks. Access to a UAV is a prerequisite for intrusion attacks against it. An attacker may gain access to a UAV by one of the following means:

- UAV battery is exhausted or UAV is damaged and the adversary gets access to the UAV when it is on the ground.
- The adversary has the capability to capture a flying UAV in its functional state.
- The adversary, being on the ground, takes control of a UAV during its flight with or without using an adversarial UAV.

Since the attacker's capabilities are dependent on the adversary model, it is necessary to characterize the adversary's attack potential before learning the security requirements for the UAV.

An attacker with a low attack potential or a layman will try to disassemble the UAV to gain access to its internal memory with the expectation that he can be read with equipment using common standard interfaces (e.g., USB, SDIO) and that the data are not ciphered. These data may include sensory data (e.g., measurements, photos) as well as mission-specific data (e.g., flight-plan for the mission, coordinates of points of interest).

An attacker with moderate attack potential, or a proficient person, will try to operate in a similar manner but he will be able to gain access to less common standard interfaces (like JTAG Test Access Port) or even proprietary interfaces with some reverse-engineering skills. Such an attacker will also certainly test if he can install his own system on the UAV (for instance, in flashing it or booting a Linux) or gain access to the embedded software.

Both types of attackers can be partially countered with the following security requirements (SR):

- (SR1) UAV should be equipped with self-destruction mechanisms [264, 461]. For instance, these mechanisms can be triggered when the UAV is flying at low altitudes.
- (SR2) Information should be stored encrypted on the UAV.

While SR1 is helpful in protecting the UAV against aforementioned attacks, it might have the unwanted side effect of destroying the UAV when it needs to land due to battery exhaustion if it is triggered by the altitude. SR2 can also help to delay the time required by an attacker to access data if the ciphering algorithm is resistant to brute-force. However the cryptographic keys have to be stored somewhere in an unencrypted form and it is not satisfactory. In addition, only the data are encrypted and an attacker may still have access to the embedded software.

The aforementioned requirements are not suitable to counter an attacker with high attack potential, such as a security expert or a group of expert hackers sponsored by state agencies. An expert attacker can perform various well-known attacks studied and applied during past decades in the world of smart cards but applicable to any systems as it will be shown below. Smart card (under its different form factors) is one of the most secure devices running successfully in the worst adversary conditions (where even its owner can be malicious), and knowledge of the attacks they have been subjected to is helpful to draw relevant security requirements of UAVs. The main categories of attacks to consider include:

- Side-channel attacks [306, 116, 334, 52, 33], which are blackbox attacks consisting in identifying some information leakages from algorithms running on the target. From these leakages, different types of information can be retrieved (e.g., cryptographic keys [261], sequence of opcodes executed [432]). The attacks can be based on timing [260], power consumption analysis (simple power analysis [261], differential power analysis [261], high-order differential power analysis [297], correlation power analysis [68]), electromagnetic analysis (simple electromagnetic analysis [344, 178], differential electromagnetic analysis [178, 19], high-order differential electromagnetic analysis, correlation electromagnetic analysis), or a combination of different sources [436, 20]. There also exist some other powerful attacks based on side-channels like template attacks [88, 349].
- Fault injection attacks [61, 263, 384, 188, 45, 182], which consist of perturbing, usually during a short time, the execution of a process for instance by using a laser or voltage glitches to reach a state the attacker can take advantage of. For instance, using fault injection at the right time on an RSA signature process, an attacker can recover very quickly the private key used [61] in exploiting the erroneous signatures delivered by the blackbox system signing the message. With differential fault analysis, secret key cryptosystems like DES [55] or AES [147] are also vulnerable.
- Physical attacks [263, 383], which encompass microprobing, circuitry modification with a Focused Ion Beam system or a laser cutter, etc.
- Software attacks, which are highly dependent on the possibility of loading applications on the target. The loading may or may not be protected by an authentication mechanism, which can still be circumvented by another attack. If a malicious application is loaded, it is feasible to launch an attack from inside the target against other hosted applications or against the platform of the target [90, 304].

- Combined attacks [433, 46], which often combine fault injection during execution of a code loaded or already present in the target to alter the application execution in order to gain additional access privileges.

These attacks are not only applicable to smart card but also to any processor as is discussed in several research articles [334, 52, 33, 389, 188] and thus to a UAV. For instance, some researchers did a correlation power analysis [300] on Virtex-4 and Virtex-5 family, that are Xilinx FPGAs widely used in UAVs (including the Predator [456]). They suggested that the encryption mechanism can be completely broken with moderate effort. Thus, a strong adversary model makes sense, especially in the context of military usage of UAVs since the opponent can be a government-controlled organization capable of performing forensic analysis of UAVs or attacks on the UAVs.

For all these reasons, the following security requirements must be considered:

- (SR3) At least a processing part of the UAV should be tamper resistant.
- (SR4) At least a part of the UAV should offer secure key management and cryptographic features (e.g., for keys used to encipher/decipher data stored in memory or exchanged over the network).
- (SR5) The UAV should provide a remote attestation mechanism if update of embedded software or installation of new applications is authorized.
- (SR6) The UAV should provide a high level of assurance for the security mechanisms provided. In short, it means the UAV must have been subjected to a security evaluation and certification to prove that it is resistant to attacks coming from a strong adversary.

Though most of these requirements can be implemented using devices like Trusted Platform Module (TPM) or smart card, a new kind of devices is proposed in [23].

An attacker with high potential can also try to modify the location information systems used by a UAV to get the position of the data collected and/or to manage its flight path. The location information system can be Global Positioning System (GPS)-based or can use radio frequency (RF) location methods like multilateration by measuring the "time difference of arrival" (TDOA). Unfortunately, both solutions are not secure and reliable. Since it was shown that GPS can be spoofed [410] and that multilateration still has some open issues [402], an additional security requirement is:

- (SR7) The UAV should use a secure location system.

SR7 will be discussed in the next section since location systems are very useful in a network setting (especially if a geographic routing protocol is used), but GPS spoofing can be used to capture UAV [256].

A location information system may use several sensors (such as a GPS) which are embedded in the UAV. As mentioned in [205], sensors may be susceptible for attacks that might compromise their integrity and availability. In order to handle these attacks, the following security requirements are necessary:

- (SR8) The UAV should have redundancy of sensors and all of them should be checked before their use to ensure that they will operate correctly in an adverse environment.

## 7.6 Security Requirements at UAV Network Level

A UAV network is a flying Mobile Ad hoc Network (MANET) if it is a standalone swarm or a flying mesh network if ground-controlled stations are used to manage its mission. Since for both kinds of networks there are generic threats, security requirements to address them are discussed below. An adversary can launch attacks similar to those that are common among MANETs, Delay/Disruptive Tolerant Networks (DTN), and Wireless Sensors Networks. An attacker can launch the easiest attacks on a wireless link, for example, a Denial-of-Service (DoS) [451]. This attack can be launched on several layers of the OSI stack. At the physical level, a jamming attack can be launched by interfering on radio frequencies used by the UAVs or the control stations. This attack can be launched from the ground or with rogue UAVs and is very efficient on most topologies. Even though some countermeasures exist, like using spread-spectrum communication (e.g., Frequency-Hopping Spread Spectrum – FHSS) or jamming detection to switch the node in a low consumption mode when such an attack is in progress, jamming is always annoying. At the link level by exploiting the medium-access control backoff and retransmission procedures, collision attacks can be achieved to create a DoS. Collision attacks are very effective since, even though only a few bits are affected, the sender should have to send the whole message again, which can lead to a depletion of its battery (an exhaustion attack). However, solutions such as the use of error correcting codes and best-effort delivery protocols can partially help to thwart these attacks. At the network level, the attacker can use routing loop attacks [271], which create loops in the routing of a message so that the packets do not reach their destinations, and drain the battery of nodes involved in the loops. At the transport level, an attacker can perform a flooding attack, which repeatedly requests new connections to a given node, leading to memory exhaustion and rejection of connection requests from legitimate nodes. A desynchronization attack can lead to resource-costly re-transmissions [128].

If communications are not ciphered, an attacker may attempt eavesdropping, packet injection, corruption, man-in-the-middle, or relay attacks.

The attacker can also build a rogue UAV to attempt some attacks on routing protocols [252, 62, 91], like the blackhole attack, selective forwarding attack, sinkhole attack, rushing attack, Sybil attack, wormhole attack, etc. UAVs are also susceptible to application specific attacks. For example, the SkyJack project [5] describes a drone engineered to autonomously seek out, hack, and take over other drones over a wireless channel. In order to protect UAVs against the aforementioned attacks, the following generic security requirements are needed:

- (SR9) All layers of the communication stack should include protection mechanisms against DoS attacks.
- (SR10) Communication channels should be protected in terms of confidentiality and integrity.
- (SR11) Communication channels should only exist with authenticated entities.
- (SR12) If a routing protocol is used, it should be secure.

Key management-related security requirements will be discussed in the context of networked UAVs.

### 7.6.1 Security Requirements for Standalone Swarms

The UAV swarm must be defined before developing the security requirements for it. A standalone UAV swarm is a set of UAVs which together perform a given mission in an autonomous way without relying on communications with the ground control station in order to be stealthy in the adversary conditions of the mission. A UAV swarm is self-organized and equipped with swarm intelligence. From the user point of view, the swarm is easy and transparent to manage in terms of both functionality and security. The user provides the mission goals and the initial security means (cryptographic keys, for instance) to the swarm and the swarm must accomplish its mission autonomously. The swarm is able to operate optimally in the adverse environment by analyzing the situation and making decisions in real-time. For instance, if a UAV with a dedicated assignment is facing a problem (e.g., failure, too low energy level to achieve the assignment), it may decide to entrust its assignment to another UAV according to the capabilities in terms of equipment (e.g., sensors and software). A standalone UAV swarm is depicted in Figure 7.4.

Major security requirements for a standalone UAV swarm include the following:

- (SR13) Security of the whole swarm should be distributed on each UAV: In short, security should not rely on a few UAVs which would act as servers for security-specific tasks (like keys distribution or renewal, authentications server, etc.).
- (SR14) Each UAV should provide a secure unique ID on which the whole swarm could rely, for its management and networking operations.

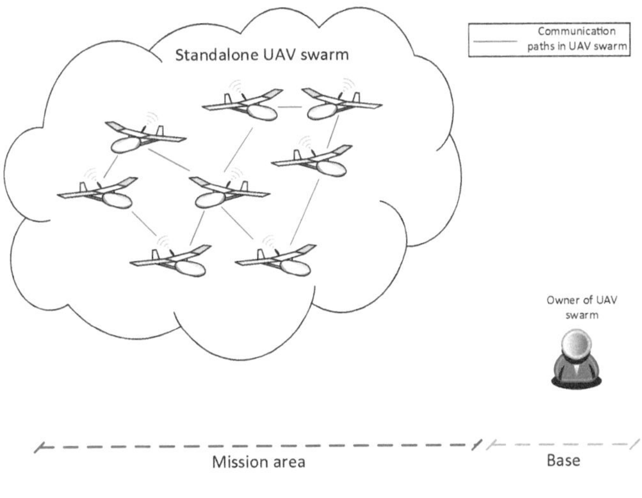


**Figure 7.4** A standalone UAV swarm

- (SR15) Security of the swarm should rely on appropriate cryptographic means that are resistant to attacks during its mission. In other words, capture and forensics of one or more UAV should not compromise the security of the UAV swarm. This requirement is also related to the fact that UAVs may join a swarm at any time and leave the swarm at any time.
- (SR16) To ensure perfect forward and backward secrecy properties, keys need to evolve dynamically throughout the mission.
- (SR17) To avoid loss of collected data (due to failure or crash of a UAV), redundancy mechanisms should exist in the swarm.

Swarm intelligence-based security methods [390] can be used in UAV swarms. However, this topic is beyond the scope of this chapter.

### 7.6.2 Security Requirements in Ground-Controlled UAV Fleets

In contrast to swarms, a ground-controlled UAV fleet appears as a mesh network. Information security in such a network may rely on certain infrastructures such as a public key infrastructure. The link for command and control (C2) between the ground station and individual UAV is the major difference with respect to the swarm. In a one-level fleet, UAVs can communicate directly among themselves according to their capabilities and they can also directly communicate with the ground. In a multi-level fleet, UAVs of the fleet can communicate between them or with special nodes, called heads, that can communicate with the ground station and with the heads of other friendly fleets. A one-level UAV fleet is depicted in Figure 7.5, whereas two multi-level UAV fleets (with only two levels – a configuration with more levels is possible) are depicted in Figure 7.6. For simplicity reasons, only communications related to head A1 are drawn. Major security requirements for UAV fleets are:

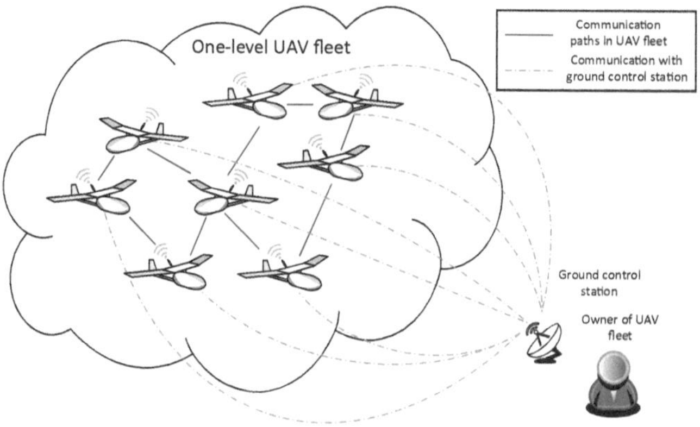


**Figure 7.5** One-level UAV fleet

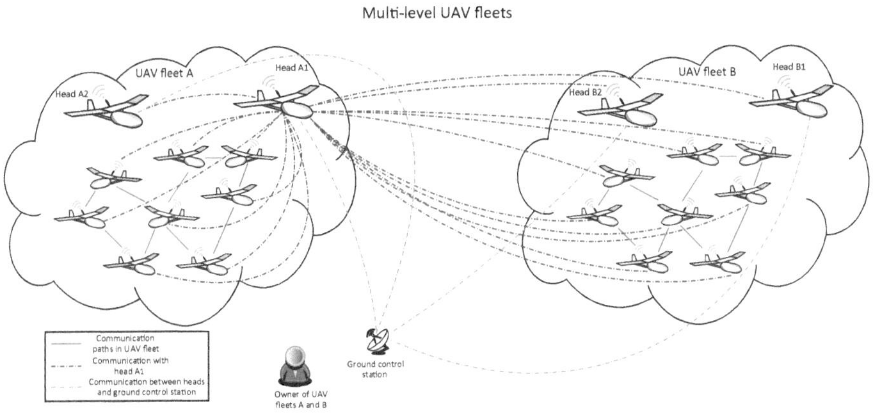


**Figure 7.6** Multi-level UAV fleets

- (SR18) The ground-controlled station should be secure to avoid compromising the whole network.
- (SR19) The C2 links should be used to renew/refresh the cryptographic means (keys, etc.) used by the UAVs of the fleet from time to time.
- (SR20) The communication channel between the UAVs should be different from the communication channel between a UAV and the ground station (for instance different RF frequencies). This requirement will make the channels more resistant to jamming attacks.
- (SR21) Data collected by UAVs should be sent to the ground station as soon as a connection is available to avoid potential loss of data (especially in the case of the heads).
- (SR22) Redundant heads are needed in multi-level fleets to avoid potential loss of collected data.

## 7.7 Ongoing Research and Products Related to UAV Security

The security issues outlined in this chapter underscore the need for significant research in UAV security. Researchers are developing security solutions and companies are working toward developing secure products. For instance, Galois Inc. claims that it has developed the most secure UAV software [3]. Xilinx Inc. claims that it has developed a tamper resistant FPGA for avionics [4].

Akram *et al.* [23] are working on a solution to secure a UAV swarm by embedding the secure elements discussed in this chapter within a UAV. This device which is under development is expected to provide a high level of security for the control network layer illustrated in Figure 7.7.

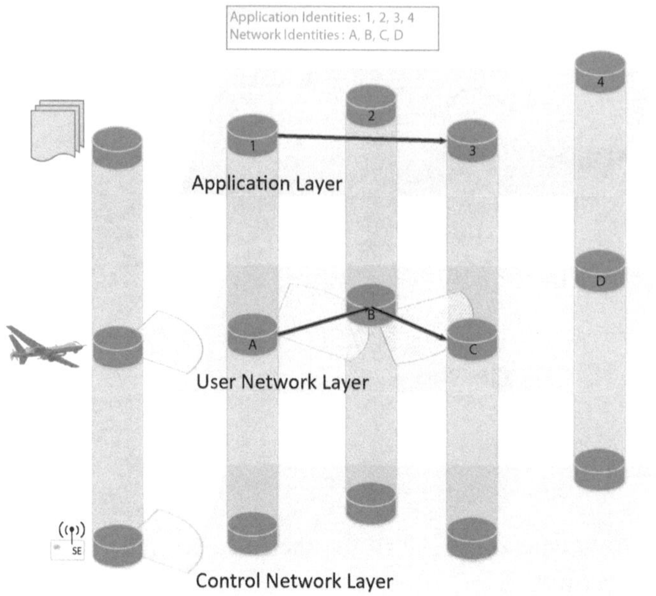


**Figure 7.7** UAV swarm using security elements proposed in [23]

## 7.8 Summary

This chapter outlined and discussed the safety, privacy, and security aspects that surround the use of UAVs and UAV networks. While safety is emphasized as the most important and critical factor among the three, the dependence of safety on employing appropriate security mechanisms is highlighted. The need for interpreting privacy in the new era of UAVs is discussed. A set of 22 security requirements are derived assuming a strong adversary model using a variety of contexts and scenarios that UAVs may encounter.

# 8 Collaboration Between Autonomous Drones and Swarming

Serge Chaumette

## 8.1 Introduction and Background

Recent technological progress, especially in robotics and mechatronics, has made it possible to create autonomous mobile vehicles such as air, ground, surface, or underwater robots. All of these vehicles are referred to as drones or unmanned systems. The focus of this chapter is on air vehicles, but all these systems share a large number of characteristics independently of the environment in which they evolve. This chapter is thus useful reading whatever type of unmanned system you are interested in. Unless stated otherwise, the term drone in what follows will refer to air vehicles. We will also use the terms unmanned aerial vehicle (UAV), unmanned aerial system (UAS), and remotely piloted aircraft system (RPAS), the latter being the now official naming. It should be noted that the wording RPAS includes the notion of a remote pilot. Still, in this chapter, the focus is on **autonomous systems** that do not have a ground pilot, and that consequently need to make their own decisions to achieve the missions that are assigned to them. Why we consider autonomous drones will be explained a bit later in this introduction.

The technological progress additionally makes it possible to embed various sensing, actuating, and communication capabilities on board drones and it is now possible for them to not only interact with their environment (by means of sensors and actuators) but also to interact with each other (and with the ground, surface, or underwater vehicles, ground sensors, and troops). It is thus possible to consider groups of UAVs flying together and **collaborating to achieve a common global mission.**[1] Such groups are referred to as **swarms**. Swarms usually assume close collaboration and close flight – even though close flight still remains a difficult problem – whereas a fleet – terminology that the reader will often find in the literature – supposes looser relationship and a notion of hierarchy between the UAVs. In this chapter, we will use both words depending on the kind of configuration we are referring to. Additionally, we are envisioning large groups of drones and since one cannot afford to have as many ground pilots as UAVs (we are talking tens or hundreds of UAVs), the UAVs we consider have to be **autonomous**: once launched, they manage themselves without any ground intervention.

Those who have already been working with or simply using a drone, know that even considered individually, these machines raise technical issues such as trajectory, altitude,

[1] We use the word *collaboration* which supposes close relationship and interaction to achieve a common goal, whereas the word *cooperation* only assumes work sharing [266].

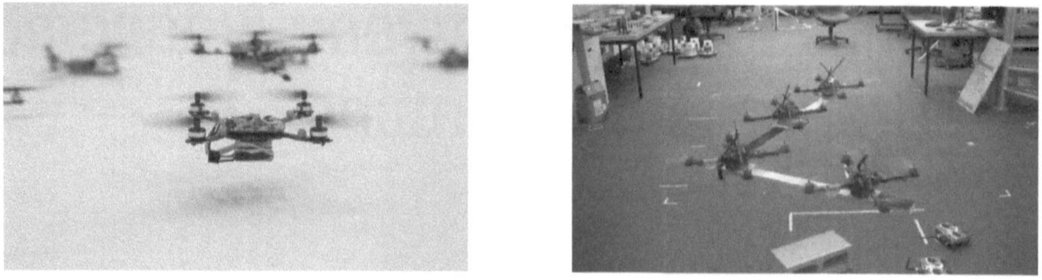

(a) Multiple quadrotors (credits: Scott Spitzer, University of Pennsylvania)

(b) Drones joining forces (credits: GRASP Lab, University of Pennsylvania)

**Figure 8.1** The power of swarms comes from collaborating and joining forces

and attitude control, endurance management, etc. Indeed, if flying UAVs in groups or swarms raises many new opportunities, it also raises a significant number of additional, complex, and almost unexplored issues that are both practical and fundamental (communication, identity management, resilience, etc.).

These issues must be addressed because there is no doubt that, should they be solved, swarms will develop in the (near) future: a swarm of unmanned vehicles is much more powerful than a single unmanned system (this will be detailed below). It can undoubtedly help in many so-called Dull, Dirty, Dangerous [137] (DDD) domains, e.g., rescue, disaster relief, and military operations, where a single unmanned system cannot support all the facets of the mission. Swarms have been envisioned for quite a few years, but the technological progress now makes them a real possibility to consider and not a future dream. The Office of Naval Research (ONR) issued a press release [320] about the Low-Cost UAV Swarming Technology (LOCUST) program announcing "demonstrations of swarming unmanned aerial vehicles." Also stated in this release is the fact that "The recent demonstrations are an important step on the way to the 2016 ship-based demonstration of 30 rapidly launched autonomous, swarming UAVs," declaration attributed to ONR program manager Lee Mastroianni.

This chapter on *Collaboration Between Autonomous Drones and Swarming* is based on our expertise and on the experience we collected from the projects we have conducted. The goal here is not to be exhaustive, but to put the stress on some of the major issues that should be solved if swarms are to be used in the field for an increasing number of missions. We first explain why swarming can significantly increase the possibilities of a mission. We then focus on a number of issues and research directions that need to be explored further. Among these, the most significant topics that we cover in this chapter are: localization, proximity detection, and positioning; man swarm interaction; degraded mode of operation; safety and legal issues; security.

## 8.2 Why Use Swarms of Unmanned Aerial Systems?

Swarms offer many advantages that are supported by close collaboration between the UAVs. This of course raises difficult to solve issues, as we will see later in this chapter, but for now we focus on the advantages. We give the main ones, those that are widely acknowledged in the literature and that emerge from the use cases that are being developed.

### 8.2.1 Continuous Flight/Mission

Continuous mission or flight is probably one of the most important features that is supported by swarms. The goal is to run a mission over the clock, or at least to be able to carry out a mission till its end without being interrupted by refueling needs or by some unexpected event that would for instance cause one (several) UAV(s) to land, crash, or to cancel its (their) mission.

There are many proposals whose goal is to offer continuous (or almost continuous) flight to a single drone. One of the domains that are explored is that of solar recharging. For instance, the AtlantikSolar 2 (AS-2) drone [43] developed at ETH Zurich's Autonomous Systems Lab flew for 81 and a half hours in July 2015. A bit earlier, in December 2014, the Orion system [324] of Aurora Flight Sciences flew for 80 hours, 2 minutes, and 52 seconds.

There are other approaches to supplying additional fuel/battery capacity during the flight. For instance, Boeing recently got a patent entitled *Autonomous Aircraft with disconnectable tether* [96] (see Figure 8.2) for a tethered refueling system that would theoretically make it possible for a drone to fly forever. Still, it should be noted that the mission would be limited in terms of the area that could be accessed, because of the need for the UAV to get back to the place where the tether and the UAV can couple.

Proposals have also been developed for many years around power beaming. In 2002, researchers from NASA's Marshall Space Flight Center, Huntsville, AL, and Dryden

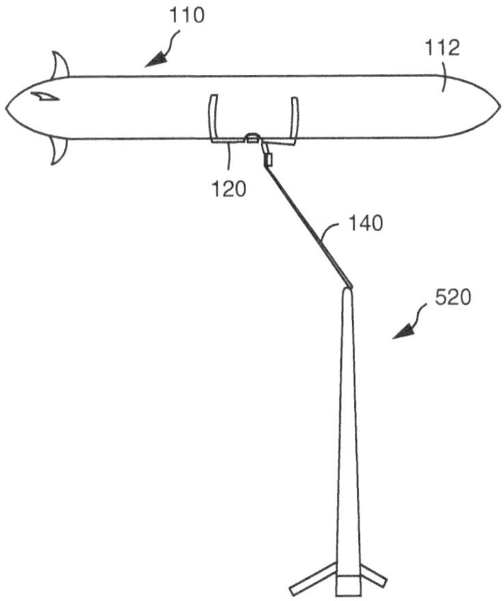


**Figure 8.2** Drone tethered to a charging tower (source: US patent US 9,045,218 B2 - June 2, 2015 [96])

Flight Research Center, Edwards, CA, and the University of Alabama in Huntsville have demonstrated a small-scale aircraft powered by a ground-based infrared laser [342].

Drone balloons are also a potential solution. They are for instance the key component of the Google Loon project [282], the goal of which is to supply Internet all over the planet. The interest of balloons is that they are lightweight and can additionally be equipped with solar panels to help increase their endurance. Funnily enough, this would be going back to the roots, that is to the unmanned balloons launched by Austria in 1849 over Venice [186].

Still, it seems that the studied approaches would not be appropriate for military or disaster relief situations for instance. The solar-based systems must be optimized in terms of weight to give significant results, and the payload they can carry is thus very limited. The other proposals described above require some form of additional infrastructure/support to be deployed or at least require some planning, which is almost impossible in many real life scenarios.

Furthermore, the above approaches only address the refueling issue, which is far from being the only reason why continuous flight might be interrupted. In a military scenario for instance, the mission can be discontinued because the UAV has been destroyed, or because it has to get back to a specific location to unload the data it has collected, etc.

Therefore, we believe that the best (if not the only) solution is to have a swarm, possibly with spare drones, the elements of which can decide on/apply a retasking strategy designed so that the mission is never discontinued.

### 8.2.2 Increased Mission Flexibility

Having a swarm makes it possible to adapt to different kinds of missions, both in terms of area covering capacity and in terms of the measurements that can be performed, and thus to better meet the expectations of situation management.

Assume a surveillance scenario at sea, the goal being to rescue people in case they cannot get back to the shore. A swarm can be used to conduct a Search And Rescue [462] (SAR) mission, and in case a distressed swimmer is detected, one of the UAVs of the swarm can get closer to the person in danger to achieve some specific operation. In case the mission requires communication between the UAVs and the ground (e.g., to provide video feedback), some UAVs of the swarm that have communication capacities can be used to set up a communication backbone.

Furthermore, the heterogeneity of swarms should not be considered only in terms of the sensors that can be embedded but also in terms of the unmanned systems themselves. Of course, drones differ in terms of endurance, size, etc. but also in terms of characteristics that can impact the deep nature of the missions they can achieve. Some of these features are unique and these drones should thus be used sparingly, only when these features are required. Consider the ATEX[2] drone of Xamen Technologies [454]

[2] The word ATEX refers to the Directive 2014/34/EU of the European Parliament and of the Council of 26 February 2014 on the harmonization of the laws of the Member States relating to equipment and protective systems intended for use in potentially explosive atmospheres [144].

**Figure 8.3** The Xamen ATEX drone in operation close to the Maersk Edgar tanker (credits: Xamen Technologies)

(Figure 8.3) with its unique capacity to fly in areas with an explosive atmosphere (note that this is the first and most likely still unique UAV certified for this capacity). It should be used sparingly and such a UAV thus only makes perfect sense when included in an accompanying swarm. Its participation in a mission in close collaboration with other UAVs clearly increases the flexibility of the operations that can be carried out in terms of the areas that can be addressed.

Additionally, even though it is out of the scope of this book, swarms can support different theaters of operation: ground, air, surface, underwater, etc. For instance, an autonomous swarm performing a mission that consists in detecting and destroying mines can adapt itself to deal with ground, surface, or underwater mines. This clearly shows the complementarity of the components of a swarm, which leads to flexibility in terms of the missions that can be addressed.

### 8.2.3 Increased Capabilities

By using a swarm, it is possible to join forces, to support and complement different capacities.

First, in terms of **brute force**. By using several UAVs that collaborate with each other, it is possible to combine forces, as was done for instance at General Robotics, Automation, Sensing & Perception (GRASP) [189] (Figure 8.1).

Second, in terms of **sensing capabilities**. There are many scenarios where embedding different capacities in different UAVs is much better than embedding all of them in the same system. Assume for instance the surveillance of a city to detect emanations of

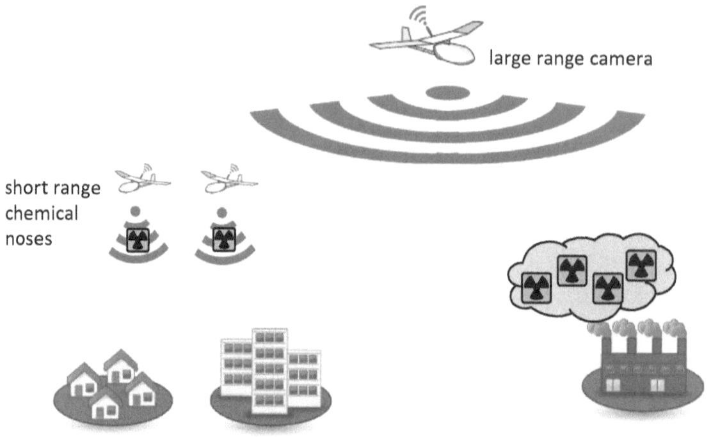


**Figure 8.4** Increasing sensing capacities by using a swarm

toxic gas (Figure 8.4). To detect toxic gas it is required to embed a chemical nose in a UAV. The problem here is that the sensing capacity of this kind of system is limited in terms of covering area. Therefore, having a single UAV covering a large area (a whole city for instance) is not a proper solution because it may take too much time for the UAV to fly over the whole area compared to the speed at which the emanation of toxic gas must be dealt with. One possible way to cope with this issue is to embed a camera in another UAV that would fly at a higher altitude (Figure 8.4). This latter UAV will achieve detection (of candidate threats) based on video capture. Of course, given the fact that this detection is only based on a video capture and not on a chemical nose, some (perhaps many) false positives will be detected. These potential threats will then be transmitted to the low level UAV(s) that embeds the chemical nose(s) so that it (they) can go and confirm or deny the risk.

### 8.2.4 Additional Features

There are a number of additional reasons (often considered as less central) why using swarms of UAVs makes sense. Among these are:

- Security: having information distributed over a number of collaborating systems makes it more difficult for an enemy to identify a critical point by analyzing, for instance, the communication traffic.
- Safety: by using UAVs working together there are many cases where smaller UAVs could be used. Thus, should one of the UAVs fall down, it would not cause such serious injuries and damage as a big drone would.
- Resilience: distributing and possibly duplicating the information over a number of systems helps make the whole system resilient. In case one of the UAVs crashes or lands (for instance for refueling), the others can take over its workload so that the mission can continue. This is closely related to the notion of a degraded mode

of operation that we consider as one of the key features of swarms and that will be detailed later in this chapter (see Section 8.3.3).

- Inter-authorities cooperation/collaboration: the capability to cooperate/collaborate between a number of systems makes it possible to consider cooperation between UAVs owned by different authorities. For instance, armies of different countries can share information and collaborate to achieve some mission of common interest. Still, this is a difficult feature to achieve because most of the time the level of collaboration (both in terms of mission and in terms of information to be shared) varies over time. The cooperation system must thus be dynamic.
- Load sharing: UAVs can embed different kinds of sensors that can collect a large amount of information. Filtering and processing such information can generate an important workload. The possibility to filter and to synthesize the information that makes sense (for situation management for instance) in a distributed manner over a large number of systems can significantly reduce the individual load and the communication needs. This is an important aspect because we are talking about systems that have limited resources.

### 8.2.5 Summary

UAVs are a solution to Dull, Dirty, and Dangerous (D3) situations and contexts. It is possible to send a sensing capacity to some given area where it is considered not acceptable to send a human.

Swarms of UAVs add a new dimension. They support continuous mission, reconfiguration, and much more adaptation capacity to the environment and to the requirements of the mission. Still they raise a world of new, difficult issues in addition to those that need to be considered when dealing with a single UAV.

## 8.3 Major Issues and Research Directions

If one single unmanned system raises issues in itself, it is of course much more complicated when a collaborative swarm is considered. Swarms are powerful because they provide a number of additional features (compared to one single UAV), as described in the previous section. These features are sustained by high-level mechanisms such as flight formation, re-tasking, dynamic flight plan management, etc., that are helpful to deal with complex missions. These high-level features rely on the collaboration that takes place within the swarm and that requires a number of mechanisms that we study in what follows. These are: localization, proximity detection, and positioning; man swarm interaction; degraded mode of operation; safety and legal issues; security.

### 8.3.1 Localization, Proximity Detection, and Positioning

**Localization** is an important feature which is necessary to ensure collaborative missions. One of the advantages of swarms is their reconfiguration and re-tasking capacity, which enables them to adapt to both external events and new missions or to updates of the current mission. In some circumstances, this requires that each UAV has both a precise localization system (as in single UAV flights) and the capacity to precisely locate its neighbors in the swarm. Localization in a single UAV is usually achieved by means of a GNSS [213] (GPS, Galileo, etc.) and this can of course also be used for swarms when considered sufficient.

But swarms furthermore raise the issue of **proximity detection**, which is mandatory to ensure a number of features, including compact flight. Many sensing technologies exist [415] and many have been and are still being experimented with to cope with this difficult issue. It should be noted that this is not only a question of physical detection: the collected information must furthermore be processed as quickly as possible so that the appropriate measures can be taken. Additionally, this requires collaboration between the members of the swarm, the moves of one of the UAVs possibly having consequences over the system as a whole. For instance, in the example illustrated in Figure 8.5, repositioning of UAV number 1 so that it avoids some obstacle impacts the positioning strategy of several of the other drones, at least those that are its closest neighbors (but the relocation information is propagated throughout the whole swarm).

In the scenario shown in Figure 8.6, four UAVs have formed a circle and maintain regular spacing between each other. They have to enforce this configuration to achieve their assigned mission. If drone 1 accelerates for some reason (configuration 1), it impacts drone 2 that must go away and drone 4 must come closer (configuration 2). This in turn impacts drone 3 (configuration 3) and so on. The system considered as a whole will be oscillating till it reaches a balance; note that the whole repositioning process should be dealt with in a totally distributed manner (this is a *sine qua non* condition to ensure resilience – see Section 8.3.3).

Even though the problem is not exactly the same, we can most likely take advantage of all the work carried out in the domain of localization for wireless sensor networks [93, 199].

It should also be noted that the **integration of UAVs in the airspace system** (i.e., with commercial flights) is a major regulation and research concern. It is of course closely

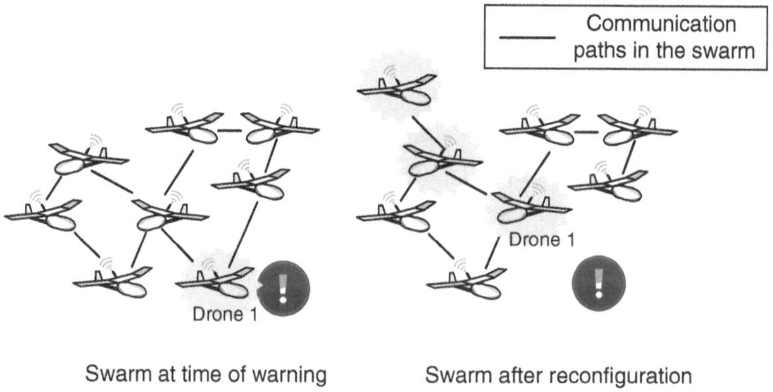


**Figure 8.5** Global impact of repositioning a single UAV in a swarm

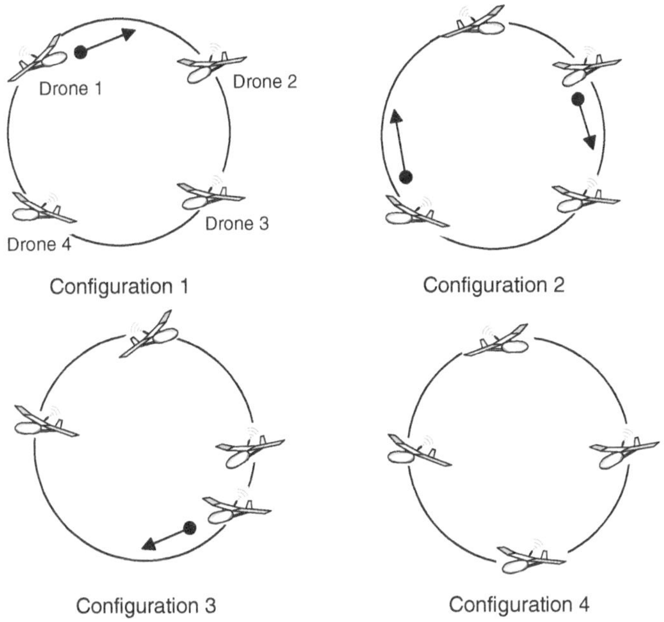


**Figure 8.6** Oscillation and distributed repositioning within a swarm

related to localization and proximity detection. The problem is known to be difficult even when a single UAV or a number of unrelated UAVs are concerned, but it definitely needs to be addressed especially with the *Next Generation Air Transportation Systems* [316] (NextGen) in mind. NASA is currently developing a cloud-based *Unmanned Aircraft System Traffic Management* [315] (UTM) system and the question of autonomy (in terms of decisions, not in terms of endurance), which is one of the keys to the complexity of the problem, is currently addressed by many projects among which is the *safe autonomous systems operations* [314] (SASO) project (also at NASA).

The question of swarms, even though still in its infancy, has been identified quite early. In a document entitled *Integration of Unmanned Aircraft Systems into the National Airspace System – Concept of Operations* [166], which is dated back to 2012, the Federal Aviation Administration (FAA) addresses the question of swarms crossing the National Airspace System (NAS). Among the requirements for UAS integration in the NAS, it is said that "Each UAS has a flight crew appropriate to fulfill the operators' responsibilities, and includes a PIC.[3] Each PIC controls only one UA.[4]" And the reader is referred to a footnote that says: *"This restriction does not preclude the possibility of a formation of UA (with multiple pilots) or a "swarm" (one pilot controlling a group of UA) from transiting the NAS to or from a restricted airspace, provided the formation or swarm is operating under a COA.[5] This constraint addresses generally only those UAS operations that will be integrated into the NAS."*

[3] Pilot In Command
[4] Unmanned Aircraft
[5] Certificate of Waiver or Authorization

This is a first and important step, but integrating autonomous swarms into the regular traffic will require more time, more experimentation, and definition of additional rules that take their specific characteristics into account.

### 8.3.2 Man Swarm Interaction

The interaction between man and swarm has several purposes, among which are safety, mission management, and situation management. It is acknowledged that the interaction between a Ground Control Station (GCS) and one single UAV is not easy to achieve but the problem is of course much more difficult when a swarm is involved, especially when it is a collaborative swarm, where the whole system is impacted by local changes and by decisions taken at the level of each individual drone.

**Health-management** [428] is not a real issue, because simple solutions exist. In the CARUS project [89] that consisted in flying an autonomous swarm of UAVs, one of the partners (the Fly-n-Sense company [171], now part of deltadrone [135]) developed a ground control station that could manage five UAVs. It consisted of an extension of their classical ground control station with additional features to give simple orders to all the members of the swarm at the same time.

But health management is relatively simple compared to **mission management**, which requires **high-level interactions**. A number of questions arise: How to manipulate the swarm as a whole and still control each UAV individually in case this is needed for instance for safety reasons? What are the appropriate devices/Man-Machine Interface (MMI) components to do so? What level of responsiveness do we need and what level are we able to achieve? What about synchronization issues, i.e., when an order is given to a group of UAVs and should be applied to all of them at the same time, or when several pieces of information are collected from different UAVs and need to be presented to the ground control?

Many research projects are addressing these issues and, most of the time, multi-modal approaches are proposed (see for instance the experiments carried out by the Army Research Laboratories [197] (ARL)). One of the questions is to offer swarm manipulation as a whole, still providing individual UAV control. This is for instance investigated in the framework of the DAISIE [123] project (Design of innovative applications for swarms of autonomous vehicles), the goal of which is to study the intuitive control of swarms (see Figure 8.7).

**Regulation** is also an issue because the law in most countries requires one remote pilot per UAV – how to have 100 pilots for 100 UAVs? In addition to the number of ground pilots, there is the issue of coordinating these pilots. For instance, if one pilot has to land his/her UAV, how to avoid crashes with the other UAVs?

Managing a swarm of UAVs thus remains a challenge and both fundamental research and experimentation should continue. It is important not only for mission management: we believe that by working on swarm control, there is a chance to impact the regulation and decrease the constraints imposed by the regulation authorities such as having one pilot per UAV, which is definitely not feasible in case large swarms are to be considered.

**Figure 8.7** Swarm interaction in the SUSIE/DAISIE project (credits: Deev Interaction & Télécom Bretagne, DGA/MRIS, REI, 2009.34.0003)

### 8.3.3 Degraded Mode of Operation

Even though not studied very often, support of the degraded mode of operation is key to the development of swarms. When considering hundreds of UAVs working together, there is almost no chance that all of them will remain available till the end of the mission to which they have been assigned.[6] There is also no doubt that the communications will be jammed at some point with possible disconnections and re-connections at any time. The question is thus the following: **can the success of the mission be guaranteed** in spite of these inevitable degradations?

Indeed, to meet this goal, the loss of a UAV or of a communication link must not be considered an error but rather be part of the nominal mode of operation of the swarm. Of course, under such conditions and constraints, the objectives of the mission and the expected quality of the result have to be revised downwards – and one should expect more resources to be consumed. One way to proceed is to rely on as little information as possible when designing a mission so as to be able to operate under any kind of restriction. Indeed, the less we assume we have at our disposal the better. The weakest hypothesis we can work with is that a UAV can only broadcast a message but it is not known if the message arrives at some target destination or not. Of course, this very constraining assumption has to be adapted depending on (the environment of) the mission that is considered because there is no reason to impose useless constraints. For instance, there are many real-world scenarios where communication is possible with the ground, or two-way communication is available, or communication is reliable, etc. Here is a theoretical scenario where we make some hypothesis on the number of UAVs that would get lost in a given interval of time. Assume a number of UAVs with short-range communication that have to fly in a so-called *follow the leader* configuration. If we can set up the communication pattern described in Figure 8.8, i.e. two neighbors at one hop and two neighbors at two hops for each UAV except at the extremities of the chain, and if we assume that only one UAV can leave the swarm at a time and that in this case we have enough time to rebuild the missing links (both one hop and two hops) before another loss arises, then it is possible to guarantee that the swarm will remain connected

[6] Symmetrically, new UAVs can attempt to join the swarm, which raises other issues such as insecure boundaries – see Section 8.3.5.

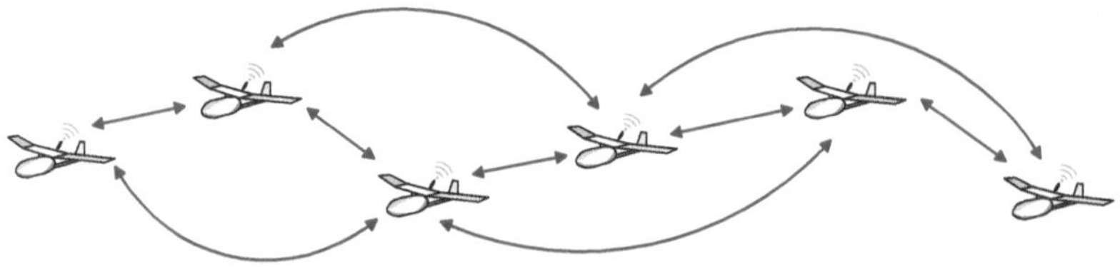

**Figure 8.8** A configuration resilient to the loss of one UAV at a time

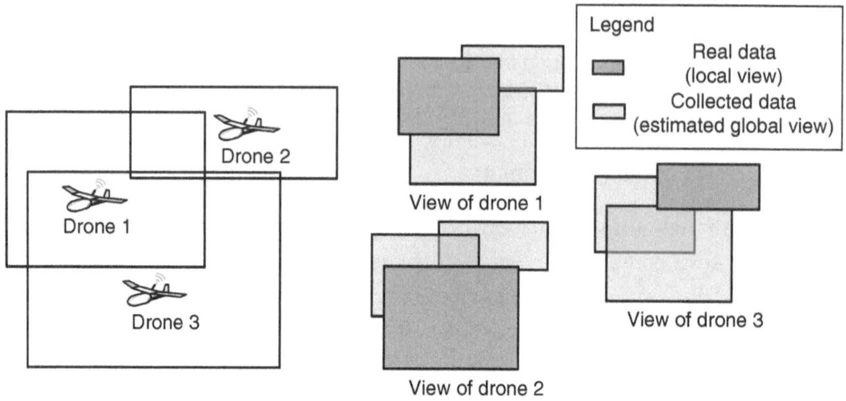


**Figure 8.9** Construction of a global view in a swarm

during the whole mission, thus being resilient to the loss of UAVs (under the conditions defined above).

Here is another example of the kind of concession one may have to make. The scenario that follows has been developed in the CARUS [89] project and we did fly the mission on a military camp. The goal was to survey a number of points of interest (POIs) using a swarm and to support a degraded mode of operation (loss of UAVs and loss of communication). We managed to guarantee that each POI to be surveilled was surveilled at regular intervals of time. Nevertheless, to achieve this goal (in spite of losses of UAVs and communication) we had to accept that from time to time several UAVs did survey the same POI, which is useless but makes it possible to guarantee the success of the mission.

In most cases, the approach that is adopted is as follows. Each UAV builds its own estimated global view of the whole situation (Figure 8.9). To do so, each UAV regularly broadcasts its estimated global view. When receiving an estimated global view coming from another UAV, it combines it with its own estimated global view and this becomes the view that it will now broadcast.

It should also be noted (this will be discussed later in Section 8.3.5) that supporting resilience also has **consequences in terms of security**. Since you can lose any UAV and/or communication link at any time, you cannot assume that there is a node that can be used to share information or to serve as a reference, for instance as a Certification Authority [16]. Thus, you also have to lower what you expect from your mission, but this time in terms of security.

### 8.3.4 Safety and Legal Issues

We encountered swarm specific safety and legal issues while conducting the CARUS project that has been introduced earlier in this chapter. We started the study as a theoretical problem, the goal of which was to set up autonomous swarming algorithms, but once we decided to fly a prototype we faced a number of real-world problems. This section is based on the data that we collected [89].

**Legal constraints** (we use the word constraints, but we acknowledge that these are for the benefit of everyone) are a very important question when considering swarms. There is a lot of work going on at the level of the International Civil Aviation Organization [218] (ICAO) and in most countries around the world, e.g., by the Federal Aviation Administration [159] (FAA) in the United States and by the *Direction Générale de l'Aviation Civile* (DGAC) [140] in France. But they have a strong constraint which is that the candidate new rules must take into account what already exists in terms of regulation for planes, because the slightest change in the current rules would require a round of agreements and signatures of all the involved countries. This also makes adding new rules a delicate operation, even though country-specific rules can be defined thus avoiding a global agreement.

We believe that by doing research on the verification of missions and on the associated formal models, we will be able to convince the regulation authorities that they can lower the constraints that are currently required for allowing UAVs to fly. A promising approach consists in considering a swarm as a single UAV. This could be a way to comply with the regulations and still be able to fly swarms.

When looking at **safety**, legal issues are definitely clear and if the problem is difficult when considering a single UAV – there is still a lot of work to do on geofencing for instance – it is much more difficult when dealing with a swarm. We began working on a failure tree within the context of one of our projects (CARUS, see Section 8.3.3), but it quickly appeared that as soon as several drones were involved the size of the failure tree did explode. The reason was not only the number of UAVs involved, but also the relationships that have to be considered between them. We thus postponed this study, but it is clear that it remains a fundamental issue. In this precise case, we did set up specific flight areas so as to avoid any crash between the UAVs. The system works as follows. The UAVs circulate at different levels and an additional level is used for the interventions over the POIs as shown Figure 8.10. We also defined intervention circles and elevator circles around the POIs (Figure 8.11).

Each UAV has its own elevator on the elevator circle (•, •, •, , •) that it can use to leave its working level to reach the intervention level. By using this elevator, there is no risk that a drone will crash into another drone during this process. Once it has reached the intervention level, it moves to its own intervention point on the intervention circle to deal with the incident detected on the POI.

It is clear that **experimentation must continue** and it is to be noted that there are some domains that are adapted to experimentation because safety is not a fundamental issue. In agriculture for instance, UAVs will fly over the fields (whatever they are) and the risk is limited (or can be limited) that human damages might be caused in case

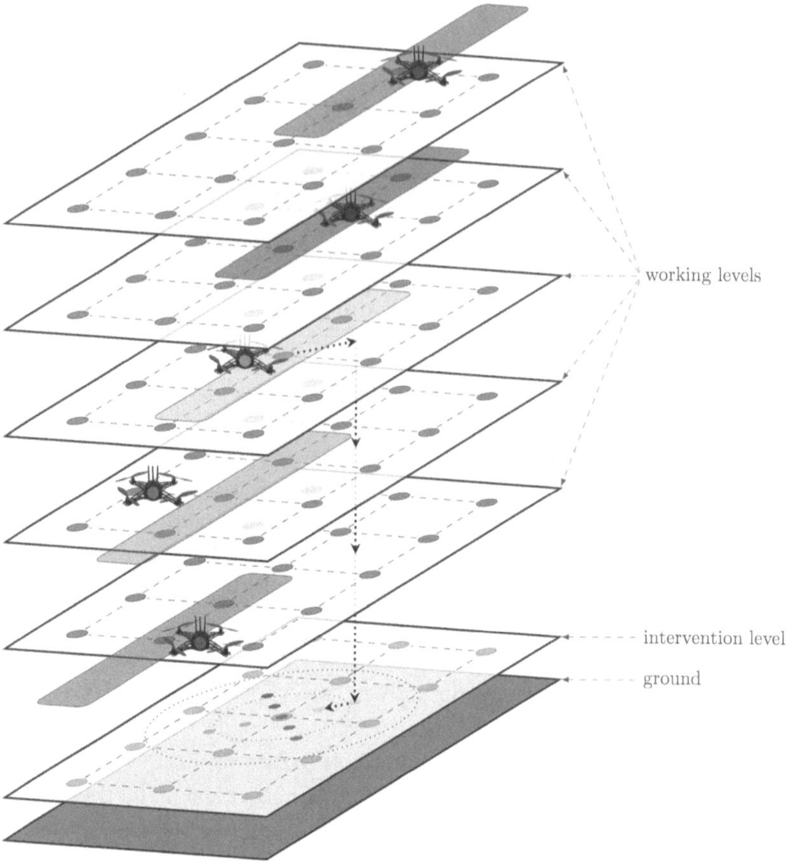


Figure 8.10 CARUS 3D swarm configuration (credits: Rémi Laplace [270])

of system failure. It is the same if we consider the surveillance of shores for potential pollution. These domains can then be used as experimentation fields while waiting for proper systems and procedures to be available.

### 8.3.5 Security

Security is also a major concern. Even though it has been dealt with in another chapter of this book, it is important to address here some specific aspects that are closely related to the collaborative model of swarms and that impact the algorithms underlying the missions. In addition to the usual ciphering and authentication concerns (inside the swarm and for swarm-GCS communication), we are facing here two main additional issues.

A swarm has, by nature, a dynamic structure and this raises the issue of **insecure boundaries**. A swarm is an evolving system: UAVs can join and leave at any time, either for planned reasons or because of the occurrence of an unexpected event. For instance, a UAV can leave to refuel or refill its battery or to bring data back to a collection point.

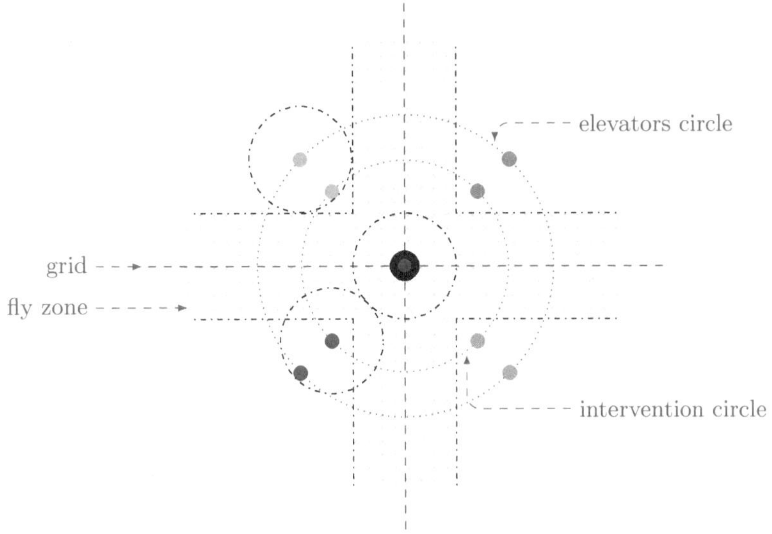


Figure 8.11 CARUS circles and elevators (credits: Rémi Laplace [270])

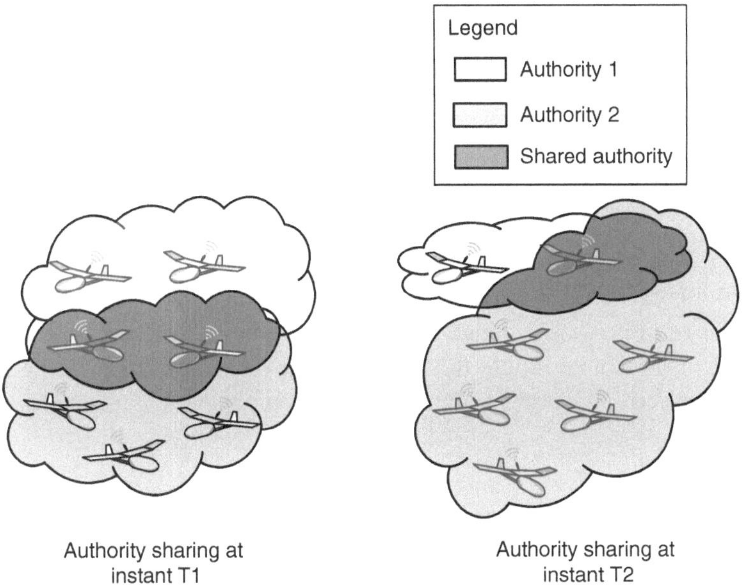


Figure 8.12 Evolving collaboration and authority sharing within a coalition

A UAV can join the swarm in place of one that is leaving or to bring in some additional capacity to the overall system. There thus should be a mechanism to prevent the intrusion of a malicious UAV.

Unfortunately, because swarms should support resilience and because they are dynamic, there is **no stable UAV** available in a swarm. Therefore, unless the network that is set up between the UAVs themselves or between the UAVs and the ground is considered stable and secure – because of some pre-planning before the start of the mission – it is not possible to rely on the existence of a hypothetical node that could be used as a Certification Authority [16] to deal with authentication, certificates, and keys. The security system must thus be implemented in a distributed manner so as to support resilience. There are some cases where identifiers/keys can be distributed on the ground while the mission is being prepared, but this is not always the case, for instance when authority sharing (which is by nature dynamic, see below) is to be supported. Even though we have not yet studied it in depth, we believe that a certificateless-based approach [25] could be a candidate to deal with this issue.

Additionally, one should take care **not to overload any specific UAV** that could thus be detected by the adversary (by some side channel attack) and become the target of attacks; this would make the system fragile. This is one additional reason for distributing the processing load over the whole swarm.

It should also be noted, but we will not detail it here, that there can be issues related to **authority sharing**. It is often the case that swarms that belong to different authorities (different NATO[7] countries plus a local authority for instance) need to collaborate, to share information. This thus requires an authentication/security system that furthermore needs to evolve over time. For instance in a military coalition, the collaboration between the different authorities (and then between their respective swarms) will most likely evolve over geographic areas and over time (see Figure 8.12). The security system should support these features.

## 8.4 Conclusion

Given the services that they can support and offer, for both civilian and military applications, collaborative swarms clearly have the potential to become key players in the domains of surveillance, rescue, construction in hard to reach areas, and also in many other domains that are still to be discovered. But the domain is still quite new (at least not mature yet) and swarms raise many unsolved problems. In this chapter we presented

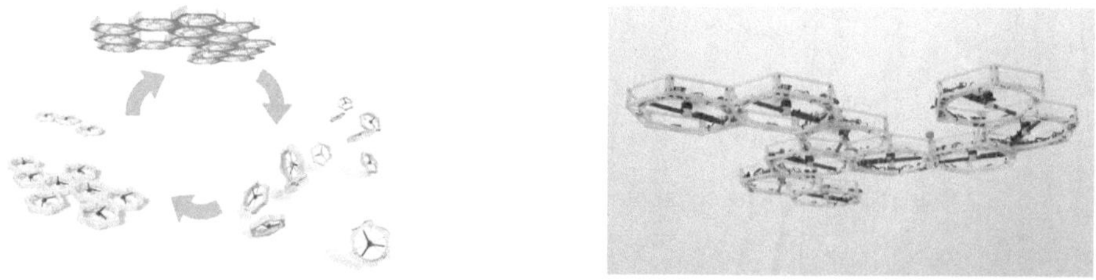

(a) Life cycle of the Distributed Flight Array (credits: Carolina Flores, ETH Zurich)

(b) Distributed Flight Array on flight (credits: Raymond Oung, ETH Zurich)

**Figure 8.13** The distributed flight array developed at ETH Zurich

[7] North Atlantic Treaty Organization

the main issues that we believe must be further investigated to make this potential a reality and to make collaborative swarms part of our future.

To conclude this chapter, I would like to leave the reader with a research direction that I definitely believe promising: combining the advantages of a single UAV (simplicity, ability to fit in the current regulation, etc.) with the advantages of a swarm. The practical idea is to have a number of small UAVs that could combine in the air to make a bigger UAV that would then be seen as a single system. By doing so, the swarm could from time to time be considered as a real collaborative swarm, or as a set of autonomous UAVs or as a single unmanned system. An instantiation of this concept is for instance the Distributed Flight Array developed at ETH Zurich (see Figure 8.13).

# 9 Real-World Applications

Leanne Hanson and Kamesh Namuduri

## 9.1 Introduction

Innovations in UAV research and technology opened up a number of new opportunities that were not imagined before. From pizza delivery to providing broadband services to rural and disadvantaged areas, UAVs are being envisioned for numerous real-world applications. Realization of such applications requires not only technological advances but also regulations, policies, and best practices for safety, privacy, and security reasons. This chapter reviews two of the many applications of UAVs and UAV networks that researchers are currently pursuing: (1) wildlife detection and (2) emergency communications. These examples showcase the unique value and innovation that UAVs can bring to real-world applications.

## 9.2 Wildlife Detection

Wildlife population counts are traditionally conducted either from ground or from fixed-wing aircraft observations; however, both of these methods have limitations (accuracy, cost, safety, and timeliness). Small unmanned aircraft systems (sUAS), and their cameras and sensors, have been tested for their suitability to provide accurate population estimates of Sandhill cranes (*Grus canadensis*) and detect Greater sage-grouse (*Centrocercus urophasianus*) in Colorado, USA. The first Federal Aviation Administration (FAA)-approved sUAS flights in the United States national airspace for the US Department of the Interior (DOI) were conducted in March 2011 by scientists with the United States Geological Survey (USGS) using the Raven RQ-11A sUAS at Monte Vista National Wildlife Refuge. These flights determined the suitability of the sUAS to enumerate sandhill cranes, by comparing ground counts to number of cranes estimated from sUAS imagery. In 2012, the first FAA-approved sUAS night flights were conducted at Monte Vista, to obtain a population estimate of the number of roosting cranes at five roosts on the refuge. In April 2013, the sUAS was flown near greater sage-grouse leks (breeding sites) in Middle Park, Colorado, to determine if the system could detect greater sage-grouse and what (if any) reaction the sage-grouse might have to the sUAS. These studies proved that the Raven sUAS provides a non-intrusive, safe, and accurate way to estimate sandhill crane population abundance on roost sites, and detect greater sage-grouse on lek sites. sUAS technology has become more widely available to natural resource managers to track wildlife population numbers, health, and trends and will be relied upon in the future as a critical source of data.

### 9.2.1 Aerial Wildlife Counts

Traditionally, aerial wildlife counts and surveys have been conducted from occupied, fixed-wing aircraft [82] or helicopters and have been applied in research for a wide range of species [472], including sage-grouse [112]. These techniques tend to be expensive and can have safety, wildlife disturbance, and logistical concerns [249] but allow natural resource managers to access areas that are difficult to reach as well as large expanses of habitat that are otherwise inaccessible [343]. As early as 2001, Efroymson and others [473] called for the need "to provide guidance for the assessment of ecological risks from low-altitude aircraft overflights." Although sUAS are being used more readily in many disciplines, their application for wildlife management has only recently been explored and has not been studied in depth [370]. Watts and others [441] identify the use of sUAS to survey large terrestrial animals and marine species and as a potential to decrease disturbance to wildlife populations by conducting sUAS surveys vs. traditional, occupied aircraft overflights [442].

Studies have been completed assessing the impacts on avian species of traditional occupied aerial overflights [134]. Military flight disturbance as a result of sonic booms has been studied [183, 34, 192, 391, 187], but these reports do not explore potential impacts on avian species from overflights with minimal noise. Studies have documented disturbance to raptor species [34, 192, 391], seabirds [69, 203], and waterfowl [113, 440, 187] with reported disturbance effects differing among species and overflight patterns. From prior noise-recording tests conducted by USGS with the Raven sUAS, noise decibels (dB) peaked at 103.1dB during the propeller check, and then decreased to 60–70dB during flights at altitudes between 60ft and 200ft (18m and 61m) above ground level (AGL). These readings, however, did not distinguish wind from aircraft noise.

There have been very few studies completed on the use and effectiveness of sUAS to monitor and survey avian populations. All the studies found in the literature referenced research conducted on water and wading birds, including Canada goose (*Branta canadensis*) and snow goose (*Chen caerulescens*) [86], black-headed gull (*Chroicocephalus ridibundus*) [370], white ibis (*Eudocimus albus*), and other wading birds [249]. There are no published reports or papers reporting the effects of unmanned aircraft surveys on either sandhill cranes or greater sage-grouse. Booth and others [474] used a manned light sport airplane to conduct greater sage-grouse lek counts. Out of 14 approaches, grouse flushed from the lek on 12 approaches when the aircraft was flown within 200 to 300m of the lek. Other grouse lek surveys using manned aerial flights (either airplanes or helicopters) for greater prairie-chicken (*Tympanuchus cupido*) [274, 291] and lesser prairie-chicken (*Tympanuchus pallidicinctus*) [378] were able to successfully identify a lek site but were not always able to achieve an accurate grouse count compared to ground surveys.

### 9.2.2 Raven RQ-11A Small Unmanned Aircraft System

The Raven RQ-11A small Unmanned Aircraft System (Raven sUAS, see Figure 9.1) is a hand-launched (Figure 9.2) reconnaissance and data-gathering tool developed for the US Department of Defense by AeroVironment, Inc. Originally designed to provide aerial observation, day or night, at line-of-site ranges up to 6.2 miles (mi) (10 kilometers [km]), the Raven sUAS has a wingspan of 4.5 feet (1.38 meters) and weighs 4.2 pounds (1.9kg). A 60-minute (min) lithium-ion rechargeable battery powers the system, which

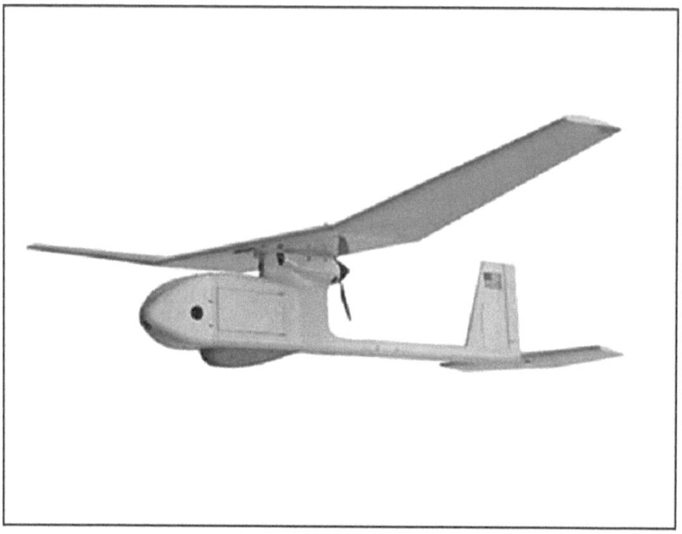

**Figure 9.1** The Raven RQ-11A small Unmanned Aircraft System (Raven sUAS). Image courtesy of AeroVironment, Inc

**Figure 9.2** Hand launch of Raven sUAS by J. Sloan, sUAS pilot/observer, Park County, Colorado, USA. USGS Image

**Figure 9.3** Raven sUAS ground control station, showing (l to r) L. Hanson, sUAS pilot/mission operator, M. McGuire, project partner, M. Cowardin, project partner, and C. Holmquist-Johnson, sUAS pilot, Park County, Colorado, USA. USGS Image

also transmits live videos (color or infrared imagery), compass headings, and location information to a ground control station. The Raven sUAS is typically operated by a three-person flight crew consisting of a pilot, a mission operator, and a trained observer (Figure 9.3).

The Raven sUAS video cameras are located in the nosecone and have fixed focal length, aperture, and depth of field settings. During flight, the aircraft altitude is used to change the camera's field of view and obtain sufficient pixel resolution to identify the item of interest (sage-grouse vs. sandhill crane). As designed, the Raven sUAS has the option of using two types of on-board cameras and two camera positions. The camera types are a visible wavelength electro-optical (EO) or a thermal infrared (IR) video camera, and are mounted for either a 45 forward-looking or 45 side-looking position. USGS staff modified a nosecone to allow for a near-vertical (NADIR) view for the Raven sUAS IR camera. The Raven sUAS has two flight-control methods to navigate the plane, either manually or autonomously, and five modes of flight (manual, altitude, home, loiter, and navigating waypoints). In the manual flight-control method, the pilot controls the plane's flight path and altitude, while in the autonomous flight-control method, the plane is flown via pre-established mapped coordinates. Both flight-control methods and all five flight modes were used during data collection flights. The image from the thermal IR camera, can be displayed in white- or black-hot modes as shown in Figure 9.4. This describes how the thermal signature is shown in the image as either white or black. Previous thermal IR flights showed that the white-hot mode provided the best thermal detection capability for wildlife data collection flights.

Approval to fly the Raven sUAS in the US national airspace requires a Certificate of Authorization (COA) from the Federal Aviation Administration (FAA). The FAA COA review can take up to two months and requires approval from (1) the land owner and (or) manager of the areas over which flights were planned, and (2) the USGS, Department of the Interior (DOI), US Army, and the US National Telecommunications and Information Administration (NTIA), which grant approval for the aircraft's radio communication

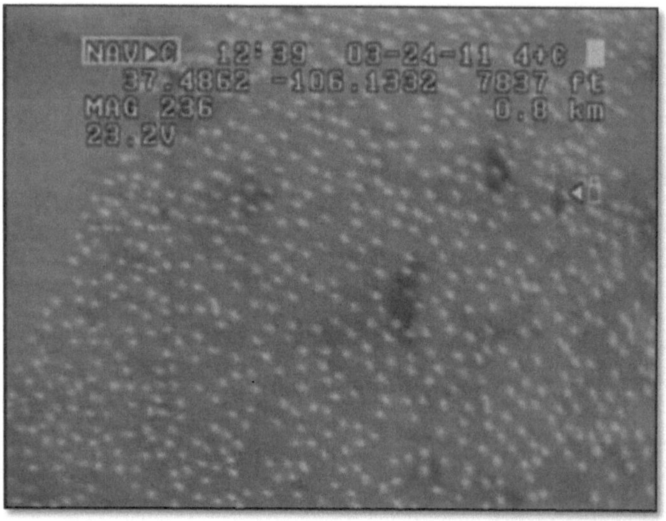


**Figure 9.4** Image from Raven sUAS taken with NADIR-IR video camera at 200 feet above ground level. USGS Image

frequency. This approval process takes a minimum of 45 days and is required because the Raven sUAS is a decommissioned military aircraft. Additionally, the FAA COA proposal package must contain information about the project, proposed flight operations, class of airspace, specific airframe description and capabilities, aircraft airworthiness certification, emergency procedures, flight crew qualifications, and any special-use permits relating to the project. Within the USGS, the COA package is reviewed by the USGS National UAS Project Office, the Bureau Aviation Manager, and the DOI Office of Aviation Services before the package is submitted to the FAA. The FAA-approved COA contains operating provisions specific to the flights, flight crew, safety, local and regional air traffic control centers notifications, notice to airmen (NOTAM), and flight reporting requirements.

### 9.2.3 Using the Raven RQ-11A sUAS to Estimate the Abundance of Sandhill Cranes (*Grus canadensis*) at Monte Vista National Wildlife Refuge, Colorado, USA

The US Fish and Wildlife Service's Migratory Bird Program Central Flyway Manager, located in Lakewood, Colorado, was interested in testing the Raven sUAS for its suitability to count sandhill cranes during their migratory stop-over at the Monte Vista National Wildlife Refuge near Monte Vista, Colorado. The first FAA approved sUAS flights in the US national airspace for the DOI were conducted between March 21 and 24, 2011 to test the Raven sUAS's capability to count sandhill cranes. Flights were conducted over feeding, loafing, and roosting sandhill cranes during daylight hours. These flights were focused over a small group of cranes on the refuge in order to monitor their reaction and minimize potential negative impacts from the Raven sUAS overflights. During these flights, sandhill cranes were not disturbed while roosting; they were slightly disturbed while loafing, and some birds flushed while feeding. Cranes were disturbed while loafing when the Raven sUAS was at or below 200 feet above ground level (AGL). Specifically, behavior included: becoming alert to overhead noise and movement, running away from the approaching Raven sUAS, and taking short flights of 20–30 feet high above before settling back to the ground. Some feeding cranes took flight when the Raven sUAS approached at altitudes between 75 and 400 feet AGL. As a result of the test flights, the best opportunities to obtain a population estimate of the cranes were while the cranes were roosting at night.

To verify the number of cranes on a roost, six wildlife biologists were positioned around a selected roost to count the number of cranes flying in to roost the evening before the Raven sUAS flights. The biologists conducting the crane count were selected by the USFWS Division of Migratory Bird Management and all had previous experience with making ground-based abundance estimates. Biologists were at their location approximately one hour before sunset and counted cranes coming onto the roost site until darkness prevented the continuation of their count. Unless disturbed, cranes have strong roost-site fidelity and do not move between roosts after nightfall. Each biologist reported the total number of cranes entering the roost. In order to verify roost fidelity, the biologists repeated their count of cranes leaving the roost the next morning, beginning before dawn until all cranes had left the roost. The crane tally was not shared with the Raven sUAS crew or data analysts until after an estimate was derived from Raven sUAS imagery. During pre-flight planning Raven sUAS pilots established transect guides using FalconView planning software for East–West flights at altitudes between 200 and 400 feet AGL. These transect guides were used during flights by the Raven sUAS crew to autonomously fly the Raven, while ensuring the imagery collected at each altitude had sufficient overlap to obtain full coverage of the roost (Figure 9.5).

On March 24, 2011, Raven sUAS pilots, observers, and support crew met at the refuge one hour before the start of civil twilight to conduct an operational briefing (including safety, weather, and flight plan review) for data collection flights over the roosting cranes. Raven sUAS pilots, observers, and support crew were positioned and the Raven sUAS was launched at the start of civil twilight (6:34 am). Eight transects were flown at 200 feet AGL within 13 minutes of launching the Raven sUAS and there was no observed disturbance to the cranes. Five additional flights over the roost were conducted for image verification at 300 and 400 feet AGL. The total flight time for all transects was 24 minutes.

Raven sUAS video imagery collected during the March 24th flights at 200 feet AGL was processed by converting the raw image data stream from analog to digital format. Still images (500 by 380 pixels) were collected from the video data and point matching was conducted using Photoshop software. Matched photos were re-sampled to 800 by 600 pixels and contrast-stretched. These photos were stitched together into a mosaic of the roost area. Feature recognition algorithms within feature analyst software were applied to the mosaic to determine the number of cranes on the roost area.

The number of cranes occupying the roost from March 23 to 24 as counted by ground observers was 2,692. The number of cranes on the roost as derived from Raven sUAS imagery was 2,567, a difference of 4.6% (125 birds). These flights demonstrated that the

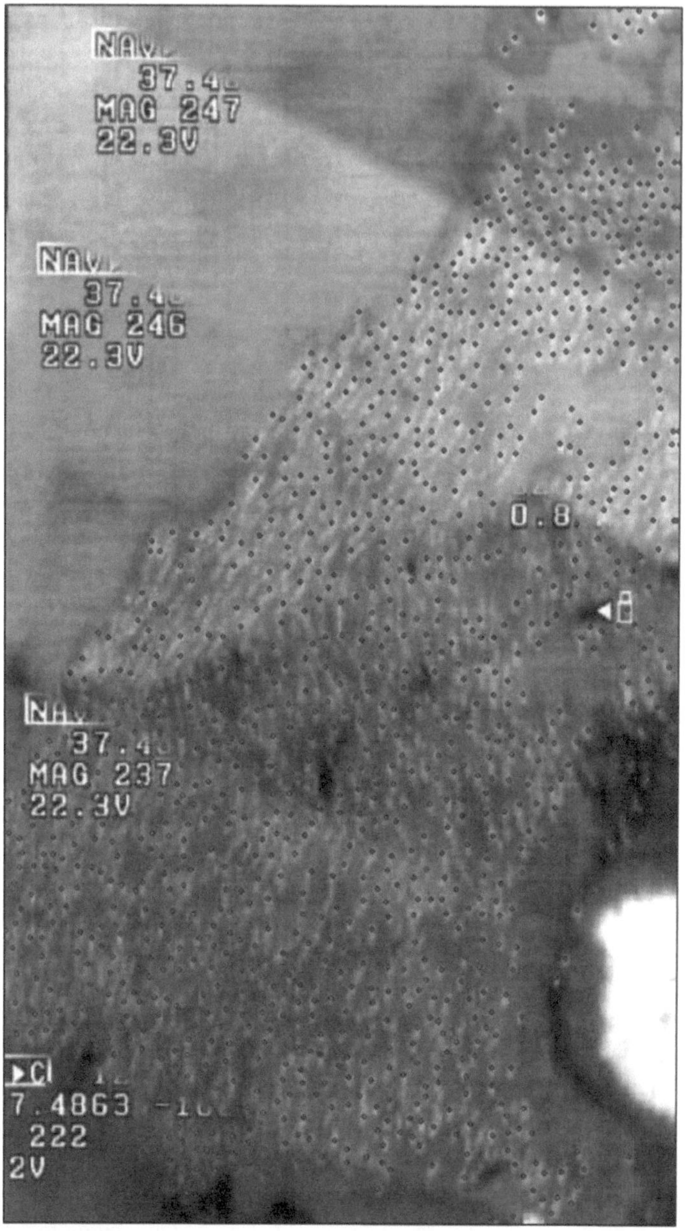


Figure 9.5 Thermal image mosaic of sandhill cranes on roost at Monte Vista National Wildlife Refuge, Monte Vista, Colorado, USA. USGS Image

Raven sUAS can be successfully used to collect abundance estimates of roosting sandhill cranes. To complete a one-night count of sandhill cranes at Monte Vista in 2012, project personnel from USGS and USFWS focused on obtaining approval for night flights from the FAA and expanding the COA boundary to include all roost sites on Monte Vista National Wildlife Refuge.

The first FAA-approved sUAS night flights in the United States national airspace were conducted on March 21, 2012 using the Raven sUAS. The focus of these flights was a complete survey of the cranes occupying five roosts on the refuge. During six hours of flight, 47 transects were flown to capture thermal IR imagery. Flights collected thermal videography, from which still images were extracted, and matched to derive flight line mosaics. The same image processing steps were followed as in 2011, above. The final count of sandhill cranes on five roost sites at Monte Vista National Wildlife Refuge on March 21, 2012 was 14,658. It is important to conduct the population survey over one night, because significant inter-roost movements occur between successive nights, and would result in biased estimates.

### 9.2.4 Evaluation of the Raven sUAS to Detect Greater Sage-Grouse (*Centrocercus urophasianus*) on Leks, Middle Park, Colorado, USA

Staff from the USGS and the Colorado Parks and Wildlife (CPW) Hot Sulphur Springs Office conducted a study testing the Raven sUAS for its suitability to detect greater sage-grouse on breeding sites (leks). During April 2013, the Raven sUAS was flown over two known lek sites to determine the reaction of the grouse to the aircraft and to determine if the technology had potential for future use of locating new leks and obtaining population counts on known, active lek sites.

The USGS and CPW staff monitored greater sage-grouse (grouse) reaction to the Raven sUAS flights and tested the Raven sensor's capability to detect and count grouse from April 11 to 13, 2013. Lek 1 was on private lands and was flown on the mornings of April 11 and April 13, 2013. Lek 2 was on Bureau of Land Management (BLM) lands and was flown only on the morning of April 12, 2013. The official names of the leks are not disclosed to protect the breeding site locations and to respect the privacy of individual landowners that granted access for the Raven sUAS flights. Test flights were conducted over the lek sites beginning at civil twilight (approximately 30 min before sunrise) and continued until the birds dispersed from the lek (approximately 8:00 am). The first flight near Lek 1 was conducted to determine the altitude needed to detect the grouse without disturbing the birds. Lek 2 has had recorded depredation by golden eagles (*Aquila chrysaetos*). Because of this, we hypothesized that the grouse response to the Raven sUAS may differ from birds displaying on Lek 1, so the initial flight near Lek 2 was to again determine the response of the grouse to Raven sUAS overflights.

Flight launch and observation locations were selected to provide an unobstructed line of sight to the lek while being far enough away from the lek not to disturb the birds during the setup and operation of the aircraft. Prior to flights, the USGS Raven sUAS, pilots, observers, and support crew met at the predetermined sUAS operation and launch site one hour before the start of civil twilight to conduct an operational briefing (including safety, weather, and flight plan review) and to prepare for the flights. In addition, trained grouse observers were deployed to the lek, within close viewing distance of the displaying grouse, to record and relay information to the operation center about the response and reaction of the grouse to the Raven sUAS flights and to collect a grouse count. Grouse observers recorded male and female grouse numbers on the lek

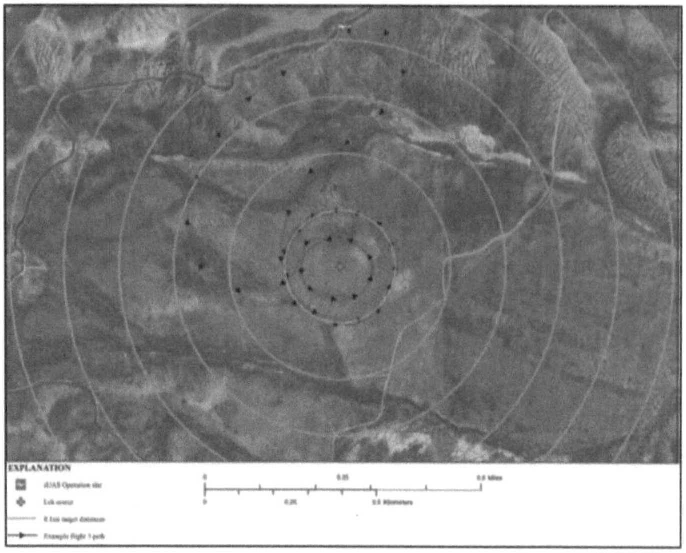

**Figure 9.6** Lek 1, Flight 3: side-looking thermal infrared sensor in a circular flight path using loiter mode. (sUAS, small unmanned aircraft system; mi, miles). USGS Image

and monitored the grouse to record any change in grouse behavior during flights. The grouse observers were in radio communication with the operation center during the Raven sUAS flights. Flights did not begin until the grouse observers had sufficient light to obtain a reliable count of all grouse on the lek. Imagery was collected over the grouse leks at various flight altitudes, using both camera types (EO and IR) and two of the camera positions (side-looking and NADIR). Previous thermal IR flights showed that the white-hot mode provided the best thermal detection capability for wildlife; therefore, thermal imagery was collected using the white-hot mode. A variety of flight paths were followed (see Figure 9.6 for example) at various flight altitudes (340ft [104m] to 100ft [30m]) and distances from the center of the lek (0.4mi [0.64km] to directly above) [202].

Greater sage-grouse response to the Raven sUAS flights was similar at both lek sites on all three mornings. There was no long-term response or reaction to the flights regardless of flight altitude or pattern, and no birds flushed from the leks. The grouse observers noted that, as the sUAS flew over the area, there was a loud buzzing noise, but that it was not obtrusive. Recorded decibel levels during previous Raven sUAS flights ranged between 60 and 70dB, which is equivalent to decibel readings for a normal conversation. Displaying males on the lek, however, reacted differently. Some males showed no reaction to the sUAS flights, while other males momentarily paused or crouched at times when the aircraft flew over or near the leks. If males showed any sign of disturbance by the Raven sUAS, they would only momentarily stop or slow their display, but never retracted their display feathers. Generally, the females also showed little to no response to sUAS flights. Females would crouch at times when the aircraft flew over but never flushed from the lek. Several females continued to forage as the Raven sUAS flew over the lek.

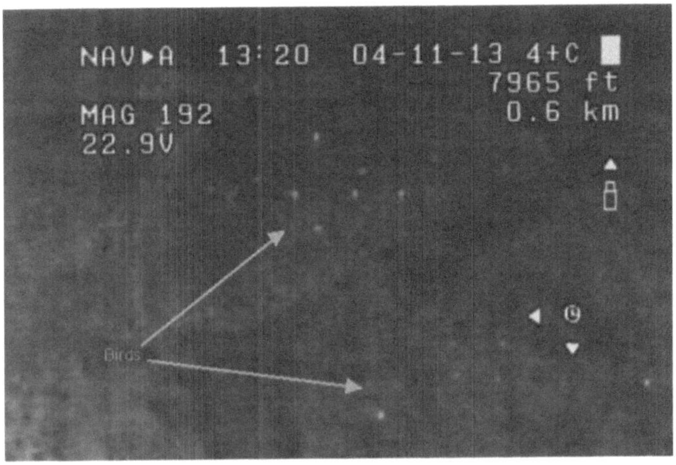


**Figure 9.7** Thermal infrared image of lek showing example of greater sage-grouse heat signature and identification. USGS Image

Video captured during the 4.5hr of data-collection flights was compiled by date, lek, and flight number. The 80GB of data were organized into folders, and each of the 15 flights was edited into flight videos to include only the footage collected over the leks. This included editing out pre-flight testing and travel to and from the launch site and lek. These videos were then viewed by project personnel who made notes regarding the flight time when birds were seen in the imagery. Imagery collected with the thermal IR sensor was viewed, with personnel taking notes of grouse, cattle, and antelope sightings recorded while watching. The grouse in the IR video appeared as white, undefined dots (Figure 9.7); as a result, we were not able to differentiate between males and females using the thermal IR camera.

This project demonstrated that sUAS technology can be successfully used to detect greater sage-grouse at known lek locations. Greater sage-grouse experts consulted prior to the sUAS flights believed that the grouse would flush from the lek if the Raven sUAS was flown directly over the lek at an altitude of 400ft [122m] AGL or less. This project demonstrated that greater sage-grouse may be more tolerant of overhead aerial disturbance than originally believed, when a battery-powered sUAS is used. sUAS overflights were conducted between 120ft [37m] and 300ft [91m] directly over the lek. The grouse observers at the leks noted that the noise from the Raven sUAS electric motor was minimal from the ground. These flights also determined that it was possible to identify grouse displaying on the lek with both the IR and EO cameras without flushing the grouse. The Raven sUAS provides a non-intrusive, safe, and potentially effective method to locate new leks and obtain count estimates of breeding greater sage-grouse. Improving the on-board camera systems with accompanying georeferencing information will improve efficiencies of using the Raven sUAS for future wildlife population estimates. The use of sUAS to conduct wildlife monitoring activities is another tool that can be utilized by wildlife managers.

## 9.3 Enabling Emergency Communications

Critical lessons learned from past disasters stress the importance of restoring communications within a few hours of a disaster, to ensure lives are not lost and communities move towards recovery quickly. However, oftentimes limited access to the impacted areas creates challenges for utility companies and restoration crew delaying the repair of critical communications infrastructure. Restoring communications among first responders is essential for planning and coordinating relief efforts, and rescuing people that are affected. In the event of a loss in terrestrial connectivity, aerial communication strategies have been offered by the industry including: (1) Loon project which proposes to provide Internet to people using high altitude platforms (HAPs) to lift communication equipment up in the stratosphere by Google [185], and (2) Deployable 4G LTE solution on a high-altitude balloon platform by Oceus [319]. These solutions are primarily targeted for commercial purposes and military applications rather than disaster response.

### 9.3.1 Aerial Base Stations

An aerial base station is one that can be airlifted by a balloon or unmanned aerial vehicle (UAV) providing the infrastructure needed to facilitate wireless and/or cellular communications during disaster response following natural and man-made disasters (see Figure 9.8). As opposed to high altitude platforms, UAV enabled aerial base stations can fly or hover at an altitude of a few thousand feet above the ground providing Wi-Fi/cellular coverage for areas affected by disaster. For example, in recent times, a tornado in Oklahoma knocked down several cell towers on its way through a 12 mile × 1 mile

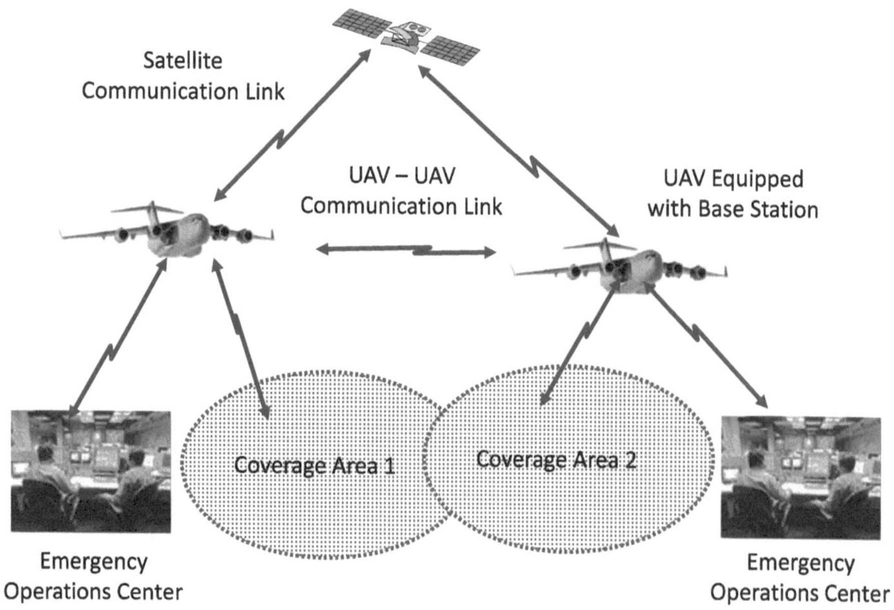


**Figure 9.8** Aerial base station: an illustration

stretch leaving the citizens in this area without cellular coverage. Deployable solutions such as cells on wheels (CoWs) that are available today take a significant amount of time to deploy in disaster areas. They may not reach their target destination in some instances, such as flooding, when the roads are impassable. Aerial base stations, on the other hand, are quicker to deploy and cost effective, and hence provide an efficient alternative to CoWs. They also help increase the capacity of an existing cellular network during public events such as Presidential inaugurations and tours, sports events, or concerts. There are, however, several obstacles that we need to overcome before aerial base stations become a reality.

### 9.3.2 Cyber Physical System Perspective

An aerial base station deployed during emergency response represents a cyber physical system (CPS) as illustrated in Figure 9.9. While the airborne network, consisting of UAVs, cellular infrastructure, sense and avoid strategies, communications with ground control towers, and associated information dissemination, forms the cyber component, the human network, involving the first responders, volunteers, decision makers engaged in the emergency operations and management, forms the physical component. During a real-world disaster deployment, urgency demands will dictate the evolution of intense interactions between the cyber and physical component systems. In order to identify these cyber physical interactions and challenges in maintaining such interactions, there is a need to mimic a real-world disaster scenario through disaster drills. This is especially important to understand the social interactions among various stakeholders during the disaster response process, that is central actors coordinating information flow, and the sub-group culture not directly evidenced in the hierarchical Incident Command System (ICS) normally used at a typical Emergency Operations Center (EOC). The concurrent analysis of airborne network and social network during emergency drills will lead to the CPS design principles that form the foundation for designing and deploying aerial base stations in a real-world scenario.

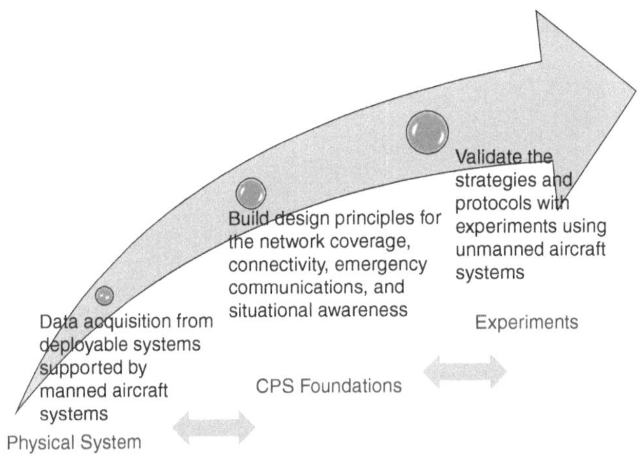


**Figure 9.9** Cyber physical system perspective of aerial deployable systems

### 9.3.3 Scientific and Engineering Challenges

In order to deploy aerial base stations in disaster recovery situations, there are several engineering as well as scientific challenges that need to be addressed. Figure 9.10 illustrates a UAV-enabled emergency communication platform. It showcases an aerial network with two UAVs equipped with base stations (Wi-Fi/cellular) in the air and a network of stakeholder organizations, including police, fire, civil air patrol, schools, utilities, transportation, hospitals, and non-profits such as the Red Cross. The aerial network needs to co-exist with the human network on the ground. This requires careful examination of the interactions that go on between the two networks – airborne network and human/social network. Below, we briefly outline these two types of networks and the interactions between them.

1. *Airborne Network Analysis:* Making the network available is an important aspect of the aerial deployable systems. This requires us to study the connectivity, coverage, scalability, and interference aspects of the aerial network. Cellular communications between an aerial base station and citizens and first responders will interfere with the existing and operational cellular communications on the ground. Scaling the airborne network in order to cover a larger area and to provide service to more users including first responders, citizens, media, and non-profit entities is another important dimension that requires exploration. Stability and reliability of the airborne network, and the factors on which the network stability depend need to be investigated as well.
2. *Human and Social Network Analysis:* An aerial base station deployed for emergency communications needs to be seamlessly integrated with the existing public safety as well as commercial communication infrastructure on the ground. The need for such integration poses its own challenges, including interoperability among various devices and spectrum sharing among different public safety

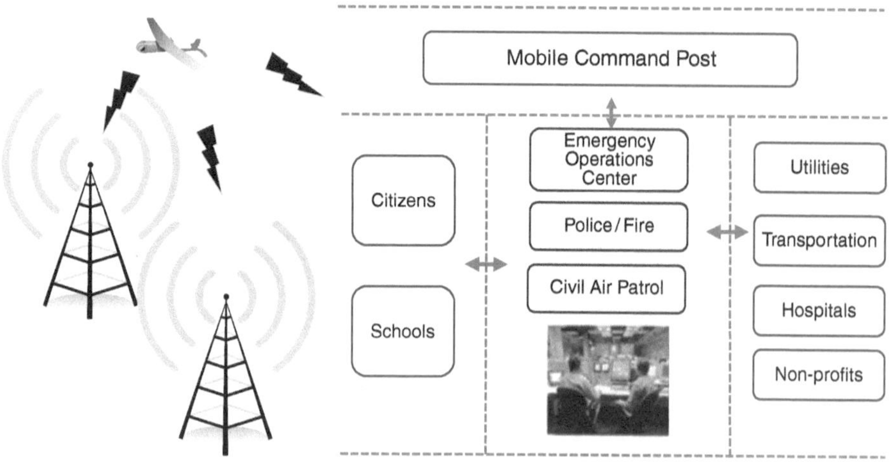


**Figure 9.10** Integrating an aerial deployable platform with existing emergency communication system

and commercial entities. Human factors, organizational challenges, and training requirements in facilitating the use of aerial base station services by citizens and emergency response stakeholders need to be investigated.

### 9.3.4 Disaster Response and Emergency Communications

Past disasters have clearly demonstrated the importance of uninterrupted communication networks to a city's ability to respond, recover, and rebound from disasters. For example, immediately after the 2013 Boston Marathon bombings, cellular networks were overwhelmed making it difficult for citizens and emergency responders to exchange information, while Hurricane Sandy in 2012 knocked out 25% of the cell towers in its path making it extremely difficult for inundated communities to reach out for assistance. Hurricane Katrina in 2005 saw major power and communication failures that disrupted recovery operations, suggesting a need to design "a low power, low weight, and low cost reliable communication solution for use in disrupted, interrupted, and low-bandwidth environments" [445]. Thus, for a viable and sustainable smart city that is prepared for natural and human induced threats to exist, it is essential to work on developing, implementing, and maintaining a robust warning system with common alert protocols [64].

**UAV-enabled Communication Platforms**

While Wi-Fi and cellular technologies have been extensively investigated in recent years [453], designing and testing for UAV network enabled emergency communication systems for enhancing emergency response are still open subjects for research. More recently, MIT Lincoln Laboratory and Penn State University have developed an Airborne Remote Communication (ARC) platform for enabling communications in disaster/disadvantaged environments [445]. However, in the context of emergency management and the vision for a smart city, the ARC platform has its limitations. While users can leverage their existing Wi-Fi enabled devices to interface with ARC, the coverage is limited in scale. Although the ARC prototype system consists of a wearable user node that communicates with other user nodes, the Internet, or Public Safety services via a UHF repeater, the system needs to be compatible with the ones that are already in use by first responders. The human and institutional dimensions are also key components of a smart city. In the emergency response context, no matter how well the systems are integrated, "we cannot build away all disaster risk. No matter how stringent, no matter how well enforced, no matter how costly, building codes cannot eliminate disaster risk" [405]. Thus, it is important to work with local and regional government agencies that are responsible for emergency management and planning to determine the applicability of the technology during emergency response.

**Federal Communication Commission's Initiatives**

In 2011, the FCC considered Deployable Aerial Communications Architectures and started seeking feedback from the Telecommunication companies [164]. Google's Loon project proposes to provide Internet to people using high altitude platforms to lift communication equipment up into the stratosphere [185]. The movements of these platforms are determined and controlled (maneuvered) using the direction of the wind flow present in this layer of the atmosphere. In October 2013, Oceus Networks Inc. successfully demonstrated the efficacy of a deployable 4G LTE solution on a high-altitude balloon platform, which can be used to rapidly provide broadband coverage and communications services to first responders within the first 72 hours following an emergency or natural disaster [319]. The exercise involved an airborne 4G LTE cellular network, allowing engineers to collect data and characterize the performance of a high altitude 4G LTE public safety system. The payload traveled nearly 200 miles and reached an altitude of 75,000 feet. While prior work has focused on a deployable 4G LTE solution on a high-altitude, aerial base stations discussed in this chapter take this research and technology to the next level by integrating UAV-enabled aerial base stations with the existing emergency communications systems. With the ongoing efforts by the FAA to integrate UAVs into the national airspace, the time is right for the design and deployment of aerial base stations. Aerial base stations will help restore communications immediately after a disaster, speeding up the recovery process.

#### Homeland Security Initiatives

Disaster simulations (drills) are widely used throughout the world and considered a fundamental tool for evaluating local disaster response capacity and suggesting improvements [190]. Currently, the Homeland Security Exercise and Evaluation Program (HSEEP) doctrine provides a standardized policy, methodology, and terminology for exercise design, development, conduct, evaluation, and improvement planning. HSEEP helps achieve the objectives of the National Preparedness Goal [169] and the National Preparedness System through the development of the HSEEP toolkit, a web-based collection of tools for exercise scheduling, design, development, evaluation, and improvement planning. Consistent with HSEEP, the proposed integrated aerial communication system would allow local response agencies to test response communication capabilities (one of the Presidential Policy Directive 8: National Preparedness goals). Moreover, the DHS–Science and Technology Directorate (S&T) has partnered with FEMA to investigate and assess new technologies and processes for an updated HSEEP Enterprise Platform (EP) for operational use [243]. The current HSEEP toolkit does not facilitate information sharing or provide a seamlessly integrated exercise support system. DHS (S&T), in partnership with FEMA, and in collaboration with Sandia National Laboratories, has developed the Standard Unified Modeling, Mapping, and Integration Toolkit (SUMMIT) as a geo-agile platform that enables users to discover and reuse models, integrate them quickly and economically, and apply them in analyses to improve the planning and execution of large and small exercises, planning efforts, and eventually response operations [290].

### 9.3.5 Research Challenges

The proposed research includes investigation of airborne networks analysis and ad hoc social networks that are formed among the stakeholders involved in disaster recovery operations. Airborne network analysis includes developing theoretical models for network coverage, connectivity, and scalability. Social network analysis involves abstracting the critical communication links among the first responders and the decisions involving disaster recovery operations. The intent of this integrated approach is to investigate the synergy between the cyber and physical components and abstract the CPS design principles behind the design and deployment of aerial base stations. Detailed descriptions of the research tasks are outlined below.

### Mobility-driven Airborne Network Analysis

Through appropriate modeling and analysis of mobility of airborne nodes, one can predict aircraft body blockage, area coverage, link duration, path duration, throughput, and delay, which can significantly contribute to the design and selection of networking algorithms for airborne networks. We will explore mobility-aware adaptive routing strategies for airborne networks based upon (1) the prediction of mobility of neighboring agents, and (2) the mobility statistics of the network. For instance, if a relay node is predicted to be moving out of the transmission range, this node will not be chosen even if it results in a reduced number of hops. Also, the updates from the routing table benefit from understanding the performance statistics of the network, such as path duration and link lifetime. For example, one can design a routing protocol that selects a route among the available routes based on the principle of "longest residual life," which will enhance network connectivity.

### Prototype of an Aerial Base Station

The University of North Texas has been working on the design of a UAV network-enabled emergency communication system in order to study the fundamental science and engineering aspects of aerial base stations. A prototype of the aerial base station airlifted by a Helium balloon was demonstrated during the Global City Team Challenge hosted by the National Institute of Standards and Technology on June 1, 2015 in Washington DC. This demonstration is illustrated in Figure 9.11.

This prototype is used to investigate the coverage, connectivity, scalability, reliability, situational awareness, and safety and security issues of the UAV platform as well as communication range, bandwidth, data rate, and quality of service aspects of the air-to-air and air-to-ground communication links. Preliminary results of the investigation suggest that a UAV-enabled communication platform is advantageous compared to Cells on Wheels. Experiments with balloons also provide benchmark data for comparative purposes. Experiments with UAVs will provide results that can be more tuned for real-world scenarios. Based on the laboratory design and testing of the aerial communication platform, we will use commercially available UAVs and equip them with Wi-Fi hotspots/cellular base stations. We will begin with one aerial node and extend the network to include several nodes. The coverage area can be scaled as a function of the number of nodes. With multiple nodes in the air forming different configurations, we will be able to measure the network characteristics, including the extent of coverage, scalability, network connectivity, signal strength, data/voice quality, and compare these measurements with the theoretical estimates. The prototype platform consists of the

**Figure 9.11** Prototype of a balloon-supported Aerial Base Station demonstrated in Global City Teams Challenge, 2015

following components: (1) a Wi-Fi hotspot/cellular base station allowing communication between first responders and citizens, (2) interface between the cellular base station and emergency first responders communication network, and citizens' cell phones, and (3) software tools and services such as location service and walkie-talkie service.

## 9.3.6 Deriving Theoretical Models

Experiments to measure connectivity, coverage, scalability, and interference with terrestrial communications are important for the design and development of aerial base stations. The purpose of these experiments is to develop the fundamental knowledge that is needed to build the theoretical models that explain the principles of aerial base stations in terms of networking and communications. At UNT, we are conducting the following sets of experiments.

### Signal Strength Measurements

In the first set, a pure sinusoidal signal is transmitted from an aerial node to a receiver placed on the ground. The goal is to measure the received signal strength on the ground at various distances such as the altitude at which the signal is transmitted. In order to conduct the experiment, the receiver is placed at different horizontal distances and the transmitter is airlifted using a balloon or UAV to various heights as shown in Figure 9.12.

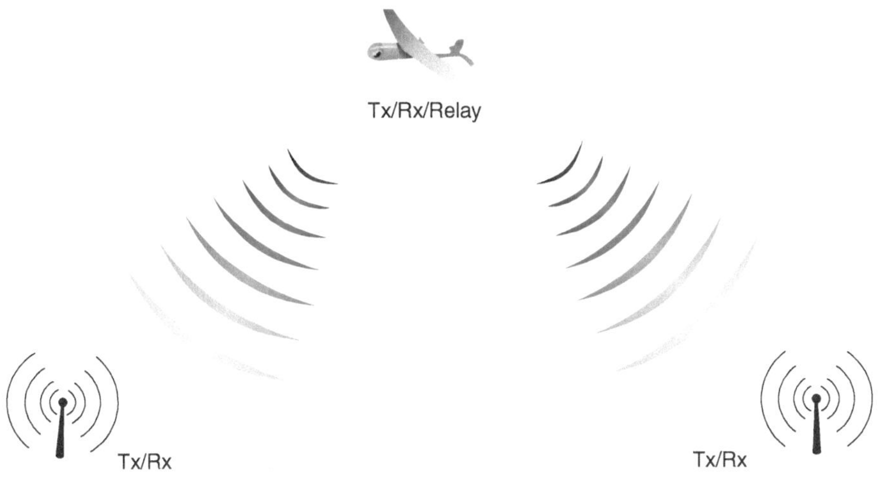


**Figure 9.12** Aerial deployable system: relay

The novelty of this experiment is that it allows us to transmit the signal in its analog form and measure the signal strength at the receiver. Students have already designed the transmitter with an oscillator to generate the sinusoidal signal at different frequencies. A receiver with a circuit to measure the average signal strength is designed as well. Preliminary experiments demonstrate that due to direct line of sight, free-space propagation can be used to model radio frequency signal propagation. This seems to be a good justification for the use of aerial nodes. However, when there are several aerial nodes, the strong signals coming from other aerial nodes create strong interference. Thus, modeling the signal quality when several aerial nodes are communicating with ground nodes will need further investigation.

## Connectivity, Coverage, and Scalability

The second set of experiments is aimed at modeling the performance of an ad hoc network created using a set of aerial nodes. This is more challenging not only because of multiple aerial nodes being deployed, but also due to the need for coordination among the aerial nodes. In order to generate the experimental datasets for bench marking as well as to create theoretical models for connectivity, coverage, and scalability, we intend to conduct the experiments using balloons. This will simplify the process of deployment of aerial nodes and facilitates for longer duration for conducting the experiments. The nodes will be airlifted using balloons filled with helium. The balloons will be tethered appropriately to keep them at given altitudes. Each balloon will be equipped with a Raspberry PI and a Wi-Fi transceiver. Adding a GPS and an altitude sensor to each balloon will also give us the precise measurement of its location and altitude. This platform will also be useful to test routing protocols such as disruption/delay tolerant networking. We are currently working towards developing such protocols for space applications with the support and mentoring from NASA engineers. The concept behind the experiment is illustrated in Figure 9.13.

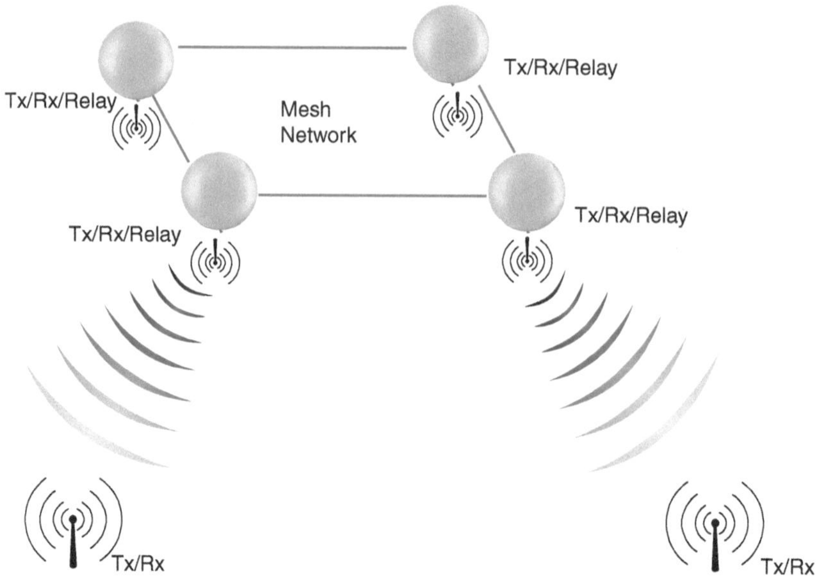


**Figure 9.13** A network of aerial base stations: connectivity, coverage, and scalability

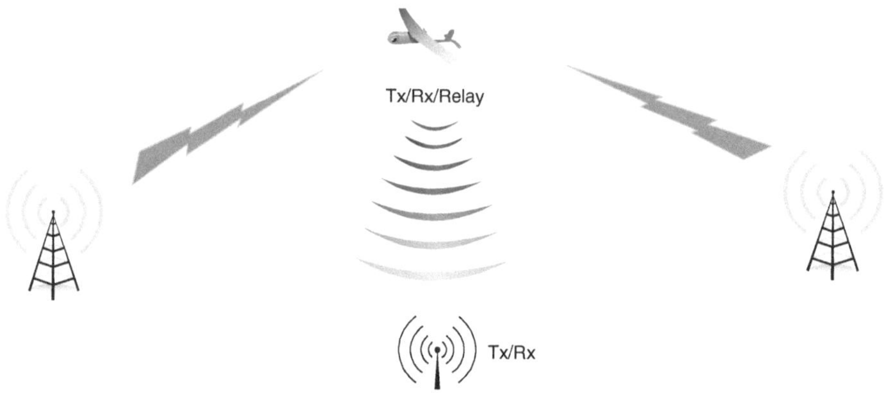


**Figure 9.14** Illustration of an aerial base station serving as a bridge between an end user and the nearest functional cell tower

## Interference

When there are several aerial nodes, the strong signals coming from other aerial nodes create strong interference. Thus, modeling the signal quality when several aerial nodes are communicating with ground nodes will need further investigation. In the third set of experiments involving airlifting a communicating node that serves as a bridge between an end user without cellular service to the nearest functional cell tower. The experimental set-up is illustrated in Figure 9.14. As shown here, the aerial node bridges the gap in the cellular coverage by providing coverage to the users in locations where cellular service is not available. The aerial node connects the users to the nearest cellular tower, thus providing the cellular service. The purpose of this set of experiments is to measure the performance of the base station in terms of area of coverage, bandwidth, and number of simultaneous calls that can be made with the aerial base station. In geographic areas where there is cellular coverage, this aerial base station offers the capability to provide coverage. At the same time, the strong signal from the aerial node may interfere with the ongoing terrestrial communication. Thus, experiments need to be conducted to collect data to model the signal quality and interference between aerial and terrestrial communications. This will lead to efficient strategies to share the spectrum among various stakeholders including citizens and first responders, as well as between commercial and public safety entities during the disaster recovery process.

## 9.4 Summary

This chapter illustrated two real-world applications that immensely benefit from using UAVs. In the first application, small unmanned aircraft systems (sUAS) with cameras and sensors were utilized to provide accurate population estimates of wildlife. In the second application, an aerial base station that can provide wireless and/or cellular communications during disaster response and recovery is explained.

# References


[1] Albatross. www.kickstarter.com/projects/164361630/the-albatross-uav. [Accessed 28 August 2015].

[2] DTN research group. sites.google.com/site/dtnresgroup/.

[3] Made in Oregon: The Most Secure UAV Software in the World -> 60 Minutes. www.softwarediligence.com/made-in-oregon-the-most-secure-uav-software-in-the-world-60-minutes-video/. [Accessed 28 August 2015].

[4] Secure Solutions. www.xilinx.com/applications/aerospace-and-defense/secure-solutions.html. [Accessed 28 August 2015].

[5] SkyJack Project. samy.pl/skyjack/. [Accessed 28 August 2015].

[6] SkyScanner. www.laas.fr/projects/skyscanner/. [Accessed 30 December 2015].

[7] Solar-Storm. wiki.paparazziuav.org/wiki/Solar-Storm. [Accessed 28 August 2015].

[8] Global information grid (GIG) overarching policy. Department of Defense Directive, November 2003.

[9] iNET Needs Discernment Report, version 1.0. Central Test and Evaluation Investment Program (CTEIP), May 2004.

[10] The ns-3 Network Simulator. www.nsnam.org, July 2009.

[11] 112th Congress, 2nd Session, U.S. House of Representatives. FAA Modernization and Reform Act of 2012, February 2012. www.gpo.gov/fdsys/pkg/CRPT-112hrpt381/pdf/CRPT-112hrpt381.pdf.

[12] M. Abolhasan, T. Wysocki, and E. Dutkiewicz. A review of routing protocols for mobile ad hoc networks. *Ad Hoc Networks*, 2(1):1–22, January 2004.

[13] M.C. Achtelik, J. Stumpf, D. Gurdan, and K.-M. Doth. Design of a flexible high performance quadcopter platform breaking the mav endurance record with laser power beaming. In *Proc. IEEE/RSJ Intl. Conf. Intelligent Robots and Systems (IROS)*, pages 5166–5172, September 2011.

[14] E. Adamey and U. Ozguner. A decentralized approach for multi-UAV multitarget tracking and surveillance. In *Society of Photo-Optical Instrumentation Engineers (SPIE) Conference Series*, volume 8389, May 2012.

[15] A. Adams, J. Nicholas, and W. Siadak. Protocol independent multicast, dense mode protocol specification (PIM-DM). *Internet RFC 3973*, February 2005.

[16] Carlisle Adams and Steve Lloyd. *Understanding PKI: Concepts, Standards, and Deployment Considerations*. Addison-Wesley Longman Publishing Co., Boston, MA, USA, 2nd edition, 2002.

[17] B. Adamson, C. Bormann, M. Handley, and J. Macker. NACK-oriented reliable multicast (NORM) transport protocol. *RFC 5740*, November 2009.

[18] Agencia Nacional de Aviacao Civil. *Regulamentos Brasileiros de Aviacao Civil (RBAC)*, 2012. www2.anac.gov.br/biblioteca/rbha.asp.

[19] Dakshi Agrawal, Bruce Archambeault, Josyula R. Rao, and Pankaj Rohatgi. The EM Side—Channel(s). In Burton S. Kaliski, Çetin K. Koç, and Christof Paar, editors, *Cryptographic Hardware and Embedded Systems – CHES 2002*, volume 2523 of *Lecture Notes in Computer Science*, pages 29–45. Springer, Berlin, Heidelberg, August 2003.

[20] Dakshi Agrawal, Josyula R. Rao, and Pankaj Rohatgi. Multi-channel attacks. In Colin D. Walter, Çetin K. Koç, and Christof Paar, editors, *Cryptographic Hardware and Embedded Systems – CHES 2003*, volume 2779 of *Lecture Notes in Computer Science*, pages 2–16. Springer, Berlin, Heidelberg, September 2003.

[21] R. Ahlswede, Ning Cai, S.-Y.R. Li, and R.W. Yeung. Network information flow. *IEEE Transactions on Information Theory*, 46(4):1204–1216, July 2000.

[22] Jeff Ahrenholz, Claudiu Danilov, Thomas R. Henderson, and Jae H. Kim, CORE: A real-time network emulator. *IEEE MILCOM 2008*, San Diego, CA, November 2008.

[23] Raja Naeem Akram, Pierre-François Bonnefoi, Serge Chaumette, Konstantinos Markantonakis, and Damien Sauveron. Improving security of autonomous UAVs fleets by using new specific embedded secure elements – a position paper. In *2nd AETOS International Conference on "Research Challenges for Future RPAS/UAV Systems,"* Bordeaux, France, September 2014.

[24] Ian F. Akyildiz, Xudong Wang, and Weilin Wang. Wireless mesh networks: a survey. *Computer Networks*, 47(4):445–487, 2005.

[25] Sattam S. Al-Riyami and Kenneth G. Paterson. *Advances in Cryptology – ASIACRYPT 2003: Proceedings of the 9th International Conference on the Theory and Application of Cryptology and Information Security*, Taipei, Taiwan, November 30–December 4, 2003, Certificateless Public Key Cryptography, pages 452–473. Springer, Berlin, Heidelberg, 2003.

[26] Roberto Albertani, Bret Stanford, James Hubner, and Peter Ifju. Characterization of flexible wing mavs: Aeroelastic and propulsion effects on flying qualities. In *Proceedings of AIAA Atmospheric Flight Mechanics Conference and Exhibit*, pages 15–18, August 2005.

[27] Mohammed J. F. Alenazi, Santosh Ajith Gogi, Dongsheng Zhang, Egemen K. Çetinkaya, Justin P. Rohrer, and James P. G. Sterbenz. Implementation of aeronautical network protocols. In *Proceedings of the AIAA Infotech@Aerospace Conference*, Boston, MA, August 2013.

[28] M. Allman, S. Dawkins, D. Glover, J. Griner, D. Tran, T. Henderson, J. Heidemann, J. Touch, H. Kruse, S. Ostermann, K. Scott, and J. Semke. Ongoing TCP research related to satellites. RFC 2760 (Informational), February 2000.

[29] M. Allman, D. Glover, and L. Sanchez. Enhancing TCP over satellite channels using standard mechanisms. RFC 2488 (Best Current Practice), January 1999.

[30] M. Allman, V. Paxson, and W. Stevens. *TCP Congestion Control*. RFC 2581 (Proposed Standard), April 1999. Obsoleted by RFC 5681, updated by RFC 3390.

[31] J. Allred, A. Hasan, S. Panichsakul, W. Pisano, P. Gray, J. Huang, R. Han, D. Lawrence, and K. Mohseni. SensorFlock: An airborne wireless sensor network of micro-air vehicles. In *Proceedings ACM International Conference Embedded Networked Sensor Systems*, pages 117–129, 2007.

[32] A.I. Alshbatat and D. Liang. Adaptive MAC protocol for UAV communication networks using directional antennas. In *Proc. Intl. Conf. Networking, Sensing and Control (ICNSC)*, pages 598–603, April 2010.

[33] Hassan Aly and Mohammed ElGayyar. Attacking AES using Bernsteins attack on modern processors. In Amr Youssef, Abderrahmane Nitaj, and AboulElla Hassanien, editors, *Progress in Cryptology – AFRICACRYPT 2013*, volume 7918 of *Lecture Notes in Computer Science*, pages 127–139. Springer, Berlin, Heidelberg, June 2013.

[34] David E. Andersen, Orrin J. Rongstad, and William R. Mytton. Response of nesting red-tailed hawks to helicopter overflights. *Condor*, pages 296–299, 1989.

[35] T. Andre, K. Hummel, A. Schoellig, E. Yanmaz, M. Asadpour, C. Bettstetter, P. Grippa, H. Hellwagner, S. Sand, and S. Zhang. Application-driven design of aerial communication networks. *IEEE Communications Magazine*, 52(5):129–137, May 2014.

[36] J.G. Andrews, S. Buzzi, W. Choi, S.V. Hanly, A. Lozano, A.C.K. Soong, and J.C. Zhang. What will 5G be? *IEEE Journal on Selected Areas in Communications*, 32(6):1065–1082, June 2014.

[37] ARINC. 618-6: Air. *Ground Character-Oriented Protocol Specification.*

[38] M. Asadpour, D. Giustiniano, and K. Hummel. From ground to aerial communication: dissecting WLAN 802.11n for the drones. In *Proc. ACM MobiCom – WinTech Wksp*, pages 25–32, 2013.

[39] M. Asadpour, D. Giustiniano, K. Hummel, and S. Heimlicher. Characterizing 802.11n aerial communication. In *Proc. ACM MobiHoc Wksp on Airborne Networks and Commun.*, pages 7–12, July 2013.

[40] M. Asadpour, D. Giustiniano, K. Hummel, S. Heimlicher, and S. Egli. Now or later? Delaying data transfer in time-critical aerial communication. In *Proc. ACM CoNEXT*, pages 127–132, December 2013.

[41] M. Asadpour, B. Van den Bergh, D. Giustiniano, K. Hummel, S. Pollin, and B. Plattner. Micro aerial vehicle networks: an experimental analysis of challenges and opportunities. *Communications Magazine, IEEE*, 52(7):141–149, July 2014.

[42] G.R. Asha, V. Vaidhehi, Toa Bi Irie Guy Cedric, and S. Balaji. *A Review of Routing Protocols for Airborne Networks. IJIRAE*, 3(2), March 2015.

[43] Atlantik Solar. http://www.atlantiksolar.ethz.ch/.

[44] Hari Balakrishnan, Randy H. Katz, and Venkata N. Padmanbhan. The effects of asymmetry on TCP performance. *Mob. Netw. Appl.*, 4(3):219–241, 1999.

[45] Hagai Bar-El, Hamid Choukri, David Naccache, Michael Tunstall, and Claire Whelan. The sorcerer's apprentice guide to fault attacks. *Proceedings of the IEEE*, 94(2):370–382, February 2006.

[46] Guillaume Barbu and Christophe Giraud. New countermeasures against fault and software type confusion attacks on java cards. In David Naccache and Damien Sauveron, editors, *Information Security Theory and Practice. Securing the Internet of Things*, volume 8501 of *Lecture Notes in Computer Science*, pages 57–75, July 2014.

[47] Stefano Basagni, Imrich Chlamtac, Violet R. Syrotiuk, and Barry A. Woodward. A distance routing effect algorithm for mobility (DREAM). In *4th Annual ACM Mobile Computing and Networking*, MobiCom 1998, pages 76–84, New York, NY, USA, 1998. ACM.

[48] R.-W. Beard. Multiple UAV cooperative search under collision avoidance and limited range communication constraints. In *Proc. IEEE CDC*, pages 25–30, 2003.

[49] İlker Bekmezci, Ozgur Koray Sahingoz, and Şamil Temel. Flying ad-hoc networks (FANET): A survey. *Ad Hoc Networks*, 11(3):1254–1270, May 2013.

[50] Wells C. Bennet. Civilian drones, privacy, and the federal–state balance. Technical report, Brookings Institute, 09 2014. www.brookings.edu/~/media/Research/Files/Reports/2014/09/civilian-drones-privacy/civilian_drones_privacy_bennett_NEW.pdf?la=en.

[51] J. Berger and J. Happe. Co-evolutionary search path planning under constrained information-sharing for a cooperative unmanned aerial vehicle team. In *IEEE Congress on Evolutionary Computation (CEC)*, pages 1–8, July 2010.

[52] Daniel J. Bernstein. Cache-timing attacks on AES. Technical report, University of Illinois at Chicago, 2005.

[53] K. Bhamidipati, D. Uhlig, and N. Neogi. Engineering safety and reliability into UAV Systems: Mitigating the ground impact hazard. *AIAA Guidance, Navigation and Control Conference and Exhibit*, Hilton Head, SC, August 2007.

[54] K. Bhamidipati, Mitigating the UAV ground impact hazard using fault-tree analysis and a fuel consumption model. M.S. Thesis, Department of Aerospace Engineering, University of Illinois at Urbana-Champaign, 2007.

[55] Eli Biham and Adi Shamir. Differential fault analysis of secret key cryptosystems. In Burton S. Kaliski, editor, *Advances in Cryptology CRYPTO 1997*, volume 1294 of *Lecture Notes in Computer Science*, pages 513–525. Springer, Berlin, Heidelberg, August 1997.

[56] P. Bilaye, V.N. Gawande, U.B. Desai, A.A. Raina, and R.S. Pant. Low cost wireless internet access for rural areas using tethered aerostats. In *2008 IEEE Region 10 and the Third International Conference on Industrial and Information Systems*, pages 1–5, December 2008.

[57] Sanjit Biswas and Robert Morris. Opportunistic routing in multi-hop wireless networks. *SIGCOMM Computer Communication Review*, 34(1):69–74, 2004.

[58] B. Blum, T. He, S. Son, and J. Stankovic. IGF: A state-free robust communication protocol for wireless sensor networks. Technical Report CS-2003-11, Department of Computer Science, University of Virginia, USA, 2003.

[59] Y. Bo, Q. Yongyuan, and C. Yan, A method for fault detection and isolation in the integrated navigation system for UAV. *Measurement Science and Technology*, 17:1522–1528, 2006.

[60] F. Boehm and A. Schulte. Air to ground sensor data distribution using IEEE 802.11n Wi-Fi network. In *Proc. IEEE/AIAA Digital Avionics Sys. Conf. (DASC)*, pages 4B2–1–4B2–10, October 2013.

[61] Dan Boneh, Richard A. DeMillo, and Richard J. Lipton. On the importance of checking cryptographic protocols for faults. In Walter Fumy, editor, *Advances in Cryptology EUROCRYPT 97*, volume 1233 of *Lecture Notes in Computer Science*, pages 37–51. Springer, Berlin, Heidelberg, May 1997.

[62] Pierre François Bonnefoi, Damien Sauveron, and Jong Hyuk Park. MANETS: an exclusive choice between use and security? *Computing and Informatics. Special Issue on Interactive Multimedia and Intelligent Services in Mobile and Ubiquitous Computing (MUC)*, 27(5):799–821, 2008.

[63] J. Border, M. Kojo, J. Griner, G. Montenegro, and Z. Shelby. Performance enhancing proxies intended to mitigate link-related degradations. RFC 3135 (Informational), June 2001.

[64] Botterell. The common alerting protocol: an open standard for alerting, warning and notification. In *Proceedings of the 3rd International ISCRAM Conference*, pages 497–503, 2006.

[65] Ouns Bouachir, Alinoe Abrassart, Francisco Garcia, and Nicolas Larrieu. A mobility model for UAV ad hoc network. In *International Conference on Unmanned Aircraft Systems (ICUAS)*, 2014, pages 383–388. IEEE, 2014.

[66] R. Braden. Requirements for internet hosts – communication layers. RFC 1122 (Standard), October 1989. Updated by RFCs 1349, 4379.

[67] R. Braden. T/TCP – TCP extensions for transactions functional specification. *RFC 1644* (Experimental), July 1994.

[68] Eric Brier, Christophe Clavier, and Francis Olivier. Correlation power analysis with a leakage model. In Marc Joye and Jean-Jacques Quisquater, editors, *Cryptographic Hardware and Embedded Systems – CHES 2004*, volume 3156 of *Lecture Notes in Computer Science*, pages 16–29. Springer, Berlin, Heidelberg, August 2004.

[69] A.L. Brown. Measuring the effect of aircraft noise on sea birds. *Environment International*, 16(4):587–592, 1990.

[70] T. Brown, B. Argrow, C. Dixon, S. Doshi, R. Thekkekunnel, and D. Henkel. Ad hoc UAV ground network (Augnet). In *Proc. AIAA Unmanned Unlimited Tech. Conf.*, September 2004.

[71] Timothy X. Brown, Brian Argrow, Cory Dixon, Sheetalkumar Doshi, Roshan-George Thekkekunnel, and Daniel Henkel. Ad hoc UAV ground network (augnet). In *AIAA 3rd Unmanned Unlimited Technical Conference*, pages 1–11, 2004.

[72] Dan Broyles, Abdul Jabbar, and James P.G. Sterbenz. Design and analysis of a 3–D Gauss-Markov mobility model for highly dynamic airborne networks. In *Proceedings of the International Telemetering Conference (ITC)*, San Diego, CA, October 2010.

[73] I. Bucaille, S. Hethuin, T. Rasheed, A. Munari, R. Hermenier, and S. Allsopp. Rapidly deployable network for tactical applications: Aerial base station with opportunistic links for unattended and temporary events absolute example. In *Proc. IEEE Military Commun. Conf. (MILCOM 2013)*, pages 1116–1120, November 2013.

[74] Jack L. Burbank, Philip F. Chimento, Brian K. Haberman, and William T. Kasch. Key challenges of military tactical networking and the elusive promise of manet technology. *Communications Magazine, IEEE*, 44(11):39–45, 2006.

[75] O. Burdakov, P. Doherty, K. Holmberg, J. Kvarnstrom, and P.-M. Olsson. Positioning unmanned aerial vehicles as communication relays for surveillance tasks. In J. Trinkle, Y. Matsuoka, and J.A. Castellanos, editors, *Robotics: Science and Systems V*, pages 257–264. MIT Press, 2010.

[76] John Burgess, Brian Gallagher, David Jensen, and Brian Neil Levine. Maxprop: Routing for vehicle-based disruption-tolerant networks. In *Proceedings of the 25th IEEE International Conference on Computer Communications (INFOCOM)*, April 2006.

[77] A. Burkle, F. Segor, and M. Kollmann. Towards autonomous micro UAV swarms. *Journal of Intelligent and Robotic Systems*, 61(1-4):339–353, 2011.

[78] S. Burleigh, A. Hooke, L. Torgerson, K. Fall, V. Cerf, B. Durst, K. Scott, and H. Weiss. Delay-tolerant networking: An approach to interplanetary internet. *IEEE Communications Magazine*, 41(6):128–136, June 2003.

[79] S. Burleigh, M. Ramadas, and S. Farrell. Licklider transmission protocol – motivation. RFC 5325 (Informational), September 2008.

[80] A. Capone, L. Pizziniaco, I. Filippini, and M.A.G. de la Fuente. A SiFT: An efficient method for trajectory based forwarding. In *2nd International Symposium on Wireless Communication Systems*, 2005, pages 135–139, 2005.

[81] Eric Baldwin Carr. Unmanned aerial vehicles: Examining the safety, security, privacy and regulatory issues of integration into US airspace. *National Center for Policy Analysis (NCPA)*. Retrieved September 2013.

[82] Graeme Caughley. Sampling in aerial survey. *The Journal of Wildlife Management*, pages 605–615, 1977.

[83] Ann Cavoukian. Privacy and drones: Unmanned aerial vehicles. Technical report, Information and Privacy Commissioner, Ontario, Canada, August 2012.

[84] V. Cerf, S. Burleigh, A. Hooke, L. Torgerson, R. Durst, K. Scott, K. Fall, and H. Weiss. Delay-tolerant networking architecture. RFC 4838 (Informational), April 2007.

[85] Egemen K. Çetinkaya and James P.G. Sterbenz. Aeronautical gateways: supporting TCP/IP-based devices and applications over modern telemetry networks. In *Proceedings of the International Telemetering Conference (ITC)*, Las Vegas, NV, October 2009.

[86] Dominique Chabot and David M. Bird. Evaluation of an off-the-shelf unmanned aircraft system for surveying flocks of geese. *Waterbirds*, 35(1):170–174, 2012.

[87] H.Y. Chao, Y.C. Cao, and Y.Q. Chen. Autopilots for small unmanned aerial vehicles: A survey. *Intl. Journal of Control, Automation and Systems*, 8(1):36–44, 2010.

[88] Suresh Chari, Josyula R. Rao, and Pankaj Rohatgi. Template attacks. In Burton S. Kaliski, Çetin K. Koç, and Christof Paar, editors, *Cryptographic Hardware and Embedded Systems – CHES 2002*, volume 2523 of *Lecture Notes in Computer Science*, pages 13–28. Springer, Berlin, Heidelberg, August 2003.

[89] S. Chaumette, R. Laplace, C. Mazel, R. Mirault, A. Dunand, Y. Lecoutre, and J. N. Perbet. CARUS, an operational retasking application for a swarm of autonomous UAVs: First return on experience. In *2011 – MILCOM 2011 Military Communications Conference*, pages 2003–2010, November 2011.

[90] Serge Chaumette and Damien Sauveron. Some security problems raised by open multiapplication smart cards. In *Proceedings of the 10th Nordic Workshop on Secure IT-systems (NordSec 2005)*, pages 1–12, Tartu, Estonia, October 2005.

[91] Serge Chaumette and Damien Sauveron. Wireless sensor nodes. In Konstantinos Markantonakis and Keith Mayes, editors, *Secure Smart Embedded Devices, Platforms and Applications*, pages 335–350. Springer, New York, 2014.

[92] B. N. Cheng and S. Moore. A comparison of MANET routing protocols on airborne tactical networks. In *MILCOM 2012 – 2012 IEEE Military Communications Conference*, pages 1–6, October 2012.

[93] Long Cheng, Chengdong Wu, Yunzhou Zhang, Hao Wu, Mengxin Li, and Carsten Maple. A survey of localization in wireless sensor network. *International Journal of Distributed Sensor Networks*, 2012.

[94] Yufei Cheng, M. Todd Gardner, Junyan Li, Rebecca May, Deep Medhi, and James P.G. Sterbenz. Optimised heuristics for a geodiverse routing protocol. In *Proceedings of the IEEE 10th International Workshop on the Design of Reliable Communication Networks (DRCN)*, pages 1–9, Ghent, Belgium, April 2014.

[95] Mung Chiang, S.H. Low, A.R. Calderbank, and J.C. Doyle. Layering as optimization decomposition: A mathematical theory of network architectures. *Proceedings of the IEEE*, 95(1):255–312, January 2007.

[96] J.J. Childress and J.J. Viniotis. Autonomous aircraft with disconnectable tether, June 2, 2015. US Patent 9,045,218.

[97] Grzegorz Chmaj and Henry Selvaraj. Distributed processing applications for UAV/drones: A survey. In H. Selvaraj, D. Zydek, and G. Chmaj, editors, *Progress in Systems Engineering*, volume 1089 of *Advances in Intelligent Systems and Computing*, pages 449–454. Springer International Publishing, 2015.

[98] K. Choi and I. Lee. A UAV-based close-range rapid aerial monitoring system for emergency responses. In *Proc. Intl. Conf. Unmanned Aerial Vehicle in Geomatics UAV-g*, pages 247–252, 2011.

[99] Cisco 3845 Integrated Services Router. Specifications and technical documentation. http://www.cisco.com/en/US/products/ps5856/.

[100] Civil Aviation Authority, United Kingdom. Advisory circular: Unmanned aircraft and aircraft systems, 2013. www.caa.co.uk/default.aspx?CATID=1995.

[101] Civil Aviation Flight Safety Authority, Australia. Advisory circular: Unmanned aerial vehicle operations, design specification, maintenance and training of human resources, 2002. www.casa.gov.au/sites/g/files/net351/f/_assets/main/rules/1998casr/101/101c01.pdf?v=1378480833.

[102] Civil Aviation Flight Safety Authority, Australia. Civil aviation safety regulation part 101, 2002. www.comlaw.gov.au/Details/F2015C00567.

[103] Civil Aviation Flight Safety Authority, Australia. Notice of proposed rulemaking: Remotely piloted aircraft systems, May 2014. aussiewebsite.pdf.

[104] Civil Aviation Safety Authority, Australia. Casr part 23 - airworthiness standards for aeroplanes in the normal, utility, acrobatic or commuter category, 2013. www.casa.gov.au/standard-page/casr-part-23-airworthiness-standards-aeroplanes-normal-utility-acrobatic-or-commuter?WCMS%3APWA%3A%3Apc=PARTS023f.

[105] Judy Clapp, Anita King, and Audrey Taub. COTS Commercial off-the-shelf benefits and burdens. The MITRE Corporation Report, March 2001.

[106] David D. Clark. *Protocol Design and Performance*, April 1995.

[107] David D. Clark, Mark L. Lambert, and Lixia Zhang. NETBLT: A high throughput transport protocol. *ACM SIGCOMM Computer Communication Review*, 17(5):353–359, October/November 1987.

[108] T. Clausen and P. Jacquet. Optimized link state routing protocol (OLSR). RFC 3626 (Experimental), October 2003.

[109] T. De Cola and M. Marchese. Performance analysis of data transfer protocols over space communications. *IEEE Transactions on Aerospace and Electronic Systems*, 41(4):1200–1223, October 2005.

[110] D.T. Cole, A.H. Goktogan, P. Thompson, and S. Sukkarieh. Mapping and tracking. *IEEE Robotics Automation Magazine*, 16(2):22–34, June 2009.

[111] R. Coltun, D. Ferguson, and J. Moy, OSPF for IPv6. *Internet RFC 2740*, December 1999.

[112] John W. Connelly, W. John Arthur, and O. Doyle Markham. Sage grouse leks on recently disturbed sites. *Journal of Range Management Archives*, 34(2):153–154, 1981.

[113] John T. Conomy, James A. Dubovsky, Jaime A. Collazo, and W. James Fleming. Do black ducks and wood ducks habituate to aircraft disturbance? *The Journal of Wildlife Management*, pages 1135–1142, 1998.

[114] Marco Conti and Silvia Giordano. Multihop ad hoc networking: The reality. *Communications Magazine, IEEE*, 45(4):88–95, 2007.

[115] Thomas H. Cormen, Charles E. Leiserson, Ronald L. Rivest and Clifford Stein, Introduction to Algorithms. MIT Press, 5(3):55, 2001.

[116] Jean-Sébastien Coron, Paul Kocher, and David Naccache. Statistics and secret leakage. In Yair Frankel, editor, *Financial Cryptography*, volume 1962 of *Lecture Notes in Computer Science*, pages 157–173. Springer, Berlin, Heidelberg, February 2001.

[117] Council of the European Union. Protection of individuals with regard to processing of personal data and on the free movement of such data, 1995. http://eur-lex.europa.eu/legal-content/EN/TXT/PDF/?uri=CELEX:31995L0046&from=en.

[118] Council of the European Union. Regulation (EU) no 216/2008, 2008. eur-lex.europa.eu/LexUriServ/LexUriServ.do?uri=OJ%3AL%3A2008%3A079%3A0001%3A0049%3AEN%3APDF.

[119] Council of the European Union. Commission regulation (EU) no 1178/2011: Aircrew regulation – annexes i to iv – flight crew licensing (FCL) and medical (med) requirements, 2011. eur-lex.europa.eu/LexUriServ/LexUriServ.do?uri=OJ:L:2011:311:0001:0193:EL:PDF.

[120] Council of the European Union. *Airworthiness and Environmental Certification*, 2012. Commission regulation (EU) no. 748/2012 ISBN Number: 978-92-9210-148-0.

[121] Council of the European Union. Air Operations, 2012. Commission regulation (EU) no. 965/2012. eur-lex.europa.eu/legal-content/EN/TXT/PDF/?uri=CELEX:32012R0965&from=EN.

[122] Council of the European Union. Part 21 Acceptable Means of Compliance and Guidance Material, 2014. Commission regulation (EU) no. 748/2012. easa.europa.eu/system/files/dfu/Change%20Information%20to%20AMC-GM%20Part%2021%20-%20Issue%202.pdf.

[123] The DAISIE project. www.agence-nationale-recherche.fr/?Project=ANR-13-ASMA-0004.

[124] Claudiu Danilov, Thomas R. Henderson, Thomas Goff, Jae H. Kim, Joseph Macker, Jeff Weston, Natasha Neogi, Andres Ortiz, and Daniel Uhlig. Experiment and field demonstration of a 802.11-based ground-UAV mobile ad-hoc network. In *MILCOM 2009–2009 IEEE Military Communications Conference*, pages 1–7. IEEE, 2009.

[125] Claudiu B. Danilov, Thomas R. Henderson, Phillip A. Spagnolo, Thomas Goff, and Jae H. Kim, MANET Multicast with Multiple Gateways. *IEEE MILCOM Conference*, November 2008.

[126] Claudiu Danilov, Thomas R. Henderson, Thomas Goff, Orlie Brewer, Jae H. Kim, Joseph Macker and Brian Adamson, Adaptive Routing for Tactical Communications, *IEEE MILCOM 2012*, Orlando, FL, November 2012.

[127] P. Danzig, Z. Liu, and L. Yan. *An Evaluation of TCP Vegas by Live Emulation*, 1995.

[128] W. Dargie and C. Poellabauer. *Fundamentals of Wireless Sensor Networks: Theory and Practice*. Wireless Communications and Mobile Computing. Wiley, 2010.

[129] N. Darwin, A. Ahmad, and O. Zainon. The potential of unmanned aerial vehicle for large scale mapping of coastal area. *IOP Conference Series: Earth and Environmental Science*, 18(1):012031, 2014.

[130] M.G. de la Fuente and H. Ladiod. A performance comparison of position-based routing approaches for mobile ad hoc networks. *Vehicular Technology Conference (VTC)*, pages 1–5, October 2007.

[131] S. Deering and R. Hinden. Internet Protocol, Version 6 (IPv6) Specification. RFC 2460 (Draft Standard), December 1998. Updated by RFCs 5095, 5722.

[132] M.T. DeGarmo. Issues concerning integration of unmanned aerial vehicles in civil airspace. Technical report, MITRE, Center for Advanced Aviation System Development, McLean, Virginia, 2004.

[133] Delair-tech. Dt-18. www.avyon.com/dt-18/. [Accessed 28 August 2015].

[134] David K. Delaney, Teryl G. Grubb, Paul Beier, Larry L. Pater, and M. Hildegard Reiser. Effects of helicopter noise on Mexican spotted owls. *The Journal of Wildlife Management*, pages 60–76, 1999.

[135] deltadrone. deltadrone.com/.

[136] M. Demmer, J. Ott, and S. Perreault. Delay-tolerant networking TCP convergence-layer protocol. RFC 7242 (Experimental), 2014.

[137] Department of Defense. *Unmanned Aerial Vehicle Roadmap 2000-2025*. Washington: Office of the Secretary of Defense, 2001.

[138] A. Detti, P. Loreti, N. Blefari-Melazzi, and F. Fedi. Streaming H.264 scalable video over data distribution service in a wireless environment. In *Proc. IEEE Intl. Symp. World of Wireless Mobile and Multimedia Networks (WoWMoM)*, pages 1–3, June 2010.

[139] Deutsches Zentrum für Luft- und Raumfahrt. LDACS1 website. www.ldacs.com, 2015.

[140] DGAC. www.dgac.org/.

[141] Reinhard Diestel. Graph theory. 2005. *Grad. Texts in Math*, 2005.

[142] Direction Generale de l'Aviation Civile. Arrêté du 11 avril 2012 relatif a la conception des aéronefs civils qui circulent sans aucune personne à bord, aux conditions de leur emploi et sur les capacités requises des personnes qui les utilisent, May 2012. www.legifrance.gouv.fr/eli/arrete/2012/4/11/DEVA1206042A/jo/texte.

[143] Direction Generale de l'Aviation Civile. Arrêté du 11 avril 2012 relatif à l'utilisation de l'espace aérien par les aéronefs qui circulent sans personne à bord, May 2012. www.legifrance.gouv.fr/eli/arrete/2012/4/11/DEVA1207595A/jo/texte.

[144] Directive 2014/34/EU of the European Parliament and of the Council of 26 February 2014 on the harmonisation of the laws of the member states relating to equipment and protective systems intended for use in potentially explosive atmospheres. *Official Journal of the European Union*, 2013. pages 309–356.

[145] F. Ducatelle, G.A. Di Caro, C. Pinciroli, F. Mondada, and L. Gambardella. Communication assisted navigation in robotic swarms: Self-organization and cooperation. In *Proc. IEEE/RSJ Intl. Conf. Intelligent Robots and Systems (IROS)*, pages 4981–4988, September 2011.

[146] Robert C. Durst, Gregory J. Miller, and Eric J. Travis. TCP extensions for space communications. In *MobiCom '96: Proceedings of the 2nd Annual International Conference on Mobile Computing and Networking*, pages 15–26, New York, NY, USA, November 1996. ACM Press.

[147] Pierre Dusart, Gilles Letourneux, and Olivier Vivolo. Differential fault analysis on A.E.S. In Jianying Zhou, Moti Yung, and Yongfei Han, editors, *Applied Cryptography and Network Security*, volume 2846 of *Lecture Notes in Computer Science*, pages 293–306. Springer, Berlin, Heidelberg, October 2003.

[148] H. Eisenbeiss. The potential of unmanned aerial vehicles for mapping. In Dieter Fritsch, editor, *Photogrammetrische Woche 2011*, pages 135–145. Wichmann Verlag, Heidelberg, 2004.

[149] M. E. Elaasar, M. Barbeau, E. Kranakis, and Zheyin Li. Satellite transport protocol handling bit corruption, handoff and limited connectivity. *IEEE Transactions on Aerospace and Electronic Systems*, 41(2):489–502, April 2005.

[150] Bracha Epstein and Vineet Mehta. Free space optical communications routing performance in highly dynamic airspace environments. In *Proceedings of the 2004 IEEE Aerospace Conference*. volume 2, pages 1398–1406. IEEE, 2004.

[151] Ozgur Ercetin, Michael O. Ball, and Leandros Tassiulas. Next generation satellite systems for aeronautical communications. *International Journal of Satellite Communications and Networking*, 22(2):157–179, 2004.

[152] C. Eschmann, C.-M. Kuo, C.-H. Kuo, and C. Boller. Unmanned aircraft systems for remote building inspection and monitoring. In *Proc. Intl. Wksp. Structural Health Monitoring*, pages 1179–1186, 2012.

[153] H. Eugster and S. Nebiker. UAV-based augmented monitoring: Real-time georeferencing and integration of video imagery with virtual globes. In *Intl. Archives of the Photogrammetry, Remote Sensing and Spatial Information Sciences*, pages 1229–1235, 2008.

[154] Eurocontrol. Challenges of growth 2013. www.eurocontrol.int/sites/default/files/article/content/documents/official-documents/reports/201306-challenges-of-growth-2013-task-4.pdf, 2013.

[155] European Aviation Safety Agency. Airworthiness certification of unmanned aircraft systems (UAS), 2009. easa.europa.eu/system/files/dfu/E.Y013-01_%20UAS_%20Policy.pdf.

[156] European Aviation Safety Agency. Annex to ed decision 2012/012/r part cs-23 normal, utility, aerobatic and commuter aeroplanes, 2012. easa.europa.eu/system/files/dfu/CS-23%20Amdt%203.pdf.

[157] European Aviation Safety Agency. Concept of operations for drones: A risk based approach to regulation of unmanned aircraft, 2015. www.easa.europa.eu/system/files/dfu/204696_EASA_concept_drone_brochure_web.pdf.

[158] Shimon Even. *Graph Algorithms*. W. H. Freeman & Co., New York, NY, USA, 1979.

[159] FAA (website). www.faa.gov/.

[160] FAA. Nextgen. www.faa.gov/nextgen, 2013.

[161] Kevin Fall. A delay-tolerant network architecture for challenged internets. In *Proceedings of the 2003 Conference on Applications, Technologies, Architectures, and Protocols for Computer Communications*, SIGCOMM '03, pages 27–34, New York, NY, USA, 2003. ACM.

[162] J. Fang, T. Goff, and G. Pei. Comparison studies of OSPF-MDR, OLSR and composite routing. In *Military Communications Conference, 2010 - MILCOM 2010*, pages 989–994, October 2010.

[163] FCC. Role of deployable aerial communications architecture in emergency communications and recommended next steps. Technical report, Federal Communications Commission (FCC), September 2011.

[164] FCC. Deployable aerial communications architectures (daca). www.fcc.gov/, 2013.

[165] Federal Aviation Administration. Systems and equipment guide for certification of part 23 airplanes and airships, 2011. www.faa.gov/documentLibrary/media/Advisory_Circular/AC%2023-17C.pdf.

[166] Federal Aviation Administration. Integration of unmanned aircraft systems into the national airspace system – concept of operations, September 2012. www.suasnews.com/wp-content/uploads/2012/10/FAA-UAS-Conops-Version-2-0-1.pdf. (retrieved May 2016)

[167] Federal Aviation Administration. Operation and certification of small unmanned aircraft systems, February 2015. www.faa.gov/regulations_policies/rulemaking/recently_published/media/2120-AJ60_NPRM_2-15-2015_joint_signature.pdf.

[168] David Feldmeier. An overview of the TP++ transport protocol project. In Ahmed N. Tantawy, editor, *High Performance Networks: Frontiers and Experience*, volume 238 of *Kluwer International Series in Engineering and Computer Science*, Chapter 8. Kluwer Academic Publishers, Boston, MA, 1993.

[169] FEMA. Department of homeland security: National preparedness system. www.fema.gov/pdf/prepared/nps-description.pdf, 2011.

[170] Sally Floyd. TCP and explicit congestion notification. *SIGCOMM Comput. Commun. Rev.*, 24(5):8–23, 1994.

[171] Fly-n-Sense fly-n-sense.com/.

[172] Robert J. Fowler, Michael S. Paterson, and Steven L. Tanimoto. Optimal packing and covering in the plane are NP-complete. *Information Processing Letters*, 12(3):133–137, 1981.

[173] R. Fox. TCP big window and NAK options. RFC 1106, June 1989.

[174] Miguel Frontera. Personal communications, 2006/07. Pilot comments during personal communications, March/April 2007. Emails between March 20 and April 20 of 2007.

[175] Michael Fudge, Thomas Stagliano, and Sunny Tsiao. Non-traditional flight safety systems and integrated vehicle health management systems. Final report, ITT Industries, Advanced Engineering and Sciences Division, 2560 Huntington Avenue, Alexandria, VA 22303, August 2003.

[176] Blake Fuller, Jonathan Kok, Neil A. Kelson, and Luis F. Gonzalez. Hardware design and implementation of a mavlink interface for an FPGA-based autonomous UAV flight control system. In *Proceedings of the 2014 Australasian Conference on Robotics and Automation*, pages 1–6. Australian Robotics & Automation Association ARAA, 2014.

[177] L. Galluccio, A. Leonardi, G. Morabito, and S. Palazzo. A MAC/routing cross-layer approach to geographic forwarding in wireless sensor networks. *Ad Hoc Netw.*, 5(6): 872–884, 2007.

[178] Karine Gandolfi, Christophe Mourtel, and Francis Olivier. Electromagnetic analysis: Concrete results. In Çetin K. Koç, David Naccache, and Christof Paar, editors, *Cryptographic Hardware and Embedded Systems — CHES 2001*, volume 2162 of *Lecture Notes in Computer Science*, pages 251–261. Springer, Berlin, Heidelberg, May 2001.

[179] A. Gaszczak, T. Breckon, and J. Han. Real-time people and vehicle detection from UAV imagery. *Proc. SPIE*, 7878:78780B–78780B–13, 2011.

[180] A. Gil, K. Passino, and J.B. Cruz. Stable cooperative surveillance with information flow constraints. *IEEE Trans. Contr. Sys. Techn.*, 16(5):856–868.

[181] John Gillom. *Overseers of the Poor: Surveillance, Resistance, and the Limits of Privacy*. The University of Chicago Press, Chicago, IL, 12 2001.

[182] Christophe Giraud and Hugues Thiebeauld. A survey on fault attacks. In Jean-Jacques Quisquater, Pierre Paradinas, Yves Deswarte, and Anas Abou Kalam, editors, *Smart Card Research and Advanced Applications VI*, volume 153 of *IFIP International Federation for Information Processing*, pages 159–176. Springer, US, 2004.

[183] Douglas N. Gladwin, Duane A. Asherin, and Karen M. Manci. Effects of aircraft noise and sonic booms on fish and wildlife: Results of a survey of US fish and wildlife service endangered species and ecological services field offices, refuges, hatcheries, and research centers. Technical report, DTIC Document, 1988.

[184] M. Goodrich, B. Morse, D. Gerhardt, J. Cooper, M. Quigley, J. Adams, and C. Humphrey. Supporting wilderness search and rescue using a camera-equipped mini UAV. *Journal of Field Robotics*, 25(1–2):89–110, 2008.

[185] Google. Balloon powered internet for everyone. www.google.com/loon/how/, 2013.

[186] A. Gordon. *The American Heritage History of Flight*. American Heritage, 1962.

[187] R. Goudie and Ian L. Jones. Dose-response relationships of harlequin duck behaviour to noise from low-level military jet over-flights in central Labrador. *Environmental Conservation*, 31(04):289–298, 2004.

[188] Sudhakar Govindavajhala and Andrew W. Appel. Using memory errors to attack a virtual machine. In *Proceedings of the 2003 IEEE Symposium on Security and Privacy*, SP '03, pages 154–165, Washington, DC, USA, May 2003. IEEE Computer Society.

[189] GRASP Laboratory www.grasp.upenn.edu/.

[190] Gary B. Green, Surbhi Modi, Kevin Lunney, and Tamara L. Thomas. Generic evaluation methods for disaster drills in developing countries. *Annals of Emergency Medicine*, 41(5):689–699, 2003.

[191] Frank M. Grimsley. Equivalent safety analysis using casualty expectation approach. *AIAA 3rd Unmanned Unlimited Technical Conference, Workshop, and Exhibit*, September 2004.

[192] Teryl G. Grubb and William W. Bowerman. Variations in breeding bald eagle responses to jets, light planes and helicopters. *Journal of Raptor Research*, 31(3):213–222, 1997.

[193] D.L. Gu, G. Pei, H. Ly, M. Gerla, B. Zhang, and X. Hong. UAV aided intelligent routing for ad-hoc wireless network in single-area theater. In *Proc. IEEE Wireless Commun. and Net. Conf.*, volume 3, 2000.

[194] Bill Gunston. *Janes Aerospace Dictionary*. Janes Information Group, London, England, 3 edition, 6 1988.

[195] A. Gupta and R. K. Jha. A survey of 5G network: Architecture and emerging technologies. *IEEE Access*, 3:1206–1232, 2015.

[196] J. Haas. A new routing protocol for the reconfigurable wireless networks. *Proc. of IEEE 6th International Conference on Universal Personal Communications* 97, pp. 562–566, 1997.

[197] E.C. Haas, K. Pillalamarri, C.C. Stachowiak, and M. Fields. Multimodal controls for soldier/swarm interaction. In *2011 RO-MAN*, pages 223–228, July 2011.

[198] D. Hague, H.T. Kung, and B. Suter. Field experimentation of cots-based UAV networking. In *MILCOM 2006–2006 IEEE Military Communications Conference*, pages 1–7, October 2006.

[199] Guangjie Han, Huihui Xu, Trung Q. Duong, Jinfang Jiang, and Takahiro Hara. Localization algorithms of wireless sensor networks: A survey. *Telecommunication Systems*, 52(4):2419–2436, 2013.

[200] A. Handa, R.A. Newcombe, A. Angeli, and A.J. Davison. Real-time camera tracking: When is high frame-rate best? In Andrew W. Fitzgibbon, Svetlana Lazebnik, Pietro Perona, Yoichi Sato, and Cordelia Schmid, editors, *ECCV (7)*, volume 7578 of *Lecture Notes in Computer Science*, pages 222–235. Springer, 2012.

[201] M. Handley, S. Floyd, J. Padhye, and J. Widmer. TCP friendly rate control (TFRC): Protocol specification. RFC 3448 (Proposed Standard), January 2003. Obsoleted by RFC 5348.

[202] Leanne Hanson, Christopher Holmquist-Johnson, and Michelle L. Cowardin. Evaluation of the Raven SUAS to detect and monitor greater sage-grouse leks within the middle park population. Open File Report, US Geological Survey, 2014.

[203] Colin M. Harris. Aircraft operations near concentrations of birds in Antarctica: The development of practical guidelines. *Biological Conservation*, 125(3):309–322, 2005.

[204] Hannes Hartenstein and Kenneth P. Laberteaux. A tutorial survey on vehicular ad hoc networks. *Communications Magazine, IEEE*, 46(6):164–171, 2008.

[205] K. Hartmann and C. Steup. The vulnerability of UAVs to cyber attacks – An approach to the risk assessment. In *5th International Conference on Cyber Conflict (CyCon)*, pages 1–23, June 2013.

[206] S. Hauert, J.-C. Zufferey, and D. Floreano. Evolved swarming without positioning information: an application in aerial communication relay. *Autonomous Robots*, 26(1), 2009.

[207] S. Hayat, E. Yanmaz, and C. Bettstetter. Experimental analysis of multipoint-to-point UAV communications with IEEE 802.11n and 802.11ac. In *Proc. IEEE Intl. Symp. Pers. Indoor, Mobile Radio Commun. (PIMRC)*, August 2015.

[208] S. Hayat, E. Yanmaz, and R. Muzaffar. Survey on unmanned aerial vehicle networks for civil applications: A communications viewpoint. *IEEE Communications Surveys Tutorials*, 18(4):2624–2661.

[209] Simon Heimlicher, Merkourios Karaliopoulos, Hanoch Levy, and Thrasyvoulos Spyropoulos. On leveraging partial paths in partially-connected networks. In *Proceedings of the 28th IEEE Conference on Computer Communications (INFOCOM)*, Rio de Janeiro, Brazil, April 2009.

[210] F. Heintz, P. Rudol, and P. Doherty. From images to traffic behavior – a UAV tracking and monitoring application. In *Proc. Intl. Conf. Information Fusion*, pages 1–8, July 2007.

[211] M. Heissenbüttel, T. Braun, T. Bernoulli, and M. Wälchli. BLR: Beaconless routing algorithm for mobile ad hoc networks. *Computer Communications*, 27(11):1076–1086, 2004.

[212] Janey C. Hoe. Improving the start-up behavior of a congestion control scheme for TCP. In *SIGCOMM '96: Conference Proceedings on Applications, Technologies, Architectures, and Protocols for Computer Communications*, pages 270–280, New York, NY, USA, 1996. ACM.

[213] B. Hofmann-Wellenhof, H. Lichtenegger, and E. Wasle. *GNSS – Global Navigation Satellite Systems: GPS, GLONASS, Galileo, and More*. Springer, Vienna, 2007.

[214] J.F. Horn. Aircraft and rotorcraft system identification: Engineering methods with flight test examples (M.B. Tischler *et al.*, 2006) [bookshelf]. *IEEE Control Systems*, 28(4):101–103, Aug 2008.

[215] Vikram Hrishikeshavan, Christopher Bogdanowicz, and Inderjit Chopra. Design, performance and testing of a quad rotor biplane micro air vehicle for multi role missions. *International Journal of Micro Air Vehicles*, 6(3):155–173, 2014.

[216] Hsin-Yuan Huang, A. Khendry, and T.G. Robertazzi. Layernet: A self-organizing protocol for small ad hoc networks. *IEEE Transactions on Aerospace and Electronic Systems*, 38(2):378–387, April 2002.

[217] L. Iannone, R. Khalili, K. Salamatian, and S. Fdida. Cross-layer routing in wireless mesh networks. *1st International Symposium on Wireless Communication Systems*, pages 319–323, 2004.

[218] ICAO (website). www.icao.int.

[219] ICAO. Technical provisions for mode s services and extended squitter. 2012.

[220] ICAO. Integration of remotely piloted aircraft systems in civil controlled airspace and self-organizing airborne networks. www.icao.int/Meetings/a38/Documents/WP/wp337_en.pdf, 2013.

[221] Integration of the CORE and EMANE Network Emulators, by Jeff Ahrenholz, Tom Goff, and Brian Adamson, *IEEE MILCOM 2011*, Baltimore, MD, November 2011.

[222] ICAO. Use of self organizing airborne networks to monitor commercial aircraft globally. www.icao.int/Meetings/GTM/Documents/WP.10.Russian.Use%20of%20self%20organizing%20airborne%20networks.Revised.pdf, 2014.

[223] Richard M. Thompson III. Civilian drones, privacy, and the federal–state balance. Technical report, Congressional Research Service, 03 2015. fas.org/sgp/crs/misc/R43965.pdf.

[224] International Civil Aviation Organization. Document 9713: International civil aviation vocabulary, 2007. www1.atmb.net.cn/CD_web/UploadFile/2013052810430696.pdf.

[225] International Civil Aviation Organization. Annex 8: Airworthiness of aircraft, July 2010. www.icao.int/safety/airnavigation/nationalitymarks/annexes_booklet_en.pdf.

[226] International Civil Aviation Organization. Advisory circular: Unmanned aircraft systems, 2011. http://ICAOwebsite.pdf.

[227] International Civil Aviation Organization. Amendment 43 to ICAO Annex 2, Rules of the Air, 2012. www.icao.int/Meetings/UAS/Documents/019e.pdf.

[228] M. Iordanakis, D. Yannis, K. Karras, G. Bogdos, G. Dilintas, M. Amirfeiz, G. Colangelo, and S. Baiotti. Ad-hoc routing protocol for aeronautical mobile ad-hoc networks. In *Proceedings of the Fifth International Symposium on Communication Systems, Networks and Digital Signal Processing (CSNDSP)*, 2006.

[229] Michael Iordanakis, Dimitrios Yannis, Kimon Karras, Georgios Bogdos, Georgios Dilintas, Massimiliano Amirfeiz, Giorgio Colangelo, and Stefano Baiotti. Ad-hoc routing protocol for aeronautical mobile ad-hoc networks. In *Fifth International Symposium on Communication Systems, Networks and Digital Signal Processing (CSNDSP)*. Citeseer, 2006.

[230] Michael Iordanakis, Dimitrios Yannis, Kimon Karras, Georgios Bogdos, Georgios Dilintas, Massimiliano Amirfeiz, Giorgio Colangelo, and Stefano Baiotti. Ad-hoc routing protocol for aeronautical mobile ad-hoc networks. In *Fifth International Symposium on Communication Systems, Networks and Digital Signal Processing (CSNDSP)*. Citeseer, 2006.

[231] Abdul Jabbar, Erik Perrins, and James P.G. Sterbenz. A cross-layered protocol architecture for highly-dynamic multihop airborne telemetry networks. In *Proceedings of the International Telemetering Conference (ITC)*, San Diego, CA, October 2008.

[232] Abdul Jabbar, Justin P. Rohrer, Victor S. Frost, and James P.G. Sterbenz. Survivable millimeter-wave mesh networks. *Computer Communications*, 34(16):1942–1955, October 2011.

[233] Abdul Jabbar, Justin P. Rohrer, Andrew Oberthaler, Egemen K. Çetinkaya, Victor Frost, and James P.G. Sterbenz. Performance comparison of weather disruption-tolerant cross-layer routing algorithms. In *Proceedings of the 28th IEEE Conference on Computer Communications (INFOCOM)*, pages 1143–1151, Rio de Janeiro, April 2009.

[234] V. Jacobson. Compressing TCP/IP Headers for low-speed serial links. RFC 1144 (Proposed Standard), February 1990.

[235] V. Jacobson, R. Braden, and D. Borman. TCP Extensions for high performance. RFC 1323 (Proposed Standard), May 1992.

[236] V. Jacobson and R.T. Braden. TCP extensions for long-delay paths. RFC 1072, October 1988. Obsoleted by RFCs 1323, 2018.

[237] A. Jahn, M. Holzbock, J. Muller, R. Kebel, M. de Sanctis, A. Rogoyski, E. Trachtman, O. Franzrahe, M. Werner, and Fun Hu. Evolution of aeronautical communications for personal and multimedia services. *IEEE Communications Magazine*, 41(7):36–43, July 2003.

[238] R. Jain, F. Templin, and Kwong-Sang Yin. Analysis of l-band digital aeronautical communication systems: L-DACS1 and L-DACS2. In *Aerospace Conference, 2011 IEEE*, pages 1–10, March 2011.

[239] Japan UAV Association. Safety standards for commercial use unmanned aerial vehicle, for fixed wing aircraft, used in uninhabited areas, 2007. JUAVwebsite.pdf.

[240] Japan UAV Association. Safety Standards for Commercial Use Unmanned Aerial Vehicle, for Fixed Wing Small Aerial Vehicle, Used in Uninhabited Areas [for electrically powered], 2011. JUAVwebsite.pdf.

[241] A. Jimenez-Pacheco, D. Bouhired, Y. Gasser, J. Zufferey, D. Floreano, and B. Rimoldi. Implementation of a wireless mesh network of ultra light MAVs with dynamic routing. In *IEEE Globecom Wkshps*, pages 1591–1596, December 2012.

[242] Liu Jiong, Cao Zhigang, and K.M. Junaid. TP-satellite: A new transport protocol for satellite IP networks. *IEEE Transactions on Aerospace and Electronic Systems*, 45(2):502–515, April 2009.

[243] Charles J. John, Carmen M. Pancerella, Lynn Yang, Karim Mahrous, Gabriel R. Elkin, Adam S. Norige, Keith Holtermann, and Jalal Mapar. New technologies and processes for the homeland security exercise and evaluation program toolkit. In *IEEE International Conference on Technologies for Homeland Security (HST)*, 2011, pages 446–450. IEEE, 2011.

[244] D. Johnson, Y. Hu, and D. Maltz. The dynamic source routing protocol (DSR) for mobile ad hoc networks for IPv4. RFC 4728 (Experimental), February 2007.

[245] Eric N. Johnson and Daniel P. Schrage. System integration and operation of a research unmanned aerial vehicle. *AIAA Journal of Aerospace Computing, Information, and Communication*, 1(1), January 2004.

[246] Joint Authorities for Rulemaking of Unmanned Systems WG-3 Airworthiness. Certification specification for light unmanned rotorcraft systems, October 2013. www.nlr.nl/downloads/jarus_cs-lurs.pdf.

[247] Joint Authorities for Rulemaking of Unmanned Systems WG-5. RPAS C2 link. Required communication performance (C2 link RCP) concept, 2014. jarus-rpas.org/index.php/deliverable/category/12-external-consultation-on-jarus-c2-rcp.

[248] Joint Aviation Authority, Administrative and Guidance Material. Airworthiness and operational approval for precision rnav operations in designated European airspace, 2000. www.eurocontrol.int/sites/default/files/content/documents/navigation/tgl-10-jaa.pdf.

[249] George Pierce Jones, Leonard G. Pearlstine, and H. Franklin Percival. An assessment of small unmanned aerial vehicles for wildlife research. *Wildlife Society Bulletin*, 34(3):750–758, 2006.

[250] Tore B. Kallevig. Current status of work in the field of UAS related standards in Europe, 2012. www.eurocae.net/wgs/active/?wg=WG-73.

[251] A. Kandhalu, A. Rowe, R. Rajkumar, C. Huang, and C.-C. Yeh. Real-time video surveillance over IEEE 802.11 mesh networks. In *Proc. IEEE Real-Time and Embedded Tech. and Appl. Symp.*, pages 205–214, 2009.

[252] C. Karlof and D. Wagner. Secure routing in wireless sensor networks: attacks and countermeasures. In *Proceedings of the First IEEE International Workshop on Sensor Network Protocols and Applications*, pages 113–127, 2003.

[253] Brad Karp and H.T. Kung. GPSR: Greedy perimeter stateless routing for wireless networks. In *Proceedings of the 6th Annual International Conference on Mobile Computing and Networking*, MobiCom 2000, pages 243–254, New York, NY, USA, 2000. ACM.

[254] Matthew Keennon, Karl Klingebiel, Henry Won, and Alexander Andriukov. Development of the nano hummingbird: A tailless flapping wing micro air vehicle. In *50th AIAA Aerospace Sciences Meeting Including the New Horizons Forum and Aerospace Exposition, Nashville, TN.* January, pages 9–12, 2012.

[255] F.P. Kelly, A.K. Maulloo, and D.K.H. Tan. Rate control for communication networks: Shadow prices, proportional fairness and stability. *The Journal of the Operational Research Society*, 49(3):237–252, March 1998.

[256] Andrew J. Kerns, Daniel P. Shepard, Jahshan A. Bhatti, and Todd E. Humphreys. Unmanned aircraft capture and control via GPS spoofing. *Journal of Field Robotics*, 31(4):617–636, 2014.

[257] M.J. Khabbaz, C.M. Assi, and W.F. Fawaz. Disruption-tolerant networking: A comprehensive survey on recent developments and persisting challenges. *IEEE Communications Surveys Tutorials*, 14(2):607–640, Second 2012.

[258] Joseph King. For or against. *Soaring Magazine*, page 47, September–October 2015.

[259] Young-Bae Ko and Nitin H. Vaidya. Location-aided routing (LAR) in mobile ad hoc networks. *Wirel. Netw.*, 6(4):307–321, July 2000.

[260] Paul C. Kocher. Timing attacks on implementations of Diffie-Hellman, RSA, DSS, and other systems. In Neal Koblitz, editor, *Advances in Cryptology CRYPTO 96*, volume 1109 of *Lecture Notes in Computer Science*, pages 104–113. Springer, Berlin, Heidelberg, August 1996.

[261] Paul C. Kocher, Joshua Jaffe, and Benjamin Jun. Differential power analysis. In *Proceedings of the 19th Annual International Cryptology Conference on Advances in Cryptology*, CRYPTO 1999, pages 388–397, London, UK, 1999. Springer-Verlag.

[262] Matthew R. Koerner. Drones and the fourth amendment: Redefining expectations of privacy. *Duke LJ*, 64:1129, 2014.

[263] Oliver Kömmerling and Markus G. Kuhn. Design principles for tamper-resistant smartcard processors. In *Proceedings of the USENIX Workshop on Smartcard Technology*, WOST'99, pages 9–20, Berkeley, CA, USA, 1999. USENIX Association.

[264] Jiejun Kong, Haiyun Luo, Kaixin Xu, Daniel Lihui Gu, Mario Gerla, and Songwu Lu. Adaptive security for multi-layer ad-hoc networks. In *Special Issue of Wireless Communications and Mobile Computing*, pages 533–547. Wiley Interscience Press, 2002.

[265] M.A. Kovacina, D. Palmer, Guang Yang, and R. Vaidyanathan. Multi-agent control algorithms for chemical cloud detection and mapping using unmanned air vehicles. In *Proc. IEEE/RSJ Intl. Conf. on Intelligent Robots and Systems*, volume 3, pages 2782–2788, October 2002.

[266] Olga Kozar. Towards better group work: Seeing the difference between cooperation and collaboration. *English Teaching Forum*, 48(2):16–23, 2010.

[267] Rajesh Krishnan, James P.G. Sterbenz, Wesley M. Eddy, Craig Partridge, and Mark Allman. Explicit transport error notification (ETEN) for error-prone wireless and satellite networks. *Computer Networks*, 46(3):343–362, 2004.

[268] Sumeet Kumar and Aneesh Aggarwal. Reducing resource redundancy for concurrent error detection techniques in high performance microprocessors. In *Twelfth International Symposium on High-Performance Computer Architecture*, pages 212–221, February 2006.

[269] Greg Kuperman, Lenny Veytser, Bow-Nan Cheng, Scott Moore, and Aradhana Narula-Tam. A comparison of OLSR and OSPF-MDR for large-scale airborne mobile ad-hoc networks. In *Proceedings of the third ACM Workshop on Airborne Networks and Communications*, pages 17–22. ACM, 2014.

[270] Rémi Laplace. Applications and DTN services for cooperative UAVs fleet. Thesis, Université Sciences et Technologies – Bordeaux I, December 2012.

[271] Andreas Larsson. Report on the state of the art of security in sensor networks, 2011.

[272] LDACS2 Project Team. LDACS2 Specification. www.eurocontrol.int/sites/default/files/article//content/documents/communications/18062009-ldacs2-design-d3-v1.2.pdf, 2009.

[273] Kevin C. Lee, Uichin Lee, and Mario Gerla. Survey of routing protocols in vehicular ad hoc networks. In Mohamed Watfa, editor, *Advances in Vehicular Ad-Hoc Networks: Developments and Challenges*, Premier Reference Source, Chapter 8. Information Science Reference, Hershey, PA, USA, 2010.

[274] Valgene W. Lehmann and R.G. Mauermann. Status of Attwater's prairie chicken. *The Journal of Wildlife Management*, pages 713–725, 1963.

[275] Nancy G. Leveson. *Safeware: System Safety and Computers*. ACM Press, New York, NY, USA, 1995.

[276] Fan Li and Yu Wang. Routing in vehicular ad hoc networks: A survey. *IEEE Vehicular Technology Magazine*, 2(2):12–22, June 2007.

[277] Fan Li and Yu Wang. Routing in vehicular ad hoc networks: A survey. *Vehicular Technology Magazine, IEEE*, 2(2):12–22, 2007.

[278] Wen-Hwa Liao, Jang-Ping Sheu, and Yu-Chee Tseng. GRID: A fully location-aware routing protocol for mobile ad hoc networks. *Telecommunication Systems*, 18(1–3):37–60, 2001.

[279] Yun-Wei Lin, Yuh-Shyan Chen, and Sing-Ling Lee. Routing protocols in vehicular ad hoc networks: A survey and future perspectives. *Journal of Information Science and Engineering*, 26(3):913–932, 2010.

[280] Q. Lindsey, D. Mellinger, and V. Kumar. Construction of cubic structures with quadrotor teams. In *Proc. of Robotics: Science and Systems*, Los Angeles, CA, USA, June 2011.

[281] Haitao Liu, Baoxian Zhang, H. Mouftah, Xiaojun Shen, and Jian Ma. Opportunistic routing for wireless ad hoc and sensor networks: Present and future directions. *IEEE Communications Magazine*, 47(12):103–109, December 2009.

[282] The Loon project. www.google.com/loon/.

[283] Luftfahrt-Bundesamt. Luftverkehrs-Ordnung vom 10. August 1963 (BGBl. I S. 652), die zuletzt durch Artikel 3 des Gesetzes vom 8. Mai 2012 (BGBl. I S. 1032) gea?ndert worden ist, May 2012. www.gesetze-im-internet.de/bundesrecht/luftvo/gesamt.pdf.

[284] J.P. Macker, ed. Simplified Multicast Forwarding. *RFC 6621*, May 2012.

[285] J.P. Macker, J. Dean, and W. Chao. Simplified multicast forwarding in mobile ad hoc networks. *IEEE Military Communications Conference (MILCOM)*, Vol. 2 (2004), pp. 744–750.

[286] Joseph Macker, Ian Downard, Justin Dean, and Brian Adamson. Evaluation of distributed cover set algorithms in mobile ad hoc network for simplified multicast forwarding. *ACM SIGMOBILE Mobile Computing and Communications Review*, 11(3):1–11, 2007.

[287] S.M. Magrabi and P.W. Gibbens. Decentralized fault detection and diagnosis in navigation systems for unmanned aerial vehicles. In *IEEE Position Locations and Navigation Symposium*, March 2000.

[288] David A. Maltz and Pravin Bhagwat. TCP splice for application layer proxy performance. *Journal of High Speed Networks*, 8(3):225–240, 1999.

[289] Md. Mamun-Or-Rashid, Muhammad Mahbub Alam, Md. Abdur Razzaque, and Choong Seon Hong. Congestion avoidance and fair event detection in wireless sensor network. *IEICE Transactions on Communications*, E90-B(12):3362–3372, 2007.

[290] Jalal Mapar, Keith Holtermann, Justin Legary, Karim Mahrous, Katherine Guzman, Zach Heath, Charles J. John, Steven A. Mier, Steffen Mueller, Carmen M. Pancerella, *et al.* The role of integrated modeling and simulation in disaster preparedness and emergency preparedness and response: the summit platform. In *2012 IEEE Conference on Technologies for Homeland Security (HST)*, pages 117–122. IEEE, 2012.

[291] Steven A. Martin and Fritz L. Knopf. Aerial survey of greater prairie chicken leks. *Wildlife Society Bulletin*, pages 219–221, 1981.

[292] M. Mathis, J. Mahdavi, S. Floyd, and A. Romanow. TCP selective acknowledgment options. RFC 2018 (Proposed Standard), October 1996.

[293] M. Mauve, A. Widmer, and H. Hartenstein. A survey on position-based routing in mobile ad hoc networks. *IEEE Network*, 15(6):30–39, 2001.

[294] David B. Mayo and J. Gordon Leishman. Comparison of the hovering efficiency of rotating wing and flapping wing micro air vehicles. *Journal of the American Helicopter Society*, 55(2):25001–25001, 2010.

[295] I. Maza, F. Caballero, J. Capitan, J.R. Martinez-de Dios, and A. Ollero. Firemen monitoring with multiple UAVs for search and rescue missions. In *Proc. IEEE Intl. Wksp. Safety Security and Rescue Robotics (SSRR)*, pages 1–6, July 2010.

[296] A. J. McAuley. Reliable broadband communication using a burst erasure correcting code. *SIGCOMM Comput. Commun. Rev.*, 20(4):297–306, 1990.

[297] Thomas S. Messerges. Using second-order power analysis to attack DPA resistant software. In Çetin K. Koç and Christof Paar, editors, *Cryptographic Hardware and Embedded Systems — CHES 2000*, volume 1965 of *Lecture Notes in Computer Science*, pages 238–251. Springer, Berlin, Heidelberg, August 2000.

[298] Stuart D. Milner, Sohil Thakkar, Karthikeyan Chandrashekar, and Wei-Lun Chen. Performance and scalability of mobile wireless base-station-oriented networks. *ACM SIGMOBILE Mobile Computing and Communications Review*, 7(2):69–79, 2003.

[299] J.C. Mogul and S.E. Deering. Path MTU discovery. RFC 1191 (Draft Standard), November 1990.

[300] Amir Moradi, Markus Kasper, and Christof Paar. Black-box side-channel attacks highlight the importance of countermeasures. In Orr Dunkelman, editor, *Topics in Cryptology – CT-RSA 2012*, volume 7178 of *Lecture Notes in Computer Science*, pages 1–18. Springer, Berlin, Heidelberg, February 2012.

[301] F. Morbidi, R.A. Freeman, and K.M. Lynch. Estimation and control of UAV swarms for distributed monitoring tasks. In *American Control Conf. (ACC)*, pages 1069–1075, June 2011.

[302] S. Morgenthaler, T. Braun, Zhongliang Zhao, T. Staub, and M. Anwander. UAVNet: A mobile wireless mesh network using unmanned aerial vehicles. In *Proc. IEEE Globecom Wkshps*, pages 1603–1608, December 2012.

[303] Simon Morgenthaler, Torsten Braun, Zhongliang Zhao, Thomas Staub, and Markus Anwander. Uavnet: A mobile wireless mesh network using unmanned aerial vehicles. In *Globecom Workshops (GC Wkshps), 2012 IEEE*, pages 1603–1608. IEEE, 2012.

[304] Wojciech Mostowski and Erik Poll. Malicious code on Java card smartcards: Attacks and countermeasures. In Gilles Grimaud and François-Xavier Standaert, editors, *Smart Card Research and Advanced Applications*, volume 5189 of *Lecture Notes in Computer Science*, pages 1–16. Springer, Berlin, Heidelberg, September 2008.

[305] J. Moy. OSPF Version 2. RFC 2328 (Standard), April 1998. Updated by RFC 5709.

[306] James Alexander Muir. Techniques of side channel cryptanalysis. 2001. Master's thesis, Master of Mathematics in Combinatorics and Optimization, University of Waterloo, Ontario, Canada.

[307] Chase C. Murray and Amanda G. Chu. The flying sidekick traveling salesman problem: Optimization of drone-assisted parcel delivery. *Transportation Research Part C: Emerging Technologies*, 54:86–109, 2015.

[308] Mostafa R.A. Nabawy and William J. Crowther. On the quasi-steady aerodynamics of normal hovering flight part ii: Model implementation and evaluation. *Journal of The Royal Society Interface*, 11(94):20131197, 2014.

[309] Multi-Generator (MGEN), NRL software. http://cs.itd.nrl.navy.mil/work/mgen/index.php.

[310] Multi-Generator (MGEN). NRL software, http://cs.itd.nrl.navy.mil/work/mgen/index.php.

[311] M. Naderan, M. Dehghan, and H. Pedram. Mobile object tracking techniques in wireless sensor networks. In *International Conference on Ultra Modern Telecommunications Workshops, ICUMT '09*, pages 1–8, October 2009.

[312] N. Nakka, Z. Kalbarczyk, R.K. Iyer, and J. Xu. An architectural framework for providing reliability and security support. In *IEEE International Conference on Dependable Systems and Networks*, pages 585–594, July 2004.

[313] Hemanth Narra, Egemen K. Çetinkaya, and James P.G. Sterbenz. Performance analysis of AeroRP with ground station advertisements. In *Proceedings of the ACM MobiHoc Workshop on Airborne Networks and Communications*, pages 43–47, Hilton Head Island, SC, June 2012.

[314] NASA Safe Autonomous Systems Operations (SASO) project www.aeronautics.nasa.gov/aosp/sasop/.

[315] NASA Unmanned Aircraft System (UAS) Traffic Management (UTM) http://utm.arc.nasa.gov/.

[316] Next Generation Air Transportation System (NextGen) www.faa.gov/nextgen/.

[317] Truc Anh N. Nguyen, Justin P. Rohrer, and James P.G. Sterbenz. ResTP–A Transport Protocol for FI Resilience. In *Proceedings of the 10th International Conference on Future Internet Technologies (CFI)*, June 2015.

[318] S. Nitinawarat, G.K. Atia, and V.V. Veeravalli. Efficient target tracking using mobile sensors. In *4th IEEE International Workshop on Computational Advances in Multi-Sensor Adaptive Processing (CAMSAP)*, pages 405–408, December 2011.

[319] Oceus. Above the clouds to support public safety. http://oceusnetworks.com/, 2013.

[320] Office of Naval Research. LOCUST: Autonomous, swarming UAVs fly into the future, April 2015. www.onr.navy.mil/Media-Center/Press-Releases/2015/LOCUST-low-cost-UAV-swarm-ONR.aspx. (Retrieved May 2016)

[321] US Department of Transportation. Unmanned aircraft system (UAS) service demand 2015–2035: Literature review and projections of future usage (dot-vntsc-dod-13-01). fas.org/irp/program/collect/service.pdf, 2013.

[322] R. Ogier and P. Spagnolo. Mobile Ad Hoc Network (MANET): Extension of OSPF using connected dominating set (CDS) flooding. RFC 5614 (Experimental), August 2009.

[323] John M.H. Olmsted and Carl G. Townsend. On the sum of two periodic functions. *Two-Year College Mathematics Journal*, pages 33–38, 1972.

[324] Orion product by Aurora. www.aurora.aero/orion/.

[325] A. Ortiz, Fault detection and diagnosis in real-time, safety critical embedded systems: A methodology for uninhabited aerial vehicles. M.S. Thesis, Department of Aerospace Engineering, University of Illinois at Urbana-Champaign, 2007.

[326] A. Ortiz and N. Neogi, Optic flow: A computer vision approach to object avoidance on UAVs. *IEEE Digital Avionics Systems Conference*, Portland, Oregon, October 16–19, 2006.

[327] A. Ortiz and N. Neogi, A dynamic threshold approach to fault detection in uninhabited aerial vehicles. *AIAA Guidance, Navigation, and Control Conference and Exhibit*, Honolulu, HI, August 2008.

[328] T. Ostermann, J. Holsten, Y. Dobrev, and D. Moormann. Control concept of a tiltwing UAV during low speed manoeuvring. In *Proceedings of the 28th International Congress of the Aeronautical Sciences: ICAS*, Brisbane, Australia, 2012.

[329] J. Ott and D. Kutscher. Drive-thru internet: IEEE 802.11b for "automobile" users. In *23rd Annual Joint Conference of the IEEE Computer and Communications Societies (INFOCOM)*, volume 1, page 373, March 2004.

[330] Cüneyt Özveren, Robert Simcoe, and George Varghese. Reliable and efficient hop-by-hop flow control. In *SIGCOMM '94: Proceedings of the Conference on Communications Architectures, Protocols, and Applications*, pages 89–100, New York, NY, USA, 1994. ACM.

[331] R.C. Palat, A. Annamalai, and J.H. Reed. Cooperative relaying for ad hoc ground networks using swarms. In *Proc. IEEE Milit. Comm. Conf. (MILCOM'05)*, volume 3, pages 1588–1594, October 2005.

[332] Richard B. Parker. Definition of privacy. *Rutgers L. Rev.*, 27:275, 1973.

[333] Craig Partridge. *Gigabit Networking*. Addison-Wesley Longman Publishing Co., Boston, MA, USA, 1994.

[334] Colin Percival. Cache missing for fun and profit. In *Proceedings of BSDCan 2005*, 2005.

[335] C. Perkins, E. Belding-Royer, and S. Das. Ad hoc on-demand distance vector (AODV) routing. RFC 3561 (Experimental), July 2003.

[336] Charles E. Perkins and Elizabeth M. Royer, Ad-hoc on-demand distance vector routing. In *Proceedings of the 2nd IEEE Workshop on Mobile Computing Systems and Applications*, 1997.

[337] Charles E. Perkins and Pravin Bhagwat. Highly dynamic destination-sequenced distance-vector routing (DSDV) for mobile computers. In *ACM SIGCOMM*, pages 234–244, London, 1994.

[338] Charles E. Perkins *et al. Ad hoc Networking*, Volume 1. Addison-Wesley, Reading, 2001.

[339] Kevin Peters, Abdul Jabbar, Egemen K. Çetinkaya, and James P.G. Sterbenz. A geographical routing protocol for highly-dynamic aeronautical networks. In *Proceedings of the IEEE Wireless Communications and Networking Conference (WCNC)*, pages 492–497, Cancun, Mexico, March 2011.

[340] J. Postel. User datagram protocol. RFC 768 (Standard), August 1980.

[341] J. Postel. Transmission control protocol. RFC 793 (Standard), September 1981. Updated by RFCs 1122, 3168.

[342] Power beaming flight demonstration. www.nasa.gov/centers/dryden/multimedia/imagegallery/Power-Beaming/EC02-0232-13.html. (Retrieved May 2016)

[343] Pham Xuan Quang and Richard B Lanctot. A line transect model for aerial surveys. *Biometrics*, pages 1089–1102, 1991.

[344] Jean-Jacques Quisquater and David Samyde. ElectroMagnetic Analysis (EMA): Measures and counter-measures for smart cards. In Isabelle Attali and Thomas Jensen, editors, *Smart Card Programming and Security*, volume 2140 of *Lecture Notes in Computer Science*, pages 200–210. Springer, Berlin, Heidelberg, September 2001.

[345] K. Ramakrishnan, S. Floyd, and D. Black. The addition of explicit congestion notification (ECN) to IP. RFC 3168 (Proposed Standard), September 2001.

[346] K. K. Ramakrishnan and Raj Jain. A binary feedback scheme for congestion avoidance in computer networks. *ACM Transactions on Computer Systems*, 8(2):158–181, 1990.

[347] S. Ramanathan and Martha Steenstrup. A survey of routing techniques for mobile communications networks. *Mob. Netw. Appl.*, 1(2):89–104, October 1996.

[348] V. Ramasubramanian, Z. Haas, and E. Sirer, SHARP: A hybrid adaptive routing protocol for mobile ad hoc networks. *Proceedings of the 4th ACM International Symposium on Mobile Ad Hoc Networking and Computing*, 2003.

[349] Christian Rechberger and Elisabeth Oswald. Practical template attacks. In ChaeHoon Lim and Moti Yung, editors, *Information Security Applications*, volume 3325 of *Lecture Notes in Computer Science*, pages 440–456. Springer, Berlin, Heidelberg, August 2004.

[350] Y. Rekhter, T. Li, and S. Hares. A border gateway protocol 4 (BGP-4). RFC 4271 (Draft Standard), January 2006.

[351] L. Reynaud and T. Rasheed. Deployable aerial communication networks: Challenges for futuristic applications. In *Proc. ACM Symp. Perf. Eval. Wireless Ad Hoc, Sensor, and Ubiquitous Networks*, pages 9–16, 2012.

[352] L. Reynaud, T. Rasheed, and S. Kandeepan. An integrated aerial telecommunications network that supports emergency traffic. In *Proc. Intl. Symp. Wireless Personal Multimedia Commun. (WPMC)*, pages 1–5, October 2011.

[353] K. Ro, J.-S. Oh, and L. Dong. Lessons learned: Application of small UAV for urban highway traffic monitoring. *American Inst. of Aeronautics and Astronautics*, pages 1–19, 2007.

[354] W. Robinson and A. Lauf. Aerial MANETs: Developing a resilient and efficient platform for search and rescue applications. *Journal of Commun.*, 8(4):216–224, 2013.

[355] Justin P. Rohrer, Egemen K. Çetinkaya, Hemmanth Narra, Dan Broyles, Kevin Peters, and James P.G. Sterbenz. AeroRP performance in highly-dynamic airborne networks using 3D Gauss–Markov mobility model. In *Proceedings of the IEEE Military Communications Conference (MILCOM)*, pages 834–841, Baltimore, MD, November 2011.

[356] Justin P. Rohrer, Abdul Jabbar, Egemen K. Çetinkaya, Erik Perrins, and James P.G. Sterbenz. Highly-dynamic cross-layered aeronautical network architecture. *IEEE Transactions on Aerospace and Electronic Systems*, 47(4):2742–2765, October 2011.

[357] Justin P. Rohrer, Abdul Jabbar, Egemen K. Çetinkaya, and James P.G. Sterbenz. Airborne telemetry networks: challenges and solutions in the ANTP Suite. In *Proceedings of the IEEE Military Communications Conference (MILCOM)*, pages 74–79, San Jose, CA, November 2010.

[358] Justin P. Rohrer, Abdul Jabbar, Erik Perrins, and James P.G. Sterbenz. Cross-layer architectural framework for highly-mobile multihop airborne telemetry networks. In *Proceedings of the IEEE Military Communications Conference (MILCOM)*, pages 1–9, San Diego, CA, November 2008.

[359] Justin P. Rohrer, Kamakshi Sirisha Pathapati, Truc Anh N. Nguyen, and James P.G. Sterbenz. Opportunistic transport for disrupted airborne networks. In *Proceedings of the IEEE Military Communications Conference (MILCOM)*, pages 737–745, Orlando, FL, November 2012.

[360] Justin P. Rohrer and James P.G. Sterbenz. Performance and disruption tolerance of transport protocols for airborne telemetry networks. In *Proceedings of the International Telemetering Conference (ITC)*, Las Vegas, NV, October 2009.

[361] Allison Ryan, Xiao Xiao, Shivakumar Rathinam, John Tisdale, Marco Zennaro, Derek Caveney, Raja Sengupta, and J. Karl Hedrick. A modular software infrastructure for distributed control of collaborating UAVs. *AIAA Conference on Guidance, Navigation, and Control*, 2006.

[362] A. Ryan, M. Zennaro, A. Howell, R. Sengupta, and J.K. Hedrick. An overview of emerging results in cooperative UAV control. In *Proc. IEEE Conf. Decision and Control*, volume 1, pages 602 –607, December 2004.

[363] Safety and Airspace Regulation Group, United Kingdom. CAP 393: Air Navigation 2009: The Order and the Regulations, 2015. www.caa.co.uk/docs/33/CAP%20393%20Fourth %20edition%20Amendment%201%20April%202015.pdf.

[364] Safety and Airspace Regulation Group, United Kingdom. CAP 722: Unmanned aircraft system operations in UK airspace guidance, 2015. www.caa.co.uk/docs/33/CAP%20722 %20Sixth%20Edition%20March%202015.pdf.

[365] Y.E. Sagduyu and A. Ephremides. On joint MAC and network coding in wireless ad hoc networks. *IEEE Transactions on Information Theory*, 53(10):3697–3713, October 2007.

[366] E. Sakhaee and A. Jamalipour. The global in-flight internet. *IEEE Journal on Selected Areas in Communications*, 24(9):1748–1757, September 2006.

[367] E. Sakhaee, A. Jamalipour, and Nei Kato. Aeronautical ad hoc networks. In *IEEE Wireless Communications and Networking Conference, 2006. WCNC 2006*, volume 1, pages 246–251, April 2006.

[368] J. Sanchez, P. Ruiz, and R. Marin-Perez. Beacon-less geographic routing made practical: Challenges, design guidelines, and protocols. *IEEE Communications Magazine*, 47(8):85–91, August 2009.

[369] J.A. Sanchez, R. Marin-Perez, and P.M. Ruiz. BOSS: Beacon-less on demand strategy for geographic routing in wireless sensor networks. In *IEEE International Conference on Mobile Adhoc and Sensor Systems (MASS)*, pages 1–10, October 2007.

[370] Francesc Sardà-Palomera, Gerard Bota, Carlos Vinolo, Oriol Pallares, Victor Sazatornil, Lluis Brotons, Spartacus Gomariz, and Francesc Sardà. Fine-scale bird monitoring from light unmanned aircraft systems. *Ibis*, 154(1):177–183, 2012.

[371] Akira Sato. Civil UAV applications in Japan and related safety and certification, 2003. http://Yamahawebsite.pdf.

[372] SC-228. Overview, 2013. http://RTCAwebsite.pdf.

[373] SC-228. Command and control link whitepaper, 2014. www.rtca.org/content.asp?contentid=178.

[374] SC-228. Detect and avoid whitepaper, 2014. www.rtca.org/content.asp?contentid=178.

[375] M. Schnell, U. Epple, D. Shutin, and N. Schneckenburger. LDACS: future aeronautical communications for air-traffic management. *Communications Magazine, IEEE*, 52(5):104–110, May 2014.

[376] Michael Schnell, Ulrich Epple, Dmitriy Shutin, and Nicolas Schneckenburger. LDACS: Future aeronautical communications for air-traffic management. *Communications Magazine, IEEE*, 52(5):104–110, 2014.

[377] S. Schoenung, S. Dunagan, D. Sullivan, J. Brass, and S. Wegener. UAV over-the-horizon disaster management demonstration projects. Technical report, University of Southampton, 2000.

[378] Michael A. Schroeder, Kenneth M. Giesen, and Clait E. Braun. Use of helicopters for estimating numbers of greater and lesser prairie-chicken leks in eastern Colorado. *Wildlife Society Bulletin*, pages 106–113, 1992.

[379] H. Schulzrinne, S. Casner, R. Frederick, and V. Jacobson. RTP: A transport protocol for real-time applications. RFC 3550 (Standard), July 2003. Updated by RFC 5506.

[380] K. Scott and S. Burleigh. Bundle protocol specification. RFC 5050 (Experimental), November 2007.

[381] I. Sher, D. Levinzon-Sher, and E. Sher. Miniaturization limitations of hcci internal combustion engines. *Applied Thermal Engineering*, 29(2):400–411, 2009.

[382] Sergey Shkarayev, Gunjan Maniar, and Alexander V. Shekhovtsov. Experimental and computational modeling of the kinematics and aerodynamics of flapping wing. *Journal of Aircraft*, 50(6):1734–1747, 2013.

[383] Sergei Skorobogatov. Tamper resistance and hardware security. www.cl.cam.ac.uk/~sps32/PartII_030214.pdf, February 2014. [Accessed 3 May 2014.]

[384] Sergei P. Skorobogatov and Ross J. Anderson. Optical fault induction attacks. In Burton S. Kaliski, Çetin K. Koç, and Christof Paar, editors, *Cryptographic Hardware and Embedded Systems – CHES 2002*, volume 2523 of *Lecture Notes in Computer Science*, pages 2–12. Springer, Berlin, Heidelberg, 2003.

[385] Therese Skrzypietz. Unmanned aircraft systems for civilian missions. Technical Report BIGS Policy Paper No. 1, Brandenburg Institute for Society and Security, February 2012.

[386] Ulrich Epple, Snjezana Gligorevic, and Michael Schnell. The ldacs1 physical layer design. In Dr. Simon Plass, editor, *Future Aeronautical Communications*. InTech, 2011.

[387] South African Civil Aviation Authority. Civil Aviation Act, 2009 (Act No. 13, of 2009), 2009. www.acts.co.za/civil-aviation-act-2009/.

[388] South African Civil Aviation Authority. Frequently asked questions for unmanned aircraft systems in South Africa, 2014. http://SouthAfricaCAwebsite.pdf.

[389] Raphael Spreitzer and Benoit Gérard. Towards more practical time-driven cache attacks. In David Naccache and Damien Sauveron, editors, *Information Security Theory and Practice. Securing the Internet of Things*, volume 8501 of *Lecture Notes in Computer Science*, pages 24–39, July 2014.

[390] N.K. Sreelaja and G.A. Vijayalakshmi Pai. Swarm intelligence based key generation for stream cipher. *Security and Communication Networks*, 4(2):181–194, 2011.

[391] Mark V. Stalmaster and James L. Kaiser. Flushing responses of wintering bald eagles to military activity. *The Journal of Wildlife Management*, pages 1307–1313, 1997.

[392] Martha Steenstrup. *Routing in Communication Networks*. Prentice Hall, 1995.

[393] M.E. Steenstrup, Routing under uncertainty: A comparative study, *IEEE WCNC 2000*, Chicago, September 2000.

[394] R. Steffen and W. Forstner. On visual real time mapping for unmanned aerial vehicles. In *Proc. Congress of the Intl. Soc. Photogrammetry and Remote Sensing (ISPRS)*, pages 57–62, 2008.

[395] James P.G. Sterbenz, David Hutchison, Egemen K. Çetinkaya, Abdul Jabbar, Justin P. Rohrer, Marcus Schöller, and Paul Smith. Resilience and survivability in communication networks: Strategies, principles, and survey of disciplines. *Computer Networks*, 54(8):1245–1265, 2010.

[396] James P.G. Sterbenz, David Hutchison, Egemen K. Çetinkaya, Abdul Jabbar, Justin P. Rohrer, Marcus Schöller, and Paul Smith. Redundancy, diversity, and connectivity to achieve multilevel network resilience, survivability, and disruption tolerance (invited paper). *Telecommunication Systems*, 56(1):17–31, 2014.

[397] James P.G. Sterbenz and Joseph D. Touch. *High-Speed Networking: A Systematic Approach to High-Bandwidth Low-Latency Communication*. Wiley, 1st edition, May 2001.

[398] James P.G. Sterbenz. Drones in the smart city and IoT: Protocols, resilience, benefits, and risks. In *Proceedings of the 2nd Workshop on Micro Aerial Vehicle Networks, Systems, and Applications for Civilian Use*.

[399] James P.G. Sterbenz, Rajesh Krishnan, Regina Rosales Hain, Alden W. Jackson, David Levin, Ram Ramanathan, and John Zao. Survivable mobile wireless networks: Issues,

challenges, and research directions. In *Proceedings of the 1st ACM Workshop on Wireless Security (WiSe)*, pages 31–40, Atlanta, GA, September 2002.

[400] James P.G. Sterbenz and Gurudatta M. Parulkar. AXON: Application-oriented lightweight transport protocol design. In *Proceedings of the Tenth International Conference on Computer Communication (ICCC)*, pages 379–387, New Delhi, India, November 1990. Narosa Publishing House.

[401] W. Stevens. TCP Slow Start, congestion avoidance, fast retransmit, and fast recovery algorithms. RFC 2001 (Proposed Standard), January 1997. Obsoleted by RFC 2581.

[402] M. Strohmeier, V. Lenders, and I. Martinovic. On the security of the automatic dependent surveillance-broadcast protocol. *Communications Surveys Tutorials, IEEE*, 17(2):1066–1087, Second quarter 2015.

[403] Martin Strohmeier, Matthias Schafer, Vincent Lenders, and Ivan Martinovic. Realities and challenges of nextgen air traffic management: The case of ads-b. *Communications Magazine, IEEE*, 52(5):111–118, 2014.

[404] P.B. Sujit and D. Ghose. Search using multiple UAVs with flight time constraints. *IEEE Trans. Aerospace and Electronic Systems*, 40(2):491–509, April 2004.

[405] Sun and Lisa Grow. Smart growth in dumb places: Sustainability, disaster, and the future of the American city. In *BYU Law Review (symposium), Forthcoming*, 2011.

[406] M. Suresh and D. Ghose. Role of information and communication in redefining unmanned aerial vehicle autonomous control levels. In *Proceedings of the Institution of Mechanical Engineers, Part G: Journal of Aerospace Engineering*, 224(2):171–197, 2010.

[407] Fabrice Tchakountio and Ram Ramanathan. Tracking Highly Mobile Endpoints. In *Proceedings of the 4th ACM International Workshop on Wireless Mobile Multimedia*, WOWMOM '01, pages 83–94, New York, NY, USA, 2001. ACM.

[408] Fabrice Tchakountio and Ram Ramanathan. Anticipatory routing for highly mobile endpoints. In *Proceedings of the 6th IEEE Workshop on Mobile Computing Systems and Applications (WMCSA)*, pages 94–101, Washington, DC, 2004.

[409] Tyson Thedinger, Abdul Jabbar, and James P.G. Sterbenz. Store and haul with repeated controlled flooding. In *Second International IEEE Workshop on Mobile Computing and Networking Technologies (WMCNT)*, pages 728–733, Moscow, October 2010.

[410] Nils Ole Tippenhauer, Christina Pöpper, Kasper Bonne Rasmussen, and Srdjan Capkun. On the requirements for successful GPS spoofing attacks. In *Proc. of the 18th ACM Conference on Computer and Communications Security*, CCS '11, pages 75–86, New York, NY, USA, 2011. ACM.

[411] Abhishek Tiwari, Anurag Ganguli, and Ashwin Sampath. Towards a mission planning toolbox for the airborne network: Optimizing ground coverage under connectivity constraints. In *Aerospace Conference, 2008 IEEE*, pages 1–9. IEEE, 2008.

[412] T. Tomic, K. Schmid, P. Lutz, A. Domel, M. Kassecker, E. Mair, I.L. Grixa, F. Ruess, M. Suppa, and D. Burschka. Toward a fully autonomous UAV: Research platform for indoor and outdoor urban search and rescue. *IEEE Robotics Automation Magazine*, 19(3):46–56, Sept 2012.

[413] Transport Canada. Advisory circular: Guidance material for operating unmanned air vehicle systems under an exemption, November 2014. www.tc.gc.ca/eng/civilaviation/opssvs/ac-600-004-2136.html.

[414] Transport Canada. Canadian civil aviation regulations (SOR/96-433), August 2015. http://laws-lois.justice.gc.ca/eng/regulations/SOR-96-433/.

[415] Tony Udelhoven. Selecting proximity sensors for diverse applications, September 2012. http://www.plantengineering.com/single-article/selecting-proximity-sensors-for-diverse-applications/2f48d26acd6c636a474390c3edaaeb11.html. (Retrieved May 2016)

[416] D. Uhlig, K. Bhamidipati, and N. Neogi, A safety-oriented approach to designing UAVs using COTS technology. *IEEE Digital Avionics Systems Conference*, Portland, Oregon, October 16–19, 2006.

[417] D. Uhlig, N. Kiyavash, and N. Neogi. Fault tolerant triangulation in distributed aircraft networks with automatic dependent surveillance broadcast. *International Journal of Systems, Control and Communications*, C. Hadjicostis, C. Langbort, N. Martins, and S. Yuksel (eds.), June 2009.

[418] United States Government. Title 14 code of federal regulations. http://14CFR.pdf.

[419] United States Government. Title 14 code of federal regulations, part 21 certification procedures for products and parts. www.ecfr.gov/cgi-bin/text-idx?rgn=div5;node=14%3A1.0.1.3.9.

[420] United States Government. Title 14 code of federal regulations, part 23 airworthiness standards: Normal, utility, acrobatic, and commuter category airplanes. www.ecfr.gov/cgi-bin/text-idx?SID=9f1208983b800adf381aec3ec240f673&mc=true&node=pt14.1.23&rgn=div5.

[421] United States Government. Title 14 code of federal regulations, part 39 airworthiness directives. www.ecfr.gov/cgi-bin/text-idx?c=ecfr;rgn=div5;view=text;node=14%3A1.0.1.3.20;idno=14;sid=df8523333854981277310 9cf75d37dfc;cc=ecfr.

[422] United States Government. Title 14 code of federal regulations, subchapter b: Airmen. www.ecfr.gov/cgi-bin/text-idx?gp=&SID=&mc=true&tpl=/ecfrbrowse/Title14/14CIsubchapD.tpl.

[423] United States Government. Title 14 code of federal regulations, subchapter g: Air carriers and operators for compensation or hire: Certification and operations. www.ecfr.gov/cgi-bin/text-idx?tpl=/ecfrbrowse/Title14/14cfr121_main_02.tpl.

[424] Bill Vaglienti. Communications for the piccolo avionics. Cloud Cap Technology Technical Document, September 2006. {ALSO ANDRES 3}

[425] Bill Vaglienti. Piccolo. Control laws for 2.0.x. Cloud Cap Technology Technical Document, March 2007.

[426] Bill Vaglienti and Ross Hoag. A highly integrated UAV avionics system. Cloud Cap Technology Technical Document, April 2003. {ALSO ANDRES 2}

[427] A. Vahdat and D. Becker. Epidemic routing for partially-connected ad hoc networks. Technical Report CS-200006, Duke University, April 2000.

[428] Kimon P. Valavanis and George J. Vachtsevanos. *Handbook of Unmanned Aerial Vehicles*. Springer Publishing Company, Inc., 2014.

[429] Alvaro Valcarce, Tinku Rasheed, Karina Gomez, Sithamparanathan Kandeepan, Laurent Reynaud, Romain Hermenier, Andrea Munari, Mihael Mohorcic, Miha Smolnikar, and Isabelle Bucaille. Airborne base stations for emergency and temporary events. In R. Dhaou, A. Beylot, M. Montpetit, D. Lucani, and L. Mucchi, editors, *Personal Satellite Services*, volume 123 of *Lecture Notes of the Institute for Computer Sciences, Social Informatics and Telecommunications Engineering*, pages 13–25. Springer International Publishing, 2013.

[430] Bertold Van den Bergh, Tom Vermeulen, and Sofie Pollin. Analysis of harmful interference to and from aerial IEEE 802.11 systems. In *Proceedings of the First Workshop on Micro Aerial Vehicle Networks, Systems, and Applications for Civilian Use*, DroNet '15, pages 15–19, New York, NY, USA, 2015. ACM.

[431] Athanasios V. Vasilakos, Yan Zhang, and Thrasyvoulos Spyropoulos. *Delay Tolerant Networks: Protocols and Applications*. CRC Press, 2012.

[432] Dennis Vermoen, Marc Witteman, and Georgi N. Gaydadjiev. Reverse engineering Java card applets using power analysis. In Damien Sauveron, Konstantinos Markantonakis, Angelos Bilas, and Jean-Jacques Quisquater, editors, *Information Security Theory and Practices. Smart Cards, Mobile and Ubiquitous Computing Systems*, volume 4462 of *Lecture Notes in Computer Science*, pages 138–149. Springer, Berlin, Heidelberg, May 2007.

[433] Éric Vétillard and Anthony Ferrari. Combined attacks and countermeasures. In *Smart Card Research and Advanced Application*, volume 6035 of *Lecture Notes in Computer Science*, pages 133–147. Springer, Berlin, Heidelberg, April 2010.

[434] P. Vincent and I. Rubin. A framework and analysis for cooperative search using UAV swarms. In *Proc. ACM Symp. Appl. Comp.*, 2004.

[435] S. Waharte, N. Trigoni, and S.J. Julier. Coordinated search with a swarm of UAVs. In *Proc. IEEE Wkshp. Sensor, Mesh and Ad Hoc Commun. and Networks (SECON Workshops '09)*, pages 1–3, June 2009.

[436] Colin D. Walter and Susan Thompson. Distinguishing exponent digits by observing modular subtractions. In David Naccache, editor, *Topics in Cryptology — CT-RSA 2001*, volume 2020 of *Lecture Notes in Computer Science*, pages 192–207. Springer, Berlin, Heidelberg, April 2001.

[437] Chieh-Yih Wan, Shane B. Eisenman, and Andrew T. Campbell. CODA: Congestion detection and avoidance in sensor networks. In *Proceedings of the 1st ACM International Conference on Embedded Networked Sensor Systems (SenSys)*, pages 266–279, November 2003.

[438] Yan Wan, Kamesh Namuduri, Yi Zhou, and Shengli Fu. A smooth-turn mobility model for airborne networks. *IEEE Transactions on Vehicular Technology*, 62(7):3359–3370, 2013.

[439] Ruhai Wang and S. Horan. Protocol testing of SCPS-TP over NASA's ACTS asymmetric links. *IEEE Transactions on Aerospace and Electronic Systems*, 45(2):790–798, April 2009.

[440] David H. Ward, Robert A. Stehn, Wallace P. Erickson, and Dirk V. Derksen. Response of fall-staging brant and Canada geese to aircraft overflights in southwestern Alaska. *The Journal of Wildlife Management*, pages 373–381, 1999.

[441] Adam C. Watts, Vincent G. Ambrosia, and Everett A. Hinkley. Unmanned aircraft systems in remote sensing and scientific research: Classification and considerations of use. *Remote Sensing*, 4(6):1671–1692, 2012.

[442] Adam C. Watts, John H. Perry, Scot E. Smith, Matthew A. Burgess, Benjamin E. Wilkinson, Zoltan Szantoi, Peter G. Ifju, and H. Franklin Percival. Small unmanned aircraft systems for low-altitude aerial surveys. *The Journal of Wildlife Management*, 74(7):1614–1619, 2010.

[443] Roland Everett Weibel. Safety considerations for operation of different classes of unmanned aerial vehicles in the national airspace system. PhD thesis, Massachusetts Institute of Technology, 2005.

[444] Roland E. Weibel and Jr. R. John Hansman. Safety considerations for operation of different classes of UAVs. In *The NAS. AIAA 3rd "Unmanned Unlimited" Technical Conference, Workshop, and Exhibit*, 1:341–351, September 2004.

[445] A. Weinert, P. Erickson, H. Reis, P. Breimyer, T. Hackett, M. Samperi, J. Huff, C. Parra, E. Stoekl, P. Zundritsch, R. Morris, I. Iakimenko, E. Petschauer, and S. Bilen. Enabling

communications in disadvantage environments: An airborne remote communication (ARC) platform. In *2013 IEEE International Conference on Technologies for Homeland Security (HST)*, pages 797–803, November 2013.

[446] Alan Westin. *Privacy and Freedom*. The Bodley Head Ltd, London, England, 4 1970.

[447] WG-93 Light Remotely Piloted Aircraft Systems Operations. WG93 on lightweight RPAS, April 2012. www.eurocae.net/wgs/active/?wg=WG-93.

[448] M. Wheeler, B. Schrick, W. Whitacre, M. Campbell, R. Rysdyk, and R. Wise. Cooperative tracking of moving targets by a team of autonomous UAVs. In *IEEE/AIAA 25th Digital Avionics Systems Conference*, pages 1–9, October 2006.

[449] Wilde Solmecke. Die rechtlichen Probleme des Einsatzes von zivilen Drohnen, January 2014. www.wbs-law.de/internetrecht/die-rechtlichen-probleme-des-einsatzes-von-zivilen-drohnen-49854/.

[450] Mark P. Witton and Michael B. Habib. On the size and flight diversity of giant pterosaurs, the use of birds as pterosaur analogues and comments on pterosaur flightlessness. *PloS One*, 5(11):e13982, 2010.

[451] Anthony D. Wood and John A. Stankovic. Denial of service in sensor networks. *Computer*, 35:54–62, October 2002.

[452] Robert J. Wood. Liftoff of a 60mg flapping-wing mav. In *IEEE/RSJ International Conference on Intelligent Robots and Systems, 2007. IROS 2007.* pages 1889–1894. IEEE, 2007.

[453] Tom Wypych, Radley Angelo, and Falko Kuester. Airgsm: An unmanned, flying gsm cellular base station for flexible field communications. In *Aerospace Conference, 2012 IEEE*, pages 1–9. IEEE, 2012.

[454] Xamen Technologies www.xamen.fr/.

[455] E. Xargay, V. Dobrokhodov, I. Kaminer, A.M. Pascoal, N. Hovakimyan, and C. Cao. Time-critical cooperative control of multiple autonomous vehicles: Robust distributed strategies for path-following control and time-coordination over dynamic communications networks. *IEEE Control Systems*, 32(5):49–73, October 2012.

[456] Jim Anderson Xilinx. Case study: Secure FPGA technology enables UAV communications and control. http://mil-embedded.com/articles/case-enables-uav-communications-control/, April 2011. [Accessed 3 May 2014.]

[457] Enyang Xu, Zhi Ding, and S. Dasgupta. Target tracking and mobile sensor navigation in wireless sensor networks. *IEEE Transactions on Mobile Computing*, 12(1):177–186, January 2013.

[458] E. Yanmaz and H. Guclu. Stationary and mobile target detection using mobile wireless sensor networks. In *Proc. IEEE Conf. Computer Commun. Wksps (INFOCOM)*, pages 1–5, 2010.

[459] E. Yanmaz, S. Hayat, J. Scherer, and C. Bettstetter. Experimental performance analysis of two-hop aerial 802.11 networks. In *Proc. IEEE Wireless Commun. and Net. Conf.*, April 2014.

[460] E. Yanmaz, R. Kuschnig, and C. Bettstetter. Achieving air–ground communications in 802.11 networks with three-dimensional aerial mobility. In *Proc. IEEE Int. Conf. Computer Commun. (INFOCOM), Mini Conference*, April 2013.

[461] Attila Altay Yavuz, Fatih Alagoz, and Emin Anarim. HIMUTSIS: Hierarchical multi-tier adaptive ad-hoc network security protocol based on signcryption type key exchange schemes. In Albert Levi, Erkay Savaş, Hüsnü Yenigün, Selim Balcısoy, and Yücel Saygın, editors, *Computer and Information Sciences – ISCIS 2006*, volume 4263 of *Lecture Notes in Computer Science*, pages 434–444. Springer, Berlin, Heidelberg, 2006.

[462] S.P. Yeong, L.M. King, and S.S. Dol. A review on marine search and rescue operations using unmanned aerial vehicles. *World Academy of Science, Engineering and Technology, International Journal of Mechanical, Aerospace, Industrial, Mechatronic and Manufacturing Engineering*, 9(2):390–393, 2015.

[463] Harold Youngren, Steve Jameson, and Brian Satterfield. Design of the samarai monowing rotorcraft nano air vehicle. In *Proceedings of the American Helicopter Society AHS 65th Annual Forum and Technology Display*, 2009.

[464] M. Yuksel, R. Pradhan, and S. Kalyanaraman. An implementation framework for trajectory-based routing in ad hoc networks. *Ad Hoc Networks*, 4(1):125–137, 2006.

[465] Pengcheng Zhan, D.W. Casbeer, and A.L. Swindlehurst. A centralized control algorithm for target tracking with UAVs. In *Conference Record of the Thirty-Ninth Asilomar Conference on Signals, Systems and Computers*, pages 1148–1152, October 2005.

[466] Zhensheng Zhang. Routing in intermittently connected mobile ad hoc networks and delay tolerant networks: overview and challenges. *Communications Surveys & Tutorials, IEEE*, 8(1):24–37, 2006.

[467] W. Zhao, M. Ammar, and E. Zegura. A message ferrying approach for data delivery in sparse mobile ad hoc networks. In *Proceedings of the 5th ACM International Symposium on Mobile Ad Hoc Networking and Computing*, MobiHoc '04, pages 187–198, New York, NY, USA, 2004. ACM.

[468] Dimitrios Zorbas, Tahiry Razafindralambo, Di Puglia Pugliese Luigi, and Francesca Guerriero. Energy efficient mobile target tracking using flying drones. *Procedia Computer Science*, 19(0):80–87, 2013.

[469] Yi Zou and K. Chakrabarty. Distributed mobility management for target tracking in mobile sensor networks. *IEEE Transactions on Mobile Computing*, 6(8):872–887, August 2007.

[470] S. Shirazipourazad, P. Ghosh, and A. Sen. On connectivity of airborne networks in presence of region-based faults, in *Proceedings of the Military Communications Conference (MILCOM)*, pp. 197–2002, IEEE 2011.

[471] D. Hay, S. Shirazipourazad, and A. Sen. Optical tracking of multiple targets using UAVs, in *International Conference on Combinatorial Optimization and Applications*, pp. 750–763, Springer 2014.

[472] W.J. Sutherland. *Ecological Census Techniques – A Handbook.* Cambridge University Press, 2005.

[473] Rebecca A. Efroymson, I.I. Suter, W. Glenn, Winifred H. Rose, and Sarah Nemeth. Ecological risk assessment framework for low-altitude aircraft overflights: I. Planning the analysis and estimating exposure. *Risk Analysis*, 21(2):251–262, 2001.

[474] D.T. Booth, S.E. Cox, G.E. Simonds, and B. Elmore. Efficacy of two variations on an aerial lekcount method for greater sage-grouse. *Western North American Naturalist*, 69:413–416, 2009.

# Index